Computer-Aided Drug Design

Dev Bukhsh Singh
Editor

Computer-Aided Drug Design

Springer

Editor
Dev Bukhsh Singh
Department of Biotechnology,
Institute of Biosciences and Biotechnology
Chhatrapati Shahu Ji Maharaj University
Kanpur, Uttar Pradesh, India

ISBN 978-981-15-6814-5 ISBN 978-981-15-6815-2 (eBook)
https://doi.org/10.1007/978-981-15-6815-2

This Springer imprint is published by the registered company Springer Nature Singapore Pte Ltd.
The registered company address is: 152 Beach Road, #21-01/04 Gateway East, Singapore 189721, Singapore

Foreword

Ever since the cracking of the human genome in the beginning of the present century, scientists have been engaged in locating the drug targets and designing and developing novel drugs through the system's approach. This has resulted in a tremendous reduction in research and production costs. Earlier, the drug design process used to take many decades and was carried out haphazardly without any direction. Already the surge in bioinformatics solutions has redefined the way drug trials are done and making a shift from in vitro to in silico. In this age of multiple drug resistance, in silico drug design could be used to shorten the time of discovery and this issue shall remain the biggest challenge for years to come. In the present fast changing scenario, it is difficult to manage expressive coherence in this rapidly growing area of drug designing.

I am happy that Dr. Dev has ventured to collect twelve well-written chapters and has brought an edited book named "Computer-Aided Drug Design" to be published by Springer Nature, Singapore. I feel that the authors are quite successful in "fusing" the otherwise diverse topics of this fast-emerging area. I am sure that this book will be exceedingly useful for not only under- and postgraduate students but also for research scholars, scientists, and pharma industries involved in developing new drugs.

I hope that the readers of this book shall contribute in the future for making the text more useful for further development of this important field of computer-aided drug design.

Hony. Professor, IIIT-Allahabad
Prayagraj, India

Krishna Misra

Preface

The computer-aided drug design uses computational approaches for analysis of target, screening, and interaction of ligands, simulation of target–ligand complex, optimization of lead compounds, QSAR analysis, and ADMET studies. In structure-based drug designing, ligand molecules are built keeping in mind the binding cavity of the target by assembling small substructures in a stepwise manner. Ligand-based drug designing involves the 2D/3D analysis and chemical modification of ligand known to interact with a drug target of the disease. A large number of computational tools have been developed to fulfill the different objectives in the way of drug designing. There are many successful stories of computer-aided drug designing. This field has attracted many researchers working in diverse fields of knowledge such as chemistry, physics, biology, mathematics, and computer science. In drug designing, systematic and sequential use of different computer-aided drug designing tools/software is required. Much advancement has taken place in the algorithms and approaches of computer-aided drug designing from time to time. The existing limitations of the tools and approaches used for drug designing have also been discussed which can motivate the readers and researchers to overcome such challenges in the future.

The present book "Computer-Aided Drug Design" has been written considering the need for researchers and students working in the domain of computer-aided drug designing. This book not only represents the discussion of recent advances in the field of computer-aided drug designing but also provides a basic knowledge of principles, approaches, and tools used for drug designing. This book includes a discussion of biological database resources used for drug discovery. One chapter is focused on the computational approaches and resources used for vaccine designing. Similarly, a basic discussion and application of machine learning approaches such as genetic algorithm, artificial neural network, and support vector machine have been included. It also explains the basics and use of different biological, physical, and chemical parameters used for modeling, simulation, and ADMET prediction. The chapters provide a summary of related case studies along with the application, merits/demerits, limitations, and future perspectives related to the title. The steps and use of different computational approaches have been explained with the help of simple, suitable, and neat sketches and illustrations. This book is full of a lot of resources that can guide and motivate a learner to proceed for drug designing.

I hope this book will be very helpful in understanding the basics and recent advances in computer-aided drug design. I tried my best effort to present a good quality creation before the readers and other scientific communities. This book will cover the need for a broad spectrum of subjects such as bioinformatics, biotechnology, biochemistry, and pharmaceutical sciences. During the review and editing process, many suggestions, corrections, and suitable addition of new topics have been included. Still, I look forward to your valuable suggestions and feedback related to the content quality of the book.

Kanpur, India Dev Bukhsh Singh

Acknowledgement

I am highly grateful to Prof. (Mrs.) Krishna Mishra (Prayagraj), a renowned scientist and educator for her continuous support, guidance, and motivation. Prof. J.V. Vaishampayan, former Vice-Chancellor of CSJM University, Kanpur has encouraged me a lot to achieve academic excellence. I will always be highly thankful to him for his inspiration, encouragement, and support. I would like to acknowledge the effort of all the authors of this book for their extensive labor, vision, and planning in writing the chapters. I thank the reviewers whose critical comments improved the book in substantial ways. I am highly thankful to Dr. Pankaj Kumar Singh, GBPUA&T, Pantnagar for his technical support and suggestions. I am highly grateful to my parents Mr. Sudhakar Singh and Smt. Radhika Singh and other family members for their wishes, valuable support, and encouragement.

I am also thankful to my colleagues Dr. Manish Kumar Gupta, Dr. Satendra Singh, Dr. P. K. Yadav, Dr. Budhayash Guatam, Dr. Prashant Ankur Jain, Dr. Anil Kumar, Dr. Durg Vijay Singh, Dr. Ajay Kumar Singh, Dr. K. K. Ojha, Dr. Pramod Katara, Dr. Prem Kumar Singh, Dr. R. K. Kesharwani, other friends and staff members for their support. I would like to appreciate the effort of Dr. Rajesh Kumar Pathak, Mr. Rohit Shukla, Mr. Apoorv Tiwari, Mr. Himanshu Avasthi, Mr. Ambuj Srivastava, and Ms. Shikha Agnihotri for their support. I am thankful to Dr. Bhavik Sawhney and the entire team of Springer Nature for their continuous support and cooperation during the entire process of publication.

Contents

About the Editor

Dev Bukhsh Singh is an Assistant Professor at the Department of Biotechnology, Chhatrapati Shahu Ji Maharaj University, Kanpur, India. He received his B.Sc. and M.Sc. degrees from the University of Allahabad, Prayagraj, and his M.Tech. from the Indian Institute of Information Technology, Prayagraj. Holding a Ph.D. in Biotechnology with specialization in Bioinformatics from Gautam Buddha University, he has been actively involved in teaching and research since 2009, and his focus areas include molecular modeling, chemoinformatics, inhibitor/drug design, and in silico evaluation. He has authored numerous research articles and book chapters in the fields of medicinal research, molecular modeling, drug design, and systems biology. He also published a book on the title "protein structure, function, and dynamics" (Springer Nature, Singapore). He is a member of various national and international academic bodies and is a reviewer for several international journals.

Computational Approaches in Drug Discovery and Design

1

Rajesh Kumar Pathak, Dev Bukhsh Singh, Mamta Sagar, Mamta Baunthiyal, and Anil Kumar

Abstract

Drug discovery is an expensive and complicated process. The drug must fulfill some criteria of being nontoxic, bioavailable, and potent. In the view of evermore stringent demands about efficacy, potency, and safety, the finding of the new drug-like molecule has become a complex and resource-intensive undertaking. Now, the availability of 3D structures of molecular drug targets and advances in computational approaches and bioinformatics speed up the application of molecular modeling in discovery. In this chapter, several molecular modeling strategies employed in modern drug discovery program are discussed. The concepts of structure- and ligand-based drug designing, protein modeling and visualization, molecular docking, virtual screening, molecular dynamics simulation, pharmacophore modeling, and QSAR approaches have been explained. Besides, we also provide important database resources and tools available for drug research. Finally, we present case studies conducted in our lab, showing how computational approaches can be implemented in reality for the discovery and designing of novel drugs from natural sources.

R. K. Pathak · M. Baunthiyal (✉)
Department of Biotechnology, Govind Ballabh Pant Institute of Engineering & Technology, Pauri Garhwal, Uttarakhand, India

D. B. Singh
Department of Biotechnology, Institute of Biosciences and Biotechnology, Chhatrapati Shahu Ji Maharaj University, Kanpur, Uttar Pradesh, India

M. Sagar
Department of Bioinformatics, University Institute of Engineering and Technology, Chhatrapati Shahu Ji Maharaj University, Kanpur, Uttar Pradesh, India

A. Kumar
Rani Lakshmi Bai Central Agricultural University, Jhansi, Uttar Pradesh, India

D. B. Singh (ed.), *Computer-Aided Drug Design*,
https://doi.org/10.1007/978-981-15-6815-2_1

Keywords

Molecular modeling · Molecular docking · Molecular dynamics simulation · QSAR · Bioinformatics

1.1 Introduction

Generally, new drugs were discovered from plants and other natural sources through accidental observations and analysis (Singh et al. 2012). The field of drug discovery is extremely challenging and requires adequate infrastructure and lab facilities. People in every nation have used drugs derived from plant or animal origin to treat and prevent disease. The quest for substances to fight sickness and to alter mood and consciousness is nearly as fundamental as search for food and shelter. Various drug molecules derived from plants or animals are highly valued, but most of the drugs in the modern medical system are synthetic chemistry and biotechnology products. Thus a drug is said to be a substance of either natural or synthetic origin that is employed in the prevention, treatment, and diagnosis of disease or modulation of the target or function of the biological systems. Thus, generally, we can say that a drug is a chemical that affects the biological systems and its processes at molecular to the cellular level (Huang et al. 2010). However, the traditional approach utilized for the discovery of novel drug molecules is time-consuming and cost-intensive. Therefore, the new approaches exceed the limitations of traditional research in the field of drug discovery. It evolved based on the following consideration, i.e. the molecular target present in the body and the potential bioactive compound are directly related to each other.

For designing a drug, understanding about the disease and molecular mechanism of infectious processes is a must. For structure-based drug design, investigating a molecular target is a first step that is essential to a disease process and an infectious pathogen (Nag and Dey 2010). The next key step is to determine the molecular structure of the target through experimental or computational approaches. The success of structure-based drug discovery depends on accurate target structure with detail information of amino acid residues present in binding sites, which are further utilized by molecular docking program for screening of small molecules database (Kesharwani et al. 2018).

Drugs have historically been developed to target a single biological object, usually a protein generally known as the target, with high selectivity to prevent any unintended effects arising from mis-targeting other biological targets. Based on this, the concept of multi-targets drugs has long been marked as undesirable, as it was naturally linked with harmful effects (Bolognesi and Cavalli 2016; Ramsay et al. 2018). Parallel to this, several evidences confirmed that molecules that strike more than one target have a safer profile compared to single targets. Therefore, the idea of multi-target drugs made rapid and dramatic progress from an evolving paradigm when first laid out in early 2000, to one of the hottest drug research topics for the year 2017 (Roth et al. 2004; Ramsay et al. 2018).

In multifactorial diseases such as cancer and Alzheimer's disease, there is an urgent need to find multi-target inhibitors. As essential for Alzheimer's development, β-amyloid cleavage enzyme (BACE-1) and acetylcholinesterase (AChE) were considered promising targets for the drug (Goyal et al. 2014). Besides, numerous pathological manifestations also contribute to cancer. In fact, the medications also lead to serious side effects with their treatments. Thus, multi-indication therapies are required, which can simultaneously inhibit multiple targets and minimize side effects (Lim et al. 2019). Some multi-target drugs are available in the market for the treatment of the diseases. The recently approved (April 2017) multi-target drug is midostaurin; it is a well-known multi-kinase inhibitor for the treatment of those newly diagnosed adult patients with acute myeloid leukemia who have a particular FLT3 gene variant. A study has shown that it can inhibit the activity of protein kinase C alpha, KIT, VEGFR2, WT and PDGFR and/or FLT3 tyrosine kinases mutant (Levis 2017; Ramsay et al. 2018).

Drug discovery is based on the screening of small molecule databases on a receptor, whereas designing is based on modification in the structure of a lead compound when the lead compound having some therapeutically undesirable side effects. In addition to the above, quantitative structure–activity relationship (QSAR) is one of another potential area in molecular modeling that has helped medicinal chemists in drug designing process. In previous years, the identification of a new drug molecule is a very complex and time-consuming process. After studying more than 5000–10,000 compounds, only a single drug molecule comes to the market. The cost of each drug is about \$156 million in the discovery phase. I, II, and III clinical trial and Food and Drug Administration (FDA) processes the cost is another \$75 million. This brings the total amount is about \$231 million for each drug that comes to the market for benefits of the society. Then, for gaining FDA approval, an extensive and expensive procedure also needs to be followed (Huang et al. 2010; de Ruyck et al. 2016).

Considering the high failure levels, considerable costs, and slow speed of new drug identification research, repurposing "existing" medicines to treat common and rare diseases is becoming increasingly desirable as it requires the use of low-risk compounds to develop drugs in a shorter time with cost-effective manner (Pushpakom et al. 2019). Generally, three kinds of approaches that are widely used in drug repositioning include computational, experimental, and mixed approaches (Xue et al. 2018; Talevi and Bellera 2020). The first case of drug repositioning, in the 1920s, was an accidental discovery. After a century of progress, further strategies for accelerating the drug repositioning cycle have been suggested. In this scenario, machine learning algorithms have been implemented to boost drug repositioning efficiency. Over the computational methods, the experimental method was established that provide clear evidence of linkages between drugs and diseases, such as target screening, cell assays, animal model, and clinical approaches. These are effective and trustworthy methods. In recent years, growing numbers of researchers have merged computational and experimental methods for identifying new drug indications, called mixed approaches. Biological experiments and clinical studies confirmed the findings of the computational methods. Mixed approaches provide incentives for the successful and rapid discovery of repositioned drugs (Xue et al. 2018). Some successful repositioned drugs are Zidovudine, Minoxidil, Sildenafil, Thalidomide, Celecoxib, Atomoxetine, Duloxetine, Rituximab, Raloxifene, Fingolimod, Dapoxetine, Topiramate, Ketoconazole, and Aspirin (Pushpakom et al. 2019). However, there are also major technical and regulatory issues that need to be tackled. It is an intricate process involving several factors including technology, business models, patents, and investment as well as consumer demands. While several medical databases have been developed, it is still a challenge to choose the best approach to make full use of vast amounts of medical data.

There is an urgent need to develop a novel approach for drug repositioning. Another highlighted issue to address is the intellectual property (IP). IP safety is limited for repositioning of the drugs. For example, some novel associations of drug-target-disease discovered by repositioning researchers were verified by publications or online databases; however, because of the law, it is difficult to obtain IP protection for these associations. The IP problem prohibits the entry of such repositioned drugs into the market. In fact, some repositioning research initiatives are forced to give up, which is wastage of money and time. Therefore, developing a new commercial model is essential because the current model is a serial model which causes problems related to funding and investment (Xue et al. 2018).

Molecular modeling is a data-driven science branch, with many of the algorithms and databases being created or adapted as a response to new data forms (Xia 2017). Today computer experiments play an increasingly important role in research. The advent of high-performance computing has allowed in silico experimentation as a tool for interpolating laboratory experiments and theory (Aminpour et al. 2019). Due to advances in computational algorithms and the development of efficient software, the time requires in the identification of lead compounds reduced dramatically. A detailed flowchart highlighting the different approaches of molecular modeling in drug discovery and design is demonstrated in Fig. 1.1.

1.2 Structure-Based Drug Designing

Structure-based designing is a multidisciplinary and iterative process that is well-established in the research institution and pharmaceutical industry. It played a tremendous role in the discovery and development of several registered drugs and clinical candidates, for example, zanamivir, nelfinavir, and aleglitazar. In contrast, structure-based designing is relatively new in the agrochemical industry and at present, no products in the market that are directly investigated with the use of this approach. However, there are several databases and software programs where structure-based design has a strong impact (Huang et al. 2010). The major database resources used in a drug discovery program are listed in Table 1.1. Different approaches used in the discovery of lead molecule through computational are discussed in the following sections.

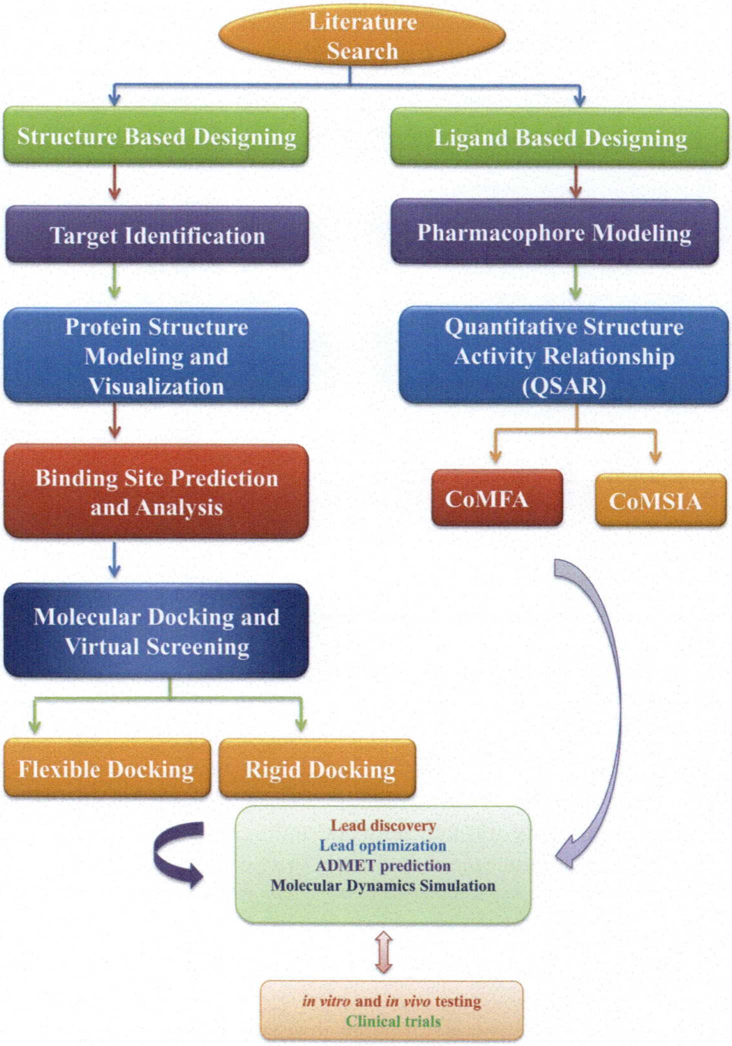

Fig. 1.1 Application of molecular modeling approaches in drug discovery and design

1.2.1 Target Identification

Drug target identification and its validation is the initial step of the drug discovery process. It is a macromolecule that has an established function in the pathophysiology of a disease. Four major drug targets are found in organisms, i.e. proteins, including receptors and enzymes, nucleic acids (DNA and RNA), carbohydrates, and lipid. The majority of drugs available in the market are addressed to proteins as a target. However, due to the decoding of several genomes of pathogens, nucleic acids could gain big importance as drug targets in the future (Gashaw et al. 2012). The

Table 1.1 Availability of major compound database resources for molecular modeling

S. No.	Database	Description	Availability	References
1	ZINC	It is a freely available database of commercially available compounds for molecular docking and virtual screening	http://zinc.docking.org/	Irwin and Shoichet (2005)
2	PubChem	It is a database of small chemical molecules, their biological activities	https://pubchem.ncbi.nlm.nih.gov/	Kim et al. (2016)
3	ChemSpider	It is a chemical structure database used for drug discovery	http://www.chemspider.com/	Pence and Williams (2010)
4	ChEMBL	It is a small molecule database that contains information about ADMET and binding for a huge number of bioactive compounds	https://www.ebi.ac.uk/chembldb/	Gaulton et al. (2012)
5	DrugBank	It is a comprehensive database resource containing information about drugs, their targets, and other useful information	https://www.drugbank.ca/	Wishart et al. (2006)

selection of potential drug targets from thousands of candidate macromolecules is a challenging task. In the post-genomic era, genomics and proteomics approaches are the most important tools for target identification (Singh et al. 2016). Besides, advances in high-throughput omics technologies generated a huge amount of data for host–pathogen interaction. These available data are also integrated and analyzed by the scientific community through network and systems biology approaches to accelerate the process of target identification in drug discovery program.

1.2.2 Modeling and Visualization of Macromolecule Structure

Determination of three-dimensional structure through experimental approaches is a costly and time taking process. Therefore, comparative modeling or homology modeling using sequence information is an accurate method for the prediction of three-dimensional structures, yielding appropriate models for a wide range of applications in the area of drug discovery (Bodade et al. 2010; Pathak et al. 2016). It is generally a choice of an algorithm when a homology among the target protein and a template structure exists (Sussman et al. 1998). This approach is based on the assumption that two identical sequences adopt similar three-dimensional structures. A higher sequence identity between the sequence of the target and template structure promises the generation of a more reliable model. Modeling the 3D structure of a protein from a sequence, in the absence of an X-ray or NMR verified structure is necessary for drug designing (Hekkelman et al. 2010; Bagaria et al. 2012). Besides, threading or fold recognition and ab initio are other methods used in modeling of 3D structure when no appropriate template detected in the PDB database (Singh and Tripathi 2020).

CASP (critical structure prediction assessment) playing a key role in protein structure prediction. It is a biennial collective project designed to evaluate the state of the art in protein structure modeling. Participants are provided with target protein amino acid sequences, and model the corresponding 3D structures. The independent assessors equate submissions with the experiment. It is a double-blinded experiment, participants do not have exposure to the experimentally determined structures, and the evaluators do not know the identity of those who apply. A variety of other aspects of protein modeling are also tested, in addition to structure models: optimization of an estimated structure similar to the experimental one, estimates of the accuracy of the overall structural model and residue, modeling of the protein oligomer structure, the ability to develop models using a range of sparse data types, and the accuracy of protein structure characteristics relevant to the deduction of functional aspects (Kryshtafovych et al. 2019). CASP studies were designed to achieve an objective for evaluation and assessment of different servers used of protein 3D structure prediction.

RasMol, PyMol, Chimera, and other visualization tools play a significant role in viewing and analyzing the predicted and experimentally determined 3D structures of macromolecules at the atomic level. Many efforts have been made in recent years to develop user-friendly simulation environments based on computer graphics for the structural biologist. It is widely used in biology for the presentation of simulation results in post-processing or experiments and by graphic editors for building models for a better understanding of atomic data of 3D co-ordinates (Seeliger and de Groot 2010; Mamgain et al. 2018).

1.2.3 Binding Site Prediction and Analysis

The determination of binding sites is not a simple task; researchers have suggested some criteria for selecting a binding site. It is investigated that the functional activity of any protein is governed by such highly conserved cluster of amino acid residues present in binding site pocket. The most available algorithms are based on similarity searches of the molecular surface for functional site databases such as PDB that contain fully reviewed and experimentally validated information of protein structures. Besides, some methods are also developed based on phylogenetic profiling of residues and several other models such as HMM, SVM, and CASP9 (Schmidt et al. 2011; Liu et al. 2014).

Generally, binding site residues are highly conserved among closely related proteins. Identification of such binding site residues is also done through the superimposition of the predicted model with their template that provided integrity for homology and assisted in the positioning of conserved active site residues (Nag and Dey 2010; Bajorath 2015). However, many protein–ligand complex structures are also available in public databases as a signature for the binding site where ligand was bound in binding site cavity of a protein. Usually, researchers separate bound ligand from protein, and this area is considered as binding site area for molecular docking studies using ligand structures because it is an experimentally determined complex structure and yielded significant outcome. Advances in the area of bioinformatics provide several computational tools that can able to predict novel binding site residues present in the cavity of the predicted protein model, which are further utilized in drug discovery research.

1.2.4 Molecular Docking

Docking intends to precisely fit the structure of a ligand inside the requirements of a receptor binding site and to accurately evaluate the strength of binding (Adrian-Scotto and Vasilescu 2008). When the binding site is not known in target protein structure, in such case, blind docking is helpful because in which whole protein structure is considered as binding site area, whereas if the binding site is known, site-specific docking is useful to predict the interacting nature of ligand molecule. Generally, the results of blind docking are less accurate and take more time and memory than site-specific docking because it targets only selected amino acid residues present in the binding site cavity. Nevertheless, the majority of available literature represents case studies where molecular docking has been used to deal with specific issues related to ligand design or target recognition (Huang et al. 2010; Yuriev and Ramsland 2013; Pathak et al. 2016; Rana et al. 2019). A summary of the highly cited molecular docking programs used in drug discovery has been listed in Table 1.2.

Table 1.2 A summary of the highly cited molecular docking programs used in drug discovery

S. No.	Programs	Description	Availability	References
1	AutoDock	Used for molecular docking. It predicts the binding affinity and poses of a small molecule to a 3D structure target protein	http://autodock.scripps.edu/	Goodsell et al. (1996)
2	AutoDock Vina	Used for virtual screening and molecular docking	http://vina.scripps.edu/index.html	Trott and Olson (2010)
3	Glide Schrodinger	A complete package for molecular modeling and computer-aided drug discovery (CADD)	https://www.schrodinger.com/	Friesner et al. (2004)
4	Hex	Used for docking studies	http://hex.loria.fr/	Ritchie (2003)
5	Molecular operating environment (MOE)	A complete package for molecular modeling and computer-aided drug discovery	https://www.chemcomp.com/	Vilar et al. (2008)

1.2.4.1 Flexible Docking

In this model, both the ligand and receptor side chain are kept flexible as well as the binding energy for different poses of the ligand fits into the receptor is calculated. For induced-fit docking, the main chain is also moved to integrate the conformational changes of the receptor upon ligand binding (Huang et al. 2010; Yuriev and Ramsland 2013). Whereas it is time taking and computationally expensive, yet this method can estimate many different probable conformations, which make it more extensive and perhaps simulate the phenomenon of real-life and hence trustworthy. Therefore, flexible docking is considered as a good approach because it yielded better prediction than conventional docking. The major drawback of other docking approaches is that it may provide poor docking scores due to incorrect ligand binding modes.

1.2.4.2 Rigid Docking

It is another method of molecular docking, in which the internal geometry of the ligand and receptor is kept fixed during docking simulation (Huang et al. 2010; Yuriev and Ramsland 2013; de Ruyck et al. 2016). The DOCK program based on rigid docking applied to the aspartic protease of HIV yielded a candidate inhibitor molecule with higher potency, and this molecule can be used as a lead for designing more powerful inhibitors. Besides, with simple bound and unbound target cases, ZDOCK correctly predicted 47% of interface contacts, demonstrating its strength in predicting binding sites. SOFTDOCK, on the other hand, predicted 66 of 83 (Pagadala et al. 2017).

1.2.5 Structure-Based Virtual Screening

Virtual screening involves the docking and screening of a compound database against the drug target, followed by scoring based on their binding free energy with the target. Many softwares are available for screening of compound databases against the selected target. Some are commercially available, and some are free for academic uses (Pathak et al. 2017). This method plays a key contribution to the drug discovery program for the identification of efficient lead compounds from small molecule databases. It also enables to boost lead identification process.

1.2.6 Validation of Molecular Docking

A variety of methods for validating the molecular docking have been published. One widely used approach is pose selection by which docking programs are used to re-dock a compound with a known conformation and orientation into the target's active site, usually from a co-crystal structure. Programs that are capable of returning poses below a pre-selected root mean square deviation (RMSD) value from the known conformation (generally 1.5 or 2 Å depending on the size of the ligand) are considered good. Pose selection is then accompanied by scoring and ranking to analyze which of the available scoring functions rank the poses most accurately in relation to their RMSD values (Hevener et al. 2009).

Another strategy of validation is to dock, a decoy set of inactive or suspected inactive, compounds that have been "seeded" against the target with compounds having known activity. Enrichment can be measured after ranking the docked decoy set by scores, and enrichment plots or receiver operating characteristic (ROC) curves can be plotted. The ROC curves map the sensitivity of a given docking/scoring combination against specificity and area under the curve for comparison can be determined (Jain 2008; Hevener et al. 2009). These approaches provide us an amazing opportunity for validation of docking results.

1.3 Ligand-Based Designing

The ligand-based drug designing is an alternative protocol, and plays a tremendous role when the structure of the target protein is unknown or cannot be predicted by available modeling methods. It uses statistical methods to correlate the activity of ligand to structural information (Huang et al. 2010). The different approaches used in ligand-based drug designing are discussed in the following sections.

1.3.1 Pharmacophore Modeling

Pharmacophore mapping is one of the real components in the drug designing program, without basic information of the target receptor. The tool at first applied to the identification of lead compounds now reaches out to lead optimization (Bauer and Stockwell 2008). Lead optimization is the mechanism by which a drug candidate is designed after having searched an initial lead compound. The method involves iterative rounds of synthesis and characterization of a putative drug to construct a picture of how chemical structure and its behavior are associated in terms of interaction by targets and its metabolism.

Pharmacophoric features can be used as a query for searching and retrieving the potential leads from chemical compound databases (discovery of lead molecules), for designing molecules with precise attributes (lead optimization). It also evaluates comparability and diversity of compounds by utilizing pharmacophore fingerprints. It can likewise be utilized to align compounds dependent on the 3D arrangement or to create prescient 3D QSAR models (Huang et al. 2010; Singh et al. 2013).

1.3.2 Quantitative Structure–Activity Relationship (QSAR)

Quantitative structure–activity relationship (QSAR) is an approach used to predict the biological activity of chemical compounds in drug designing. It uses statistical and mathematical tools to find the relationship between structures of the compound and their corresponding biological behaviors. Therefore, the QSAR model is built using structural parameters to predict the biological properties of a drug. The two-dimensional QSAR (2D-QSAR) uses 2-D structural properties of descriptors such as steric, electrostatic, hydrophobicity, and geometric behavior to interpret the molecular biological activity using multiple regression analysis. One of the foremost unremarkably used 2D QSAR strategies was given by Hansch (Clayton and Purcell 1969). 2D-QSAR techniques are not able to accurately explain the correlation between the physicochemical properties and 3D spatial arrangement as well as biological activities. Therefore, recently, 3D-QSAR approaches are introduced to overcome the limitation of 2D QSAR (Singh et al. 2013). In recent years the concept of multidimensional QSAR is introduced. It is more useful in predicting the biological properties of the chemical molecules. It includes HQSAR, G-QSAR, MIA-QSAR, and multi-target QSAR. It has made remarkable success in the drug discovery program (Wang et al. 2017). Comparative molecular field analysis (CoMFA) and comparative molecular similarity indices analysis (CoMSIA) are the two most important methods introduced for designing drug molecules (Huang et al. 2010).

1.3.2.1 CoMFA

CoMFA technique is based on a concept that the biological activity of a small molecule depends on the molecular fields. It uses Lennard-Jones and Coulombic potential to calculate steric and electrostatic fields, respectively. Both potential functions change dramatically close to the van der Waals surface of the molecule, which often requires cut-off values. Additionally, ligands must be aligned before calculating energy, but superimposed compounds orientation is correlative to the grid. It could cause key changes in the results of CoMFA. Further, a scaling factor must be added to the steric field to determine both fields in the same PLS analysis (Cramer et al. 1989; Huang et al. 2010).

1.3.2.2 CoMSIA

CoMSIA technique has recently been developed as an addition to CoMFA. This technique includes extra field properties that include steric, hydrophobic, electrostatic, hydrogen bond donors, and hydrogen bond acceptors. This technique is not sensitive to the alignment of compounds, their 3D-orientation. Besides, the improved functional algorithm is less affected by the relative distance to the surface of the van der Waals. Therefore, this model is more precise and accurate as compared to CoMFA (Klebe et al. 1994; Huang et al. 2010).

1.4 Computation of HOMO and LUMO Energy

These are the molecular orbitals playing a key role in drug discovery and design. HOMO (highest occupied molecular orbital) energy offers the small molecules area that can donate electron during the formation of the complex, whereas lowest unoccupied molecular orbital (LUMO) energy refers the ability of molecules to take electrons from the associated protein. The difference in energy of HOMO and

LUMO, referred to as HOMO–LUMO gap energy, designates the electronic excitation energy required to measure stability and chemical reactivity of the compounds (Banavath et al. 2014).

1.5 ADMET Prediction and Analysis

ADMET is one of the essential steps in any drug discovery program because it provides pharmacokinetics (ADME, i.e. Absorption, Distribution, Metabolism, and Excretion) and pharmacodynamics, i.e. toxicity (T) of lead molecules before going to wet-lab experimentation. Therefore, it reduces the risk of experimental cost, time, and drug failure. Literature studies showed that poor pharmacokinetics and pharmacodynamics are the major causes of costly and late-stage drug development failures, and it is widely recognized that computer-based ADMET prediction must be considered in the drug discovery program before going to in vitro and in vivo studies. Presently it is limited to Lipinski's rule and some other principle descriptors such as polar surface area (2D), polarizability, van der Waals surface area, refractivity, etc. Advances in combinatorial chemistry and high-throughput screening have extensively augmented the number of small molecules for which early data on ADMET are available as reference. Further, the accuracy of ADMET tools can be improved by including new principle descriptors and use of computational tools developed by the implementation of improved algorithms, the most relevant pharmacokinetics and pharmacodynamics information of any molecules can be modeled to accelerate the drug discovery process for identification of novel drug-like compounds in a cost-effective manner (Pathak et al. 2018; Singh and Pathak 2020).

1.6 Molecular Dynamics Simulation

Molecular dynamics simulation (MDS) is one of the key tools for the theoretical and computational study of biomolecules. Since molecular systems usually contain a large number of particles, the analysis of such complex systems is extremely challenging. Molecular dynamics simulation can prevent this analytical intractability by using numerical methods. The atoms and molecules can interact for a period of time during the simulation. The motion is calculated for each atom and can be used to check the overall behavior (Huang et al. 2010). It has many advantages over docking because docking gives only binding free energy of ligand with the receptor. Additionally, we can predict the actual interaction of the ligand with receptors through MDS at the atomic level. Besides, deciphering their binding mode via several bonds and amino acid residues involved in the interaction for time. During MDS, root mean square deviation (RMSD) is calculated for predicting the stability of receptor or ligand–receptor complexes, and it describes the conformational changes. Besides, root mean square fluctuation (RMSF) analysis is used to determine the flexibility with respect to time (Singh and Dwivedi 2016). It yielded novel information about the receptor or ligand–receptor complex that is further utilized for the next step in the drug discovery program. The background of the MD simulation algorithm is generally based on three steps; (1) determining the initial positions and speeds of each atom; (2) calculating the forces applied to the identified atom using inter-atomic potentials; (3) the progression of atomic positions and speeds over the short period. These new positions and speeds are then transformed into new inputs in step 2, and each repetition forms an additional time step when steps 2 and 3 are repeated (Huang et al. 2010).

1.7 Identification of New Drug-Like Molecules for Hyperuricemia from Millets: A Case Study

Xanthine oxidoreductase (XOR) plays a key role in the formation and regulation of uric acid during purine catabolism. Over expression of XOR results in the deposition of uric acid in the blood which causes damage to DNA and protein molecules. Human xanthine oxidoreductase (HsXOR) has been used as a therapeutic target against hyperuricemia. Febuxostat and allopurinol are inhibitors of HsXOR, but they have major toxic effects in the body. Millet grains contain higher levels of phenolic compounds and other phytochemicals than the major cereals (Taylor and Duodu 2015).

Bioactive compounds found in millet have enormous potential and can be used to prevent and treat hyperuricemia (Fig. 1.2). Some bioactive compounds of millets were studied of their interactions with HsXOR using docking, ADMET, and molecular dynamics simulation. The compounds derived from millet (luteolin and quercetin) showed −9.7 kcal/mol binding free energy (Fig. 1.3). Molecular dynamics simulation studies demonstrated that the luteolin forms a more stable complex with HsXOR than the quercetin. Luteolin has a high potential for testing as an HsXOR inhibitor because it can control the pathway by inhibiting HsXOR (Pathak et al. 2018).

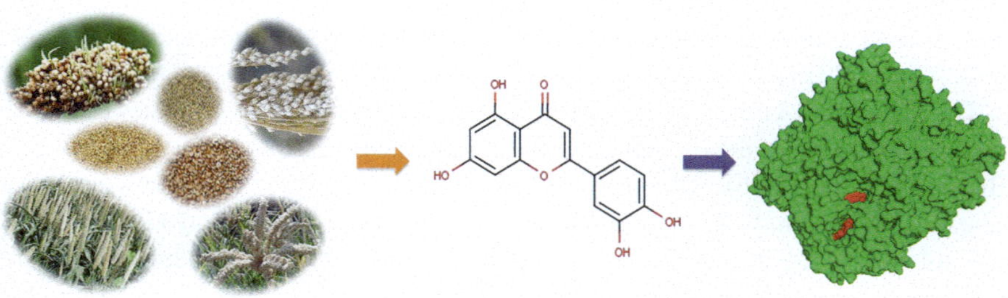

Fig. 1.2 Millets: a power house for identification of bioactive molecules through computational approaches; the figure depicted that compound luteolin is taken from millet plants and its interaction study was done through molecular docking (protein–ligand complex: HsXOR-green; luteolin-red)

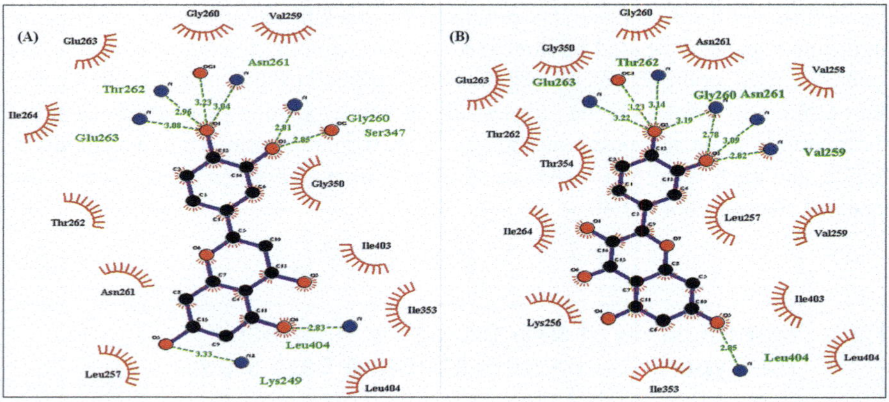

Fig. 1.3 Protein–ligand interaction diagram generated through Ligplot: (**a**) Luteolin-HsXOR docked complex showing the contribution of hydrogen bond (green line) and hydrophobic interaction (red) (**b**) Quercetin-HsXOR docked complex showing the contribution of hydrogen bond (green line) and hydrophobic interaction (red)

1.8 Discovery and Designing of Natural Lead Compounds for Liver Cancer: A Case Study

The compounds derived from natural sources such as plants, animals, microbes, etc., and having biological responses towards particular diseases are generally said to be natural lead compounds. It plays a significant role because it is a starting point of drug discovery where thousands of natural compounds are screened on a particular receptor for the identification of suitable lead compounds as new natural drugs with least or side effects. Therefore only selected compounds will be used for experimental verification. Further, lead optimizations are also done if any adverse activities are observed during wet-lab experimentation to increase the affinity, efficacy, and potency of a lead compound in which the targeted functional group of a lead compound is replaced by appropriate functional group. Molecular docking studies are also done to check its affinity with target followed by ADMET, MDS, and experimental studies to produce efficient derivatives from natural lead compounds (Pathak et al. 2014).

The hepatitis B virus x protein (HBx) of the Hepatitis B virus activates the AP-1 protein, and it causes the downregulation of tumor suppressor genes PTEN and p53 (Bouchard et al. 2006). Therefore the interactions of AP-1 with known natural anticancer compounds such as curcumin, epigallocatechin gallate (EGCG), genistein, luteolin, ellagic acid, lupeol, resveratrol, betulinic acid, and lycopene were studied using docking (Amin et al. 2009). EGCG has shown a very high affinity for its binding with AP-1 as compared to other compounds taken in the study. EGCG has shown interaction at the DNA binding domain of AP-1 that can minimize the

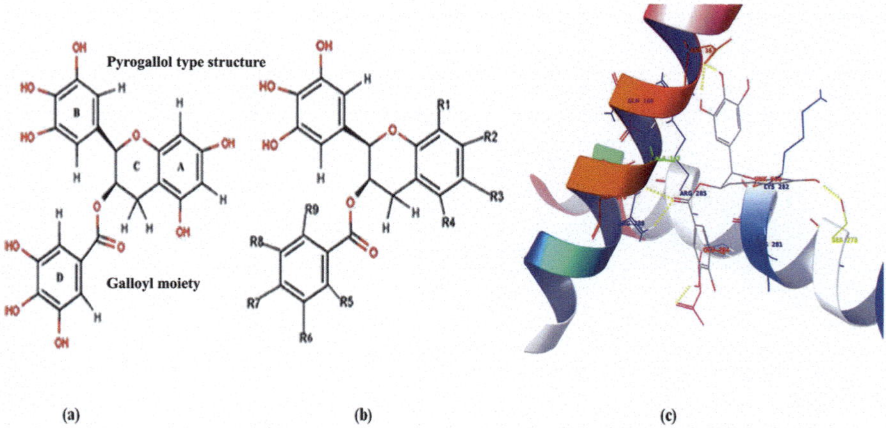

Fig. 1.4 (**a**) 2D view of EGCG: pyrogallol type structure and galloyl moiety (**b**) Position of structural modification in EGCG shown by labels (R1–R9) (**c**) Interaction view of EGCG with AP-1 protein

downregulation of the p53 and PTEN genes. Therefore derivatives were designed to improve the binding affinity and bioavailability of EGCG by imitating the positions of the H and OH groups (Fig. 1.4). One of the EGCG15 acetylates produced by replacing the OH group with OCOCH3 shows better affinity than other derivatives and binds to amino acids Asp 163(G), Ser 278(F), Lys 282(F), and Arg 288(F) with six H-bonds, and have shown −5.60 kcal/mol of binding free energy. The affinity of EGCG with AP-1 protein was greatly enhanced in the EGCG05 methoxy derivative, which forms eight H-bonds with −6.30 kcal/mol binding energy, and binds with amino acid residues Asp 163, Asp 170, Ser 278, Arg 281, and Arg 288. The substitution by OCOCH3 increases bioavailability during computational analysis after sequential addition at different positions in EGCG. Replacement by the NH2 group does not result in changes in oral absorption or bioavailability. The study is, therefore, informative to develop natural drugs against liver cancer using EGCG obtained from green tea, and its chemically synthesized derivatives for human welfare (Sagar et al. 2014).

1.9 Examples of Drugs Synthesized Using CADD

Molecular modeling has also been used in the development of drugs that have passed clinical trials and in the treatment of a number of diseases have become modern therapies. In 2003, the quest for novel transforming growth factor-β1 receptor kinase inhibitors was one of the most compelling examples of the possibilities presented by molecular modeling methods in drug discovery. One group at Eli Lilly used a conventional high-throughput screening method to investigate a lead compound, which was subsequently improved by analyzing structure–activity relationship *via*

Table 1.3 List of drugs identified through computational approaches

S. No.	Drug	Drug target	Disease	Approved year	References
1	Captopril	Angiotensin-converting enzyme (ACE)	Hypertension	1981	Talele et al. (2010)
2	Dorzolamide	Carbonic anhydrase (CA)	Glaucoma and ocular hypertension	1995	Vijayakrishnan (2009)
3	Saquinavir	HIV-1 and HIV-2 proteases	AIDS	1996	Van Drie (2007)
4	Indinavir	HIV protease	AIDS	1996	Van Drie (2007)
5	Ritonavir	HIV protease	AIDS	1996	Van Drie (2007)
6	Tirofiban	Glycoprotein IIb/IIIa receptor	Blocked coronary artery, antiplatelet drug	1998	Hartman et al. (1992)
7	Zanamivir	Neuraminidase	Influenza	1999	Kim et al. (1997)
8	Oseltamivir	Neuraminidase	Influenza	1999	An et al. (2009)
9	Raltegravir	Integrase	AIDS	2007	Schames et al. (2004)
10	Aliskiren	Renin	Hypertension, high blood pressure	2007	Cohen (2007)

in vitro assays. While the Biogen Idec group used a molecular modeling approach involving virtual high-throughput screening based on the structural interactions among the weak inhibitor and altering growth factor-β1 receptor kinase. The group at Biogen Idec found 87 hits after the virtual screening of compounds, the best hit being similar in structure with the lead compound discovered by Eli Lilly's conventional high-throughput screening approach. In this case, molecular modeling, a process with reduced costs and tons of effort, was able to investigate the same lead for drug development (Sawyer et al. 2003; Singh et al. 2003; Sliwoski et al. 2014). Some of the earliest examples of drugs synthesized after their discovery through the molecular modeling methods are listed in Table 1.3.

1.10 Success and Limitations

Discovery of new drug molecules using computational approaches has a focused research area due to advances in integrated omics, i.e. genomics, proteomics, metabolomics, and bioinformatics, it has many successful stories. Recently, the concept of pharmacogenomics is introduced to focus on personalized medicine. The key advantages of pharmacogenomics are to produce drugs based on the organizational structures of individual genomes. It is mainly used to address difficult tasks. It should not surprise that success is sometimes limited. Furthermore, several key problems related to computational complexities that have been on the agenda for decades remain to be resolved (Bajorath 2015; Hassan Baig et al. 2016; Singh 2014).

- In drug discovery practice, the potential of in silico methods should not be overestimated because it affects the credibility of serious computer work in the academic and pharmaceutical industries.
- In many cases, computational approaches can advance drug discovery projects significantly only if they are carefully selected and employed to problems, such as the selection of small molecules with a probability of displaying a specific activity, identification of novel compounds for optimization, or design of analogs that positively interact with a specified binding site in the cavity of a receptor.
- Biological systems are very complex and directed by numerous significant parameters. So there are certain restrictions, and it is not possible to copy and simulate the whole biological system on a PC using cutting edge techniques.
- One of the major challenges in drug discovery is target flexibility because most of the software provides only ligand flexibility.
- It is exceptionally hard to give total molecular flexibility to the protein as this augments the time and space complexity of the computation.
- Besides, the major limitation of pharmacophore modeling is dependent on pre-computed databases that hold a less number of low-energy conformations per compound.

1.11 Conclusion

The identification of targets for active compounds depends heavily on computational approaches for complex biological screening systems. Besides, such types of software and their applications are growth areas for molecular modeling and drug discovery. Besides, further computational research is required to reduce human bias in the creation and assessment of molecular property spaces for lead optimization and ADME analysis. For the future of computational drug discovery, further emphasis should be given on the development of new tools with more accuracy and improvement of already existing molecular modeling methods is required. Additionally, these methods have the tremendous potential to be broadly used in drug discovery practices, which is a prerequisite for success.

Competing Interest The authors declare that there are no competing interests.

References

Adrian-Scotto M, Vasilescu D (2008) Quantum molecular modeling of glycyl-adenylate. J Biomol Struct Dyn 25(6):697–708

Amin AR, Kucuk O, Khuri FR, Shin DM (2009) Perspectives for cancer prevention with natural compounds. J Clin Oncol 27(16):2712

Aminpour M, Montemagno C, Tuszynski JA (2019) An overview of molecular modeling for drug discovery with specific illustrative examples of applications. Molecules 24(9):1693

An J, Lee DC, Law AH, Yang CL, Poon LL, Lau AS, Jones SJ (2009) A novel small-molecule inhibitor of the avian influenza H5N1 virus determined through computational screening against the neuraminidase. J Med Chem 52(9):2667–2672

Bagaria A, Jaravine V, Huang YJ, Montelione GT, Güntert P (2012) Protein structure validation by generalized linear model root-mean-square deviation prediction. Protein Sci 21(2):229–238

Bajorath J (2015) Computer-aided drug discovery. F1000Res 4:630. https://doi.org/10.12688/f1000research.6653.1

Banavath HN, Sharma OP, Kumar MS, Baskaran R (2014) Identification of novel tyrosine kinase inhibitors for drug resistant T315I mutant BCR-ABL: a virtual screening and molecular dynamics simulations study. Sci Rep 4(1):1–1

Bauer AJ, Stockwell BR (2008) Neurobiological applications of small molecule screening. Chem Rev 108(5):1774–1786

Bodade RG, Beedkar SD, Manwar AV, Khobragade CN (2010) Homology modeling and docking study of xanthine oxidase of Arthrobacter sp. XL26. Int J Biol Macromol 47(2):298–303

Bolognesi ML, Cavalli A (2016) Multitarget drug discovery and polypharmacology. ChemMedChem 11(12):1190–1192

Bouchard MJ, Wang L, Schneider RJ (2006) Activation of focal adhesion kinase by hepatitis B virus HBx protein: multiple functions in viral replication. J Virol 80(9):4406–4414

Clayton JM, Purcell WP (1969) Hansch and Free-Wilson analyses of inhibitory potencies of some 1-decyl-3-carbamoylpiperidines against butyrylcholinesterase and comparison of the two methods. J Med Chem 12(6):1087–1088

Cohen NC (2007) Structure-based drug design and the discovery of aliskiren (Tekturna): perseverance and creativity to overcome a R&D pipeline challenge. Chem Biol Drug Des 70(6):557–565

Cramer RD 3rd, Patterson DE, Bunce JD (1989) Recent advances in comparative molecular field analysis (CoMFA). Prog Clin Biol Res 291:161

de Ruyck J, Brysbaert G, Blossey R, Lensink MF (2016) Molecular docking as a popular tool in drug design, an *in silico* travel. Adv Appl Bioinforma Chem 9:1–11

Friesner RA, Banks JL, Murphy RB, Halgren TA, Klicic JJ, Mainz DT, Repasky MP, Knoll EH, Shelley M, Perry JK, Shaw DE (2004) Glide: a new approach for rapid, accurate docking and scoring. 1. Method and assessment of docking accuracy. J Med Chem 47(7):1739–1749

Gashaw I, Ellinghaus P, Sommer A, Asadullah K (2012) What makes a good drug target? Drug Discov Today 17(Suppl):S24–S30

Gaulton A, Bellis LJ, Bento AP, Chambers J, Davies M, Hersey A, Light Y, McGlinchey S, Michalovich D, Al-Lazikani B, Overington JP (2012) ChEMBL: a large-scale bioactivity database for drug discovery. Nucleic Acids Res 40(D1):D1100–D1107

Goodsell DS, Morris GM, Olson AJ (1996) Automated docking of flexible ligands: applications of autodock. J Mol Recognit 9(1):1–5

Goyal M, Dhanjal JK, Goyal S, Tyagi C, Hamid R, Grover A (2014) Development of dual inhibitors against Alzheimer's disease using fragment-based QSAR and molecular docking. Biomed Res Int 2014:1. https://doi.org/10.1155/2014/979606

Hartman GD, Egbertson MS, Halczenko W, Laswell WL, Duggan ME, Smith RL, Naylor AM, Manno PD, Lynch RJ (1992) Non-peptide fibrinogen receptor antagonists. 1. Discovery and design of exosite inhibitors. J Med Chem 35(24):4640–4642

Hassan Baig M, Ahmad K, Roy S, Mohammad Ashraf J, Adil M, Haris Siddiqui M, Khan S, Amjad Kamal M, Provazník I, Choi I (2016) Computer aided drug design: success and limitations. Curr Pharm Des 22(5):572–581

Hekkelman ML, te Beek TA, Pettifer SR, Thorne D, Attwood TK, Vriend G (2010) WIWS: a protein structure bioinformatics Web service collection. Nucleic Acids Res 38(suppl_2):W719–W723

Hevener KE, Zhao W, Ball DM, Babaoglu K, Qi J, White SW, Lee RE (2009) Validation of molecular docking programs for virtual screening against dihydropteroate synthase. J Chem Inf Model 49(2):444–460

Huang HJ, Yu HW, Chen CY, Hsu CH, Chen HY, Lee KJ, Tsai FJ, Chen CY (2010) Current developments of computer-aided drug design. J Taiwan Inst Chem Eng 41(6):623–635

Irwin JJ, Shoichet BK (2005) ZINC– a free database of commercially available compounds for virtual screening. J Chem Inf Model 45(1):177–182

Jain AN (2008) Bias, reporting, and sharing: computational evaluations of docking methods. J Comput Aided Mol Des 22(3–4):201–212

Kesharwani R, Singh DB, Singh DV, Misra K (2018) Computational study of curcumin analogues by targeting DNA topoisomerase II: a structure-based drug designing approach. Netw Model Anal Health Inf Bioinform 7:15

Kim CU, Lew W, Williams MA, Liu H, Zhang L, Swaminathan S, Bischofberger N, Chen MS, Mendel DB, Tai CY, Laver WG (1997) Influenza neuraminidase inhibitors possessing a novel hydrophobic interaction in the enzyme active site: design, synthesis, and structural analysis of carbocyclic sialic acid analogues with potent anti-influenza activity. J Am Chem Soc 119 (4):681–690

Kim S, Thiessen PA, Bolton EE, Chen J, Fu G, Gindulyte A, Han L, He J, He S, Shoemaker BA, Wang J (2016) PubChem substance and compound databases. Nucleic Acids Res 44(D1): D1202–D1213

Klebe G, Abraham U, Mietzner T (1994) Molecular similarity indices in a comparative analysis (CoMSIA) of drug molecules to correlate and predict their biological activity. J Med Chem 37 (24):4130–4146

Kryshtafovych A, Schwede T, Topf M, Fidelis K, Moult J (2019) Critical assessment of methods of protein structure prediction (CASP)—round XIII. Proteins: Struct Funct Bioinf 87 (12):1011–1020

Levis M (2017) Midostaurin approved for FLT3-mutated AML. Blood 129(26):3403–3406

Lim H, Di He YQ, Krawczuk P, Sun X, Xie L (2019) Rational discovery of dual-indication multi-target PDE/Kinase inhibitor for precision anti-cancer therapy using structural systems pharmacology. PLoS Comput Biol 15(6):e1006619

Liu B, Liu B, Liu F, Wang X (2014) Protein binding site prediction by combining hidden Markov support vector machine and profile-based propensities. Sci World J 2014:1

Mamgain S, Dhiman S, Pathak RK, Baunthiyal M (2018) In 'silico' identification of agriculturally important molecule (s) for defense induction against bacterial blight disease in soybean (Glycine max). Plant Omics 11(2):98

Nag A, Dey B (2010) Computer-aided drug design and delivery systems. McGraw Hill Professional, New York

Pagadala NS, Syed K, Tuszynski J (2017) Software for molecular docking: a review. Biophys Rev 9 (2):91–102

Pathak RK, Baunthiyal M, Taj G, Kumar A (2014) Virtual screening of natural inhibitors to the predicted HBx protein structure of Hepatitis B Virus using molecular docking for identification of potential lead molecules for liver cancer. Bioinformation 10(7):428

Pathak RK, Taj G, Pandey D, Kasana VK, Baunthiyal M, Kumar A (2016) Molecular modeling and docking studies of phytoalexin (s) with pathogenic protein (s) as molecular targets for designing the derivatives with anti-fungal action on *Alternaria* spp. *of Brassica*. Plant Omics 9(3):172

Pathak RK, Baunthiyal M, Shukla R, Pandey D, Taj G, Kumar A (2017) *In silico* identification of mimicking molecules as defense inducers triggering jasmonic acid mediated immunity against alternaria blight disease in Brassica species. Front Plant Sci 8:609

Pathak RK, Gupta A, Shukla R, Baunthiyal M (2018) Identification of new drug-like compounds from millets as Xanthine oxidoreductase inhibitors for treatment of Hyperuricemia: a molecular docking and simulation study. Comput Biol Chem 76:32–41

Pence HE, Williams A (2010) ChemSpider: an online chemical information resource. J Chem Educ 87(11):1123–1124

Pushpakom S, Iorio F, Eyers PA, Escott KJ, Hopper S, Wells A, Doig A, Guilliams T, Latimer J, McNamee C, Norris A (2019) Drug repurposing: progress, challenges and recommendations. Nat Rev Drug Discov 18(1):41–58

Ramsay RR, Popovic-Nikolic MR, Nikolic K, Uliassi E, Bolognesi ML (2018) A perspective on multi-target drug discovery and design for complex diseases. Clin Transl Med 7(1):3

Rana G, Pathak RK, Shukla R, Baunthiyal M (2019) *In silico* identification of mimicking molecule (s) triggering von Willebrand factor in human: a molecular drug target for regulating coagulation pathway. J Biomol Struct Dyn 38:124–136

Ritchie DW (2003) Evaluation of protein docking predictions using Hex 3.1 in CAPRI rounds 1 and 2. Proteins: Struct Funct Bioinf 52(1):98–106

Roth BL, Sheffler DJ, Kroeze WK (2004) Magic shotguns versus magic bullets: selectively non-selective drugs for mood disorders and schizophrenia. Nat Rev Drug Discov 3(4):353–359

Sagar M, Pathak RK, Pandey RK, Singh DB, Pandey N, Gupta MK (2014) Binding affinity analysis and ADMET prediction of epigallocatechin gallate (EGCG) derivatives for AP-1 protein: a drug target for liver cancer. Netw Model Anal Health Inf Bioinform 3(1):66

Sawyer JS, Anderson BD, Beight DW, Campbell RM, Jones ML, Herron DK, Lampe JW, McCowan JR, McMillen WT, Mort N, Parsons S (2003) Synthesis and activity of new aryl- and heteroaryl-substituted pyrazole inhibitors of the transforming growth factor-β type I receptor kinase domain. J Med Chem 46(19):3953–3956

Schames JR, Henchman RH, Siegel JS, Sotriffer CA, Ni H, McCammon JA (2004) Discovery of a novel binding trench in HIV integrase. J Med Chem 47(8):1879–1881

Schmidt T, Haas J, Cassarino TG, Schwede T (2011) Assessment of ligand-binding residue predictions in CASP9. Proteins: Struct Funct Bioinf 79(S10):126–136

Seeliger D, de Groot BL (2010) Ligand docking and binding site analysis with PyMOL and Autodock/Vina. J Comput Aided Mol Des 24(5):417–422

Singh DB (2014) Success, limitation and future of computer aided drug designing. Transl Med (Sunnyvale) 4:e127

Singh DB, Dwivedi S (2016) Docking and molecular dynamics simulation study of inhibitor 2-Fluoroaristeromycin with anti-malarial drug target PfSAHH. Netw Model Anal Health Inf Bioinf 5:16

Singh DB, Pathak RK (2020) Computational approaches in drug designing and their applications. In: Experimental protocols in biotechnology. Humana, New York, NY, pp 95–117

Singh DB, Tripathi T (2020) Frontiers in protein structure, function, and dynamics. Springer, Singapore. https://doi.org/10.1007/978-981-15-5530-5

Singh J, Chuaqui CE, Boriack-Sjodin PA, Lee WC, Pontz T, Corbley MJ, Cheung HK, Arduini RM, Mead JN, Newman MN, Papadatos JL (2003) Successful shape-based virtual screening: the discovery of a potent inhibitor of the type I TGFβ receptor kinase (TβRI). Bioorg Med Chem Lett 13(24):4355–4359

Singh D, Tripathi A, Kumar G (2012) An overview of computational approaches in structure based drug design. Nepal J Biotechnol 2(1):53–61

Singh DV, Agarwal S, Kesharwani RK, Misra K (2013) 3D QSAR and pharmacophore study of curcuminoids and curcumin analogs: interaction with thioredoxin reductase. Interdiscip Sci Comput Life Sci 5(4):286–295

Singh S, Singh DB, Singh A et al (2016) An approach for identification of novel drug targets in streptococcus pyogenes SF370 through pathway analysis. Interdiscip Sci 8(4):388–394

Sliwoski G, Kothiwale S, Meiler J, Lowe EW (2014) Computational methods in drug discovery. Pharmacol Rev 66(1):334–395

Sussman JL, Lin D, Jiang J, Manning NO, Prilusky J, Ritter O, Abola EE (1998) Protein Data Bank (PDB): database of three-dimensional structural information of biological macromolecules. Acta Crystallogr Sect D 54(6):1078–1084

Talele TT, Khedkar SA, Rigby AC (2010) Successful applications of computer aided drug discovery: moving drugs from concept to the clinic. Curr Top Med Chem 10(1):127–141

Talevi A, Bellera CL (2020) Challenges and opportunities with drug repurposing: finding strategies to find alternative uses of therapeutics. Expert Opin Drug Discovery 15:397–401

Taylor JR, Duodu KG (2015) Effects of processing sorghum and millets on their phenolic phytochemicals and the implications of this to the health-enhancing properties of sorghum and millet food and beverage products. J Sci Food Agric 95(2):225–237

Trott O, Olson AJ (2010) AutoDock Vina: improving the speed and accuracy of docking with a new scoring function, efficient optimization, and multithreading. J Comput Chem 31(2):455–461

Van Drie JH (2007) Computer-aided drug design: the next 20 years. J Comput Aided Mol Des 21 (10–11):591–601

Vijayakrishnan R (2009) Structure-based drug design and modern medicine. J Postgrad Med 55 (4):301

Vilar S, Cozza G, Moro S (2008) Medicinal chemistry and the molecular operating environment (MOE): application of QSAR and molecular docking to drug discovery. Curr Top Med Chem 8 (18):1555–1572

Wang T, Yuan XS, Wu MB, Lin JP, Yang LR (2017) The advancement of multidimensional QSAR for novel drug discovery-where are we headed? Expert Opin Drug Discovery 12(8):769–784

Wishart DS, Knox C, Guo AC, Shrivastava S, Hassanali M, Stothard P, Chang Z, Woolsey J (2006) DrugBank: a comprehensive resource for in silico drug discovery and exploration. Nucleic Acids Res 34(suppl_1):D668–D672

Xia X (2017) Bioinformatics and drug discovery. Curr Top Med Chem 17(15):1709–1726

Xue H, Li J, Xie H, Wang Y (2018) Review of drug repositioning approaches and resources. Int J Biol Sci 14(10):1232

Yuriev E, Ramsland PA (2013) Latest developments in molecular docking: 2010–2011 in review. J Mol Recognit 26(5):215–239

Molecular Modeling of Proteins: Methods, Recent Advances, and Future Prospects

2

Apoorv Tiwari, Ravendra P. Chauhan, Aparna Agarwal, and P. W. Ramteke

Abstract

The three-dimensional modeling of protein structure is a reliable approach for understanding the biochemical functions along with the dynamics of protein interactions, which provides useful applications in developing drug molecules for curing diseases as well as certain other applications in biological and agricultural sciences. The conventional laboratory methods such as NMR and X-ray crystallography which are standard approaches for analysis of different proteins of interest are labor-intensive, expensive, and time-consuming. To address these challenges, the bioinformatics tools and approaches may open up new avenues for investigating the protein structures and functions. In the recent past, molecular modeling has been successfully used in various projects for 3D structure prediction of some therapeutically important proteins having applications ranging from medicine to agriculture. The approach of molecular modeling is based on the understanding of algorithms of protein structure prediction. This chapter illustrates the salient features of molecular modeling methods for a reliable and accurate structure prediction of the proteins in the field of drug designing.

A. Tiwari
Department of Computational Biology and Bioinformatics, Jacob Institute of Biotechnology and Bio-Engineering, Sam Higginbottom University of Agriculture, Technology and Sciences, Allahabad, Uttar Pradesh, India

Department of Molecular Biology and Genetic Engineering, Govind Ballabh Pant University of Agriculture and Technology, Pantnagar, Uttarakhand, India

R. P. Chauhan · A. Agarwal
Department of Molecular Biology and Genetic Engineering, Govind Ballabh Pant University of Agriculture and Technology, Pantnagar, Uttarakhand, India

P. W. Ramteke (✉)
Department of Biological Sciences, Sam Higginbottom University of Agriculture, Technology and Sciences, Allahabad, Uttar Pradesh, India

D. B. Singh (ed.), *Computer-Aided Drug Design*,
https://doi.org/10.1007/978-981-15-6815-2_2

Keywords

Molecular modeling · Structure prediction · Homology modeling · Structure validation · Drug designing · Threading

2.1 Introduction

Molecular modeling is an important approach in computational biology. For addressing the problems at the interface of biological and physical sciences, molecular modeling starting from its inception with applications of physics and computer science has been used rampantly by biological scientists. The advancements in computational biology have made molecular modeling a reality (Kumar and Chordia 2017). The current chapter discusses different methods available for protein three-dimensional structure predictions. As we know that proteins determine the biological functions of a cell and also are considered the building blocks of the cells. The dynamic processes including replication, maintenance, defense, and reproduction in living beings are encoded in the form of protein structures and functions (Berg et al. 2002).

There are 20 amino acids determined by the genetic codes. Proteins are the polypeptides that are made up of amino acids (Lodish et al. 2000). There are 20 regular amino acids in the configuration of the polypeptide chain, which determines the structure and function of the protein. Proteins are technically the end products that are decoded from cellular DNA information. Proteins are the main structural and transporter elements in a cell, which function as the workhorse of the cell unit and biocatalyst (Alberts et al. 2002). Interestingly, the functions that protein monitors in a cell are determined by genetic code. The protein structure and function depend on the genetic code encoded by DNA molecule(s), which is the building block of a gene. Different sequences of amino acids assemble to give rise to specific proteins with particular functions based on their three-dimensional configuration. The folded form or confirmation of a protein is directly dependent on the protein's linear amino acid sequence (Hooft et al. 1996, 1997).

2.1.1 Amino Acids

Amino acids are the basic units of the proteins, and each amino acid has a variable side chain (Berg et al. 2002). Multiple amino acids combine to form a long chain that is tied together with a peptide bond. The biochemical reactions govern the peptide bond formation where water (H_2O) molecule is extracted by joining the NH_2 of one and COOH of a neighboring amino acid in the polypeptide chain. A simple linear sequence of amino acids in a protein is known as the primary structure (Alberts et al. 2002). Amino acids have both polar and non-polar side chains (Lodish et al. 2000). The polarity of the side chains determines the amino acid and protein structure or conformation. The hydrogen bonding involves polar side chains, while the charged side chains of amino acid can participate in the formation of ionic bonds. Weak interactions are formed between the hydrophobic side chains to give rise to the van der Waals interactions. Cysteine is the only amino acid involved in the formation of disulfide covalent bonds, which is formed within or between two polypeptide chains that provide stability to the proteins (He et al. 2006; Neil and Bulleid Ellgaard 2011).

The 3D structure of a protein is governed by folding and intra-molecular bonding of the amino acids. The folding of the peptide chain is determined by hydrogen bonding between two neighboring amino acids. The specific patterns of this folding are termed as alpha helices and beta sheets, which are involved in the formation of a secondary structure of a protein (Voet and Vote 1990). Finally, multiple polypeptide chains join together and create the quaternary structure of a protein (Brown et al. 2003). The most energy-efficient and stable configuration of protein lies in its final form. A given protein attains its final form through a rigorous process of undergoing a variety of formations with the folding ability, which is unique and compact. A protein fold is stabilized by thousands of non-covalent bonds. The form and stability of proteins are determined by the chemical forces between a protein and its surrounding medium. Proteins inserted in the cell membranes have some hydrophobic chemical groups on their surface (Hooft et al. 1997).

2.1.2 Basic Principles of Protein Structure

The 3D structure of a protein is an atomic model of interaction of a large number of atoms (Wooley and Lin 2005). The 3D structure of protein represents a complex level of the group of atoms. In this type of arrangement, four different levels of protein structures exist, which are known as the primary, secondary, tertiary, and quaternary structures. Usually, two additional levels of intervention between secondary and tertiary structures are known as super-secondary structures and domains. The amino acid sequence itself does not directly encode disulfide bonds and other rare types of covalent bonds formed between side chains (Bailey 2018) (Fig. 2.1).

The secondary structure of protein results due to the folding of a protein sequence in a systematic form, and this fold is stabilized by the contribution of repetitive hydrogen bonding (Fig. 2.2). For the first time, Linus Pauling and Robert Corey described the chemical nature and secondary structures of proteins. The secondary structure includes alpha helix (α-helix) on the right, parallel, and antiparallel beta (β-) plated sheets and turns (Serafini 1989).

Tertiary structure is formed by a stable and compact packing of elements of secondary structure (Breda et al. 2008). The folding results in the complete three-dimensional polypeptide structure, which depends on the sequence of amino acids and atomic details (Berg et al. 2002). However, this process leads to the formation of a hydrophobic core for soluble proteins that are represented by the polar residues, which maintain its side chains inside the protein, and as a result, hydrophilic residues get exposed to the solvent. There are different types of protein folds available in nature, and two of the most common protein fold classification are SCOP and CATH (Csaba et al. 2009). The tertiary structure view for nsp10/nsp16 complex of SARS

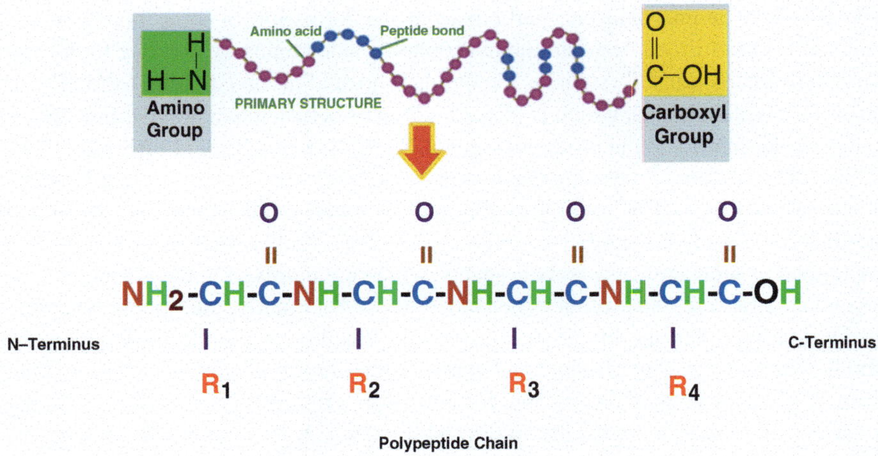

Fig. 2.1 Primary structure of a protein

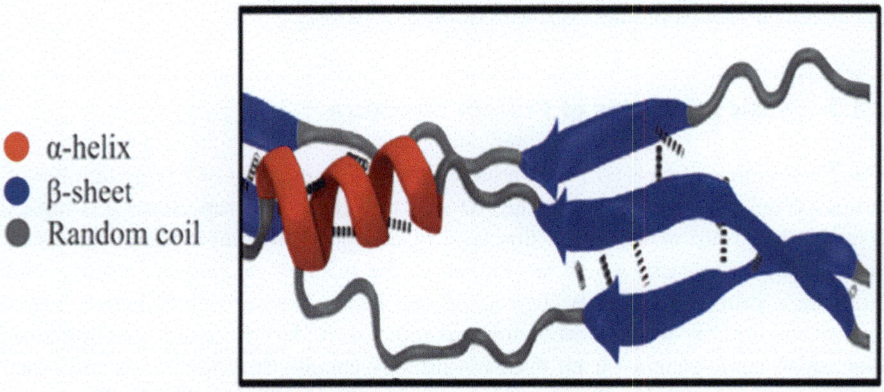

Fig. 2.2 Secondary structure of the protein

coronavirus has been shown in Fig. 2.3. The nsp16 causes sequence-dependent methylation and for successful methylation of viral mRNAs, nsp16 requires the interaction of nsp10 to initiate its methyltransferases activity (Chen et al. 2011). The nsp16 may serve as a potential drug target against corona disease. A potential inhibitor of nsp16 can serve as a therapeutic agent for the treatment of this disease.

The quaternary structure of the protein is formed by two or more identical or different polypeptide chains. Since two or more subunits are present, hence such proteins are termed as oligomers. The characterization illustrating how subunits of the native protein are arranged is based on the quaternary structure. The oligomeric subunits are held together by non-covalent forces, and therefore, can undergo rapid transformations affecting the biological activity of the protein. Hemoglobin, allosteric enzymes, actin, and tubulin are some examples of oligomeric proteins (Berg et al.

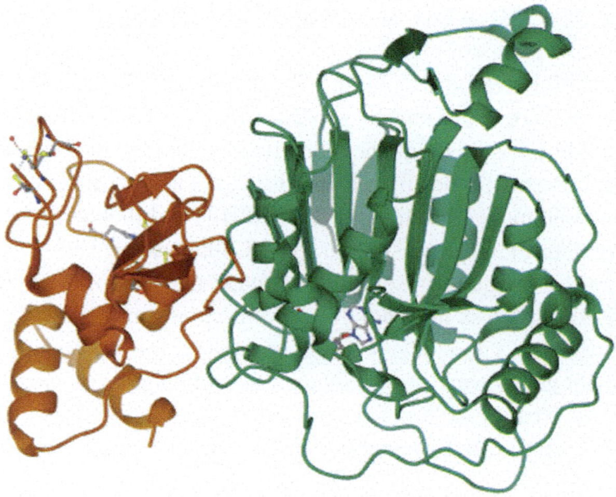

Fig. 2.3 Tertiary structure view: nsp10/nsp16 complex of SARS coronavirus (PDB: 3R24)

2002). The protein function depends on the manner how the protein surface interacts with the other molecules through bonding. Structurally similar proteins similarly interact with certain molecules and thus, represent a protein family (Alberts et al. 2002). The structural similarity of proteins within a family contributes a similar function to the family. Proteins with similar amino acid sequences belong to the same protein family, and their protein sequence remains conserved during evolution. Proteins with similar functions have a similar set of amino acid residues that interacts and binds with the substrate or signal (Cohen et al. 2009).

2.2 Explosion of Protein Related Data

Many databases are available with the sequence level information of the protein. Protein Data Bank (PDB) is an important database of protein structure. Currently, PDB has become the most popular protein databank with an archive containing more than 1,60,000 structures determined by different experimental methods like nuclear magnetic resonance (NMR) spectroscopy, crystallography, nuclear and electron cryo-microscopy (3D-EM), etc. PDB data and resources are very useful in the development of an experimental method, training, and testing of predictive models and drug discovery projects (Horiuchi et al. 2000). In recent years, different database resources have been developed which provide information about the classification and clustering of proteins, structural characterization, localization, phosphorylation, family and domain, active site, binding related information, protein disorder, conformational diversity, pathway, structure, function, etc. (Burley et al. 2017).

2.3 Protein Structure Determination

Several methods, including X-ray crystallography, NMR spectroscopy, and electron microscopy, are currently used to determine the 3D structure of a protein, and each method has its uniqueness and limitations (Kendrew et al. 1958). The user has access to several pieces of information for each of these methods to generate the final atomic model (Wang and Wang 2017) and this could be due to the X-ray diffraction pattern in X-ray crystallography (Callaway 2015). In electron microscopy, this could be the image representing the shape of the molecule. However, only experimental information is not enough to build an accurate atomic model, additional molecular structural knowledge is often required is the amino acid sequence of a protein along with geometrical features such as bond lengths and bond angles (Rankin et al. 2014). The creation of a consistent protein model requires a set of experimental data and modeling related parameters (Carroni and Saibil 2016).

2.3.1 X-Ray Crystallography

For X-ray crystallography, the protein is purified and crystallized under suitable conditions, and then subjected to an intensive X-ray beam (Burley et al. 2019). The diffraction of an X-ray beam by the protein crystals into one or other patterns is examined to determine the distribution of electron density. Finally, a map of the electron density is generated and interpreted to determine the location of each atom. The electron density map is analyzed to locate the position of each atom in 3D space. X-ray crystallography is a powerful tool that can provide coordinate information of each atom, which can illustrate the position of each atom in a protein. It is a challenging method with limitations on certain proteins but an excellent method to study and determine the structures of rigid crystals (Wang and Wang 2017). It is difficult to study the structure of a flexible protein using X-ray crystallography. The accuracy of this method depends on the quality of the crystals used for X-ray crystallography. Resolution and R-value are the two important parameters used to represent the accuracy of crystallographic structure (Haywood 1997).

X-ray crystallography is the most practical method for determining the structure of the biomolecules. Some of the salient features that this method offers include:

1. Accuracy of models for atomic resolution, also the method enables the user to solve relatively large structures and complexes.
2. Different solvents crystallize the same protein into different conformations. Thus, the method facilitates the study of the whole mechanism mediated by a single protein using XRD. Prominent examples include viral capsid structures and the ribosome, each made up of tens of thousands of atoms (Haywood 1997).

X-ray crystallography can resolve the structure of protein and protein–ligand complexes with good accuracy. But this method has some major limitations as given below:

1. The data provide only one protein position or confirmation but not dynamic behavior.
2. Contacts between molecules of crystal and the dense packaging can affect the structures.
3. The procedure is time-consuming.
4. Challenging for the analysis of highly hydrophobic or flexible proteins.
5. Difficulty in determining hydrogen positions, which requires very high resolution, thus, unfortunately, limits the reliability of this method.

2.3.2 NMR Spectroscopy

NMR spectroscopy is used to determine the 3D structure of molecules. However, during this procedure, the molecule should be pure and placed in a robust magnetic field. A distinguishable set of measured resonances can be analyzed to provide a list of nearby atomic nuclei to describe the composition of atoms that are linked together (Serdyuk et al. 2007). This list of restraints is used for model creation that indicates the location of each atom. NMR spectroscopy is used for determining the structure of proteins in solution and requires aqueous crystal. It is the first method used for the study of flexible protein structures (Snyder et al. 2005). A typical NMR structure includes a set of protein structures that are consistent with the list of experimental restraints observed.

The advantages of NMR spectroscopy are as given below:

1. The sample does not need to be in crystalline condition (which limits the applications of X-ray crystallography).
2. It provides better dynamics of the molecule.

There are some limitations associated with NMR spectroscopy that are mentioned here:

1. The technical accuracy of NMR is less as compared to X-ray crystallography.
2. MNR generates a good result for small size molecules, proteins with less than 300 residues.

2.3.3 3D Electron Microscopy

Electron microscopy is also used to determine the 3D structures of large molecular assemblies, often referred to as 3DEM. A beam of electrons and an electron lens system are used to directly image the biomolecule (Orlova and Saibil 2011). In a limited number of cases, electron diffraction from 2D or 3D crystals can be used for determining the 3D structures with an electron microscope (Rabl et al. 2010). Finally, 3DEM techniques are in advance importance for studying the biological assemblies inside the cryo-preserved cells and tissue by using electron tomography.

In terms of molecular and atomic details, both single-particle 3DEM and electron diffraction methods now provide structures with resolution limits comparable to macromolecular crystallography (i.e., enabling visualization of amino acid side chains, surface water molecules, and non-covalently bound ligands). Cryo-electron tomography provides slightly lower resolution structural information (protein domains and structural elements). In the calendar year 2016, the PDB deposition of the 3DEM structures for the first time exceeded that of NMR spectroscopy (Jonic 2016).

To investigate the very large macromolecular assemblies where lower resolution is normal, 3DEM data are increasingly being combined with information from X-ray crystallography, NMR spectroscopy, mass spectroscopy, chemical cross-linking, fluorescence resonance energy transfer, and various computational techniques to sort out the atomic details. This practice of using multiple experimental approaches is referred to as integrative or hybrid methods (I/HM) (Jonic 2016). This approach of integration has been proved much useful for investigating multimolecular structures such as complexes of ribosomes, t-RNA, protein factors, and muscle actomyosin structures, among others. A prototype data repository known as PDB-Dev, operating in parallel with the PBD, is now available for archiving of I/HM structures and data (Dever and Green 2012; Masters and Beyreuther 1998).

Structural resolution is the main limitation of EM. The EM resolution is approximately 3.5 Å, which is not enough to determine the location of side chains (Wohlgemuth et al. 2008). The advantages of electron microscopy are given below:

1. EM can solve very large biological complexes that are not accessible to crystallography with X-rays.
2. It can be used as a reference for interpreting X-ray diffraction patterns.
3. EM structures and X-ray data can be combined for determining the structure of large molecules.

2.4 Protein Structure Prediction

Three computational methods widely used for protein structure prediction are (1) homology modeling, (2) fold recognition, (3) ab initio method. Several tools for protein structure predictions are available, which utilizes different approaches and methods for modeling and refinement of the protein structure (Table 2.1).

2.4.1 Homology or Comparative Modeling

Comparative modeling is a template based modeling consists of five main steps: (a) identification of similar sequences with known structure, (b) alignment of the target sequence with template structures, (c) modeling of structurally conserved regions using templates, (d) modeling of lateral chains and loops, (e) the quality of the model being refined and evaluated by conformational sampling (Fig. 2.4). The

Table 2.1 Tools for protein structure modeling using different approaches of prediction

S. No.	Name	Description	Weblink
1	RaptorX	Remote homology detection, protein modeling, prediction of the binding site	http://raptorx.uchicago.edu/
2	ESyPred3D	Template search, better performance of alignment, used for 3D modeling based on homology	https://www.unamur.be/sciences/biologie/urbm/bioinfo/esypred
3	FoldX	Energy calculations and protein modeling integrate protein fragment libraries of different length	http://foldxsuite.crg.eu
4	Geno3D	Comparative protein modeling uses geometrical restraints (dihedral angles and distances)	https://geno3d-prabi.ibcp.fr
5	HHpred	Homology detection and template search, alignment, 3D modeling, based on hidden Markov model (HMM)	https://toolkit.tuebingen.mpg.de
6	LOMETS	Local meta-threading server for tertiary structure prediction	https://zhanglab.ccmb.med.umich.edu
7	Modeller	Comparative protein modeling, satisfaction of spatial restraints, optimization of various models of a protein	https://salilab.org
8	MOE	Molecular operating environment (MOE), template identification, use of multiple templates, loop modeling, side-chain modeling by rotamers	https://www.chemcomp.com
9	Prime	Homology modeling, assessment, and refinement of the generated model, based on the energy function	https://www.schrodinger.com/prime
10	ROBETTA	Homology modeling and ab initio fragment assembly can model multi-chain complexes	http://robetta.bakerlab.org/
11	Swiss model	Homology modeling based server, template search, local similarity/fragment assembly, model evaluation by comparison to X-ray verified structures	https://swissmodel.expasy.org/
12	I-TASSER	Combination of ab initio folding and threading methods, template-based fragment assembly	https://zhanglab.ccmb.med.umich.edu
13	NovaFold	Combination of threading and ab initio folding based on iterative assembly simulations, also predicts ligand binding site	https://www.dnastar.com/manuals/protean3d/15.3/en/topic/run-nova-applications-through-the-dnastar-website

(continued)

Table 2.1 (continued)

S. No.	Name	Description	Weblink
14	MUSTER	Threading algorithm, sequence-template alignments by profile–profile alignment	https://zhanglab.ccmb.med.umich.edu/MUSTER/
15	EVfold	Calculates evolutionarily coupled residues, evolutionary couplings calculated from correlated mutations in a protein family, ranks the models on geometric criteria and clustering	http://evfold.org/

accuracy of comparative modeling predictions depends on the degree of sequence similarity between template and target. If the target and template sequence share a high degree of similarity, then the accuracy of the predicted model is very high. For a sequence identity of 30–50%, more than 80% of C-α atom is expected to be within 3.5 Å of their true positions while significant errors are likely to occur for less than 30% of sequence identity (Rodriguez et al. 1998; Krieger et al. 2003). Comparative modeling is based on the principle that similar evolutionary sequences have similar three-dimensional folded structures.

The goal of protein modeling is to predict the structure of a target protein from its sequence information using the related known structure as a template. This would enable the rapid use of in silico protein models in all fields, such as structure-based drug prediction, protein function investigation, network analysis, antigenic behavior, and protein structure with increased soundness or novel capacities. In the case where experimental strategies are limited, one of the best ways to obtain the auxiliary data is protein modeling. Several proteins are very large and insoluble and hence cannot be studied by NMR and X-ray diffraction. Homology modeling is one of the easiest approaches to the 3D structure prediction elaborated in this chapter (Peach et al. 1994; Blundell et al. 1987).

If we have to know the structure of any protein that contains 200 amino acids, we use the BLAST tool of NCBI to compare the sequence of this protein in the PDB database and fortunately, we found a structure B with a total of 400 amino acids which aligns 50% identical residues with structure A. In this case, we can regard structure B as a template and structure A as a target, so that we can model the protein using structural information of template protein B. The homology modeling can only be used to model 3D structures of a target protein if it shares more than 30% sequence similarity with the template (Sanchez and Sali 1997; Peitsch 1997). Homology modeling is a multi-step process that includes template search, database searches, sequence alignment, structural refinement, loop search, side-chain modeling, coordinate assignment, energy minimization, and structure validation to create a quality structure (Johnson et al. 1994).

The modeler cannot make the finest protein structure, and therefore the main task of the modeling process involves genuine thinking as to how to play between

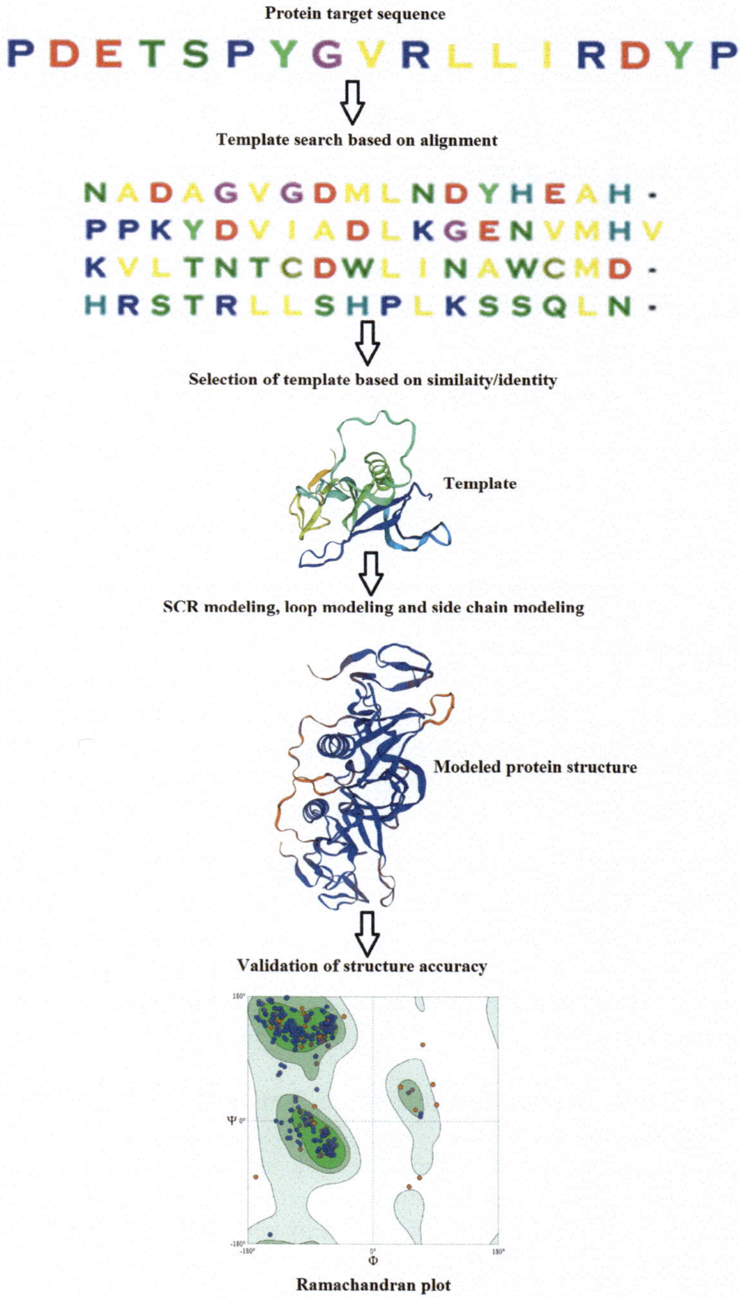

Fig. 2.4 A sequence of steps involved in homology modeling

different considerably similar choices. To construct the homology models, considerable efforts and research have been done to train the computational models to make better decisions (Peitsch et al. 2000). A sequence of steps involved in homology modeling is discussed here.

2.4.1.1 Template Recognition and Initial Alignment

The percentage identity between target protein sequence and the template is calculated using searching tools such as BLAST (Altschul et al. 1990) or FASTA (Pearson 1990). Two main matrices are used to identify these hits by comparison of the query sequence to all the known structures.

1. A matrix of residue exchange: The probability of alignment of two of the 20 amino acids is determined by the elements of this matrix.
2. A matrix of alignment: The two aligned sequences are represented by the axes of this matrix.

One needs to feed the query sequence to BLAST servers available on the web, followed by the selection of the PDB database for search. Finally, a list of templates and their alignment score with the query or target protein is received.

2.4.1.2 Alignment Correction

Several templates can be used for modeling. However, it is time to acquire better alignment using more sophisticated methods. It is difficult to model regions with a low percentage of sequence identity. Another sequence of homologous proteins can be used to find a suitable solution. Multiple sequence alignment programs such as CLUSTAL W can align several related sequences (Thompson et al. 1994) and a huge amount of data can be retrieved from the resulting alignment. If only exchanges between hydrophobic residues are observed at a certain position, then there are high chances of the residue being buried. Position-specific scoring matrices referred to as profiles are derived by multiple sequence alignments (Taylor 1986). Insertions (additional residues in the model) or deletions (missing residues in the model) can be achieved by multiple sequence alignments merely at the places where we find quite different sequences.

2.4.1.3 Modeling Structurally Conserved Region (SCR) and Backbone Generation

The important step in homology modeling is to determine the regions that are structurally conserved among the structures related to templates. Structurally conserved region (SCR) or core is determined by computing the C-alpha distance matrix for each structure and then small portions of the distance matrix are compared to find the peptide segments with lower root mean squared difference (RMSD) for related structures. In this way, all SCRs are determined and related SCRs share very high sequence and structural similarity. The generation of the model starts with the alignment of target and template protein. Template-target alignment indicates the residue blocks in the target which corresponds to SCRs of template. Coordinates of the amino acid residues of the template for structurally conserved regions between the template and target protein are taken from the template structure and assigned to the target model. There may be some varying residues within SCR region of aligned template and target, and if only their side chains differ, then the backbone (N, Cα, C, and O) coordinates of these residues are copied template and assigned to target model. Experimental protein structures are better than modeled. Choosing the template with the fewest errors is a simple way to build a good model. But what if there are two templates and each with a region that is poorly determined, but these regions are not the same, and then both templates can be used for model building using multiple templates approach. This approach is also used if there are good matches in different regions between alignments between the target sequence and templates. Multiple template modeling is done by servers like Swiss model (Peitsch et al. 2000).

2.4.1.4 Loop Modeling

During template-target alignment, gaps occur between the aligned model and the template sequence. These gaps represent the insertion/deletion between template and target. The structural fold of these gap residues or loop needs to be determined and incorporated in between the two conserved core region. It requires modification of the backbone. In the regular secondary structural elements, orientation or conformational changes are not found. Thus, it is safe to remove all the insertions or deletions within the alignment form helixes and strands and put them in turns and loops. We frequently realize different loop configurations within the template and target even while not insertions or deletions. There are the following reasons behind this problem (Krieger et al. 2003):

1. Surface loops lead to a major modification within the conformation of the template, and therefore the target.
2. Beneath the loop, the exchange of little to large side chain pushes it away.
3. The mutation of proline or glycine to the other residue in the loop.

In all cases, the residue must be placed in the loop considering the Ramachandran plot.

Two main approaches used for modeling the loop region are given here.

Knowledge-Based

Here, we search for the structure of the loops region with endpoints reminiscent from the known structures, and then the coordinates of loop structure are placed in between two cores. Most of the molecular modeling programs such as Modeller (Sali and Blundell 1993), Insight (Dayringer et al. 1986), Swiss model (Peitsch et al. 2000) or 3D-Jigsaw (Bates and Sternberg 1999) support knowledge-based approach for loop modeling.

Energy-Based

The energy function is employed to assess the loop quality and uses Monte Carlo or molecular dynamics simulation methods (Fiser et al. 2000) to generate the most accurate loop form. The energy function can be modified to generate a better loop structure that can best fit in the core (Tappura 2001). For small loops (up to 5–8 residues), the various strategies are available to predict a loop configuration that well overlaps the important structure.

2.4.1.5 Side-Chain Modeling

During the coordinate assignment in core modeling, coordinates of all amino acids are copied from template to target except those amino acids where side chain differs. At the position of the varying side chain, only the backbone coordinate of amino acid is assigned to target, and the related side chain is modeled using rotamer libraries. Rotamer libraries contain the biologically active conformation of side chains for different amino acids. As we know that all conformation of an amino acid is not biologically active, so it becomes important to determine and place the correct conformation of the side chain (Sanchez and Sali 1997).

It uses rotamers libraries derived from high-resolution X-ray structures. These rotamers are validated with a range of energy functions for their fitness (Scouras and Daggett 2010). The selection of an explicit rotamer mechanically affects the rotamers of all near residues. With a 100 residues and a median of five rotamers per residue, 5100 different mixtures would be scored already. There has been a great deal of analysis into developing strategies to create this vast search space traceable (Desmet et al. 1992). For a given backbone configuration, just one powerfully inhabited rotamer may be modeled immediately, so providing an anchor for additional versatile side chains within the surroundings. There are mainly two reasons for low prediction accuracy.

1. Flexible side chains on the surface can form several conformations.
2. Rotamers in hydrophobic packaging in the core can be easily scored, but ionic interactions on the surface, hydrogen bonds with water, and related entropic effects are challenging (Sliwoski et al. 2014).

It is vital to notice that in nearly all publications, the prediction accuracy cannot be achieved in real-world applications. The algorithms, therefore, accept the proper backbone that is not offered within the modeling of homology. The template's backbone is commonly significantly different from the target (Fiser 2010). The rotamers should be predicted based on the wrong backbone, and the predictive accuracy, in this case, tends to be lower.

2.4.1.6 Model Optimization

The right backbone is needed for the prediction of high-precision side-chain rotamers, which relies on the rotamers and their packaging. The main approach to a tangle of this kind is the re-iterative prediction of rotamers, then the ensuing backbone shifts, and the new backbone rotamers until the process converges. This method reduces the series of rotator predictions and steps of energy reduction (Hansen and Kay 2011).

The methods described above are used not only in the loop modeling but also for model optimization and should be applied for the entire protein structure (Hintze et al. 2016). At each minimization step, a few major errors, such as bumping due to atomic clashes, are eliminated, whereas several tiny mistakes are created. When small errors begin to accumulate, the model becomes less accurate. Better optimization of a model can be achieved by more accurate energy functions for force field calculation. Precision can be achieved by using the following approaches.

Quantum Force Fields

The force field calculation method should be fast and efficient to cover large protein molecules. The recent advancement in computational biology enabled methods of quantum chemistry to attain a more accurate interpretation of the charge distribution for the whole protein molecule (Liu et al. 2001).

Self-Parameterizing Force Fields

Force field accuracy depends on its variables (e.g., atomic loads and van der Waals radii). These variables are generally derived from small molecules quantum chemical calculations and fitting of these values to experimental data (Krieger et al. 2002). This method results in a rather expensive computer procedure. Take starting parameters for the force field, modify the parameter, minimize the energy of models, and save the new force field if the quality of the model improved otherwise return to the previous parameter of force field. This approach can increase the accuracy of the force field in the correct direction during energy minimization. A protein model can be optimized using molecular dynamics simulation, and it samples trajectory of the motions of the protein at a duration of 10 fs and generates the true folding dynamics of the protein (Adcock and McCammon 2006). It is therefore considered that during the simulation, the model will approach to real structure (Hospital et al. 2015).

2.4.1.7 Model Validation

Each model generated by homology contains some errors. The number of errors (for one particular method) depends primarily on two points.

1. A highly accurate protein model can be generated if the target shares very high similarity with the template. If the accuracy of the model is greater than 90%, then it can be compared accurately equivalent to an X-ray determined structure (Chothia and Lesk 1986; Sippl 1993). If the sequence identity between template and target lies below 25%, then alignment becomes meaningless for homology modeling, and the resulted model may have a high error.
2. There might be some errors in the template structure, which may result in the modeled protein. Structural errors can be estimated by the following methods:
   (a) Force field-based methods evaluate the bond angles, bond lengths, and bumps within the atoms. Lower energy models do not guarantee for the

accuracy of protein structure because sometimes the misfolded inaccurate models also achieve the low energy folds (Novotny et al. 1988).

(b) Normality indices can be used to compare the feature in the model that resembles the real structures. Many characteristics of protein structures are suitable for the analysis of normality. Many of these are based directly or indirectly on interatomic distance and contact analysis. It is important to observe the normality of torsion angles, bond angles and bond lengths and quality parameters of determined structures, but are less appropriate for model assessment (Czaplewski et al. 2000; Morris et al. 1992). Polar and non-polar residue distributions inside/outside can be used to predict misfolding in the protein model (Baumann et al. 1989). Most of the methods used to verify models can be applied to X-ray and NMR verified structures.

2.4.2 Fold Recognition or Threading Method

This method of prediction is used when there exists a low degree of similarity between template and target sequence as we cannot proceed for homology modeling due to low similarity (>30%). There is still no complete understanding of the relationship between the sequences, structure, and function. The only reliable fold prediction tools are currently the analogy based prediction algorithms. The threading approach is able to identify the most distant homologs and unrelated proteins with similar structures in some cases (Jaroszewski et al. 1998). The main challenge in the field of fold recognition is to develop tools to comply with structure, function, and analysis (Pruisner 1996) (Fig. 2.5).

The threading approach is used when the similarity between the target and template lies below 30% (Hendlich et al. 1990; Sippl 1993). In such cases, homology modeling may not generate a reliable model, so it is necessary to consider detailed structural parameters in the alignment (Jones et al. 1992). Threading methods consider structural information that is missed in the alignment process by sequence

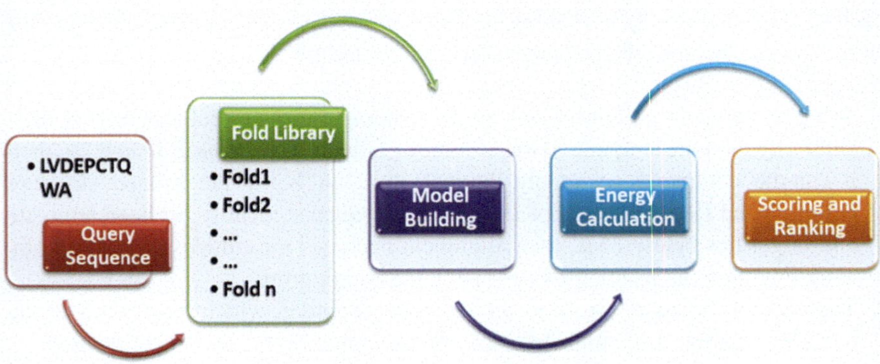

Fig. 2.5 Prediction of potential structure through threading approach

comparison. Structural details can be included in various ways (Bowie et al. 1991). The 3D profile is another method used for threading, which is based on the structural environmental class of each amino acid residue and generates a matrix for the probability of each amino acid to stay in each environmental class (Shi et al. 2001; Bowie et al. 1991). Each amino acid has a probability to reside in a particular environmental class.

Another approach calculates the contact residue potential of the pair and maximizes the hydrophobic core score. This method identifies the core of the protein structure that is essential for maintaining the structural integrity. This can be done directly, including contact potentials in pairs (Bowie et al. 1991; Jones et al. 1992). Threading is based on environmental class and uses a dynamic programming algorithm but has some limitations related to the preservation of environmental class. Contact residue potential approach considers the formation of hydrophobic core on contacts between hydrophobic residues. Threading utilizes two approaches: (1) profile of structural environmental classes (2) the contact potentials directly in pairs.

In the 3D profile method, a template structure is represented as a descriptor string that describes the structural environment. There are three basic environment classes; (1) area of the lateral chains buried by other protein atoms, (2) fraction of the lateral chains covered by polar atoms, and (3) secondary local structure. Here, a 3D protein structure is represented in the form of an ID string, which represents each residue's environmental class in the folded protein structure. The environment of a side chain is first classified as buried (*B*), partially buried (*P*) or exposed (*E*) depending on the area exposed to solvent. The buried and partially buried residue environments are further subdivided into *P* and *Pi* and *B*, *Bi*, *B*3, respectively (Peterson et al. 2014). The *E*, *P*, *Pi*, *B*, *Bi*, and *B*3 are the basic six environmental classes. In this way, there are a total of 18 environmental classes for all three secondary structures: helix, sheet, and coil. The 3D-ID scoring table where the pairing residue score *i* is given as follows with the environment *j*. *P* (*i*, *j*) represents the probability of amino acid residue *i* in environment *j*, and *Pi* is the overall probability of amino acid residue *i* in any environmental class (Ihm 2004). Here, the 3D-1D scoring table is used to generate the profile of a template structure. This is also known as sequence-structure because in this approach target and template both are represented in the form of string. The target protein is represented as a string of amino acids and the template structure represented as a string of the environmental classes (Jones 1999). The fitness score between the target and template environment class is calculated using a dynamic programming algorithm.

I-TASSER, a Yang Zhang Lab structure prediction threading method primarily identifies the structural template or fragment from the PDB subset using multiple threading approaches. Second, the initial conformations generated from the templates replica-exchange Monte Carlo simulations to produce a large number of reduced models. Third, all the models are grouped by SPICKER43 and the centroid cluster is formed by averaging the coordinates of each cluster of all decoys. Fourth, the simulation of the fragment assembly is carried out again starting from the selected cluster centroids of the cluster. Fifth, FG-MD reconstructs and refines the all-atom structures. Finally, five full-length atomic models are produced along with models of approximate accuracy. As a comprehensive process, even for new fold targets, I-TASSER performs pretty well (Yang et al. 2016).

2.4.3 Ab Initio Methods

This approach generates a protein model from sequence information due to the unavailability of structural counterparts or structural folds. The ab initio method enables us to understand the physicochemical principle related to the nature of proteins. The accuracy of ab initio modeling is low as compared to other methods of structure prediction (Simons et al. 1999). If the target sequence does not share structural similarities with structures in the database, then protein structure can be generated by determining the configuration space of atoms in amino acids. This method utilizes the knowledge of various principles of physics, chemistry, and mathematics. The use of reduced protein representations makes the computation easy. Some of the models represent a residue using only two locations, such as backbone one and side chain (Cohen et al. 2009). Others use several sites, including heavy backbone atoms and a side link. The main driving force for protein folding is known to be hydrophobic interaction, and there is some empirical energy function for the calculation of interactions in protein. For the ab initio prediction, three factors must be established: (1) reduce the representation of proteins, (2) a potential energy function for interaction, (3) method for searching the conformation space.

Simulated annealing is used for searching the configuration space of the fragment structures. A move is taken to replace the torsion angles of a randomly selected neighbor in a randomly selected position with the current configuration. Movements that bring two atoms closer within 2.5 Å are discarded, and other movements are evaluated. Baker and colleagues predicted the tertiary structure of a protein using the sequence information of amino acid, and no template details were considered.

Most structure prediction methods currently depend on the information provided by the structures predicted by experimental methods, which is not much supportive in exploring the basic law of protein folding. Template-free methods are guided by the practical application as well also consider the fundamental principles of protein folding. Template-free methods are based on information from known structures, their development may better reflect the prediction of the theoretical and technical level of protein structure than template-based methods. ROSETTA one of the efficient ab initio *modeling* approaches (Han and Baker 1995; Shortle et al. 1998) is a template-free method created by the David Baker Lab, assembling a complete structure based on fragments of 3–9 residues from PDB. Similar to the template-based methods, the selection of fragments is based on the similarity between known and predicted secondary structure. Monte Carlo method simulates the assembly process with the annealing search technique. QUARK is an incredible Yang Zhang Lab fragment assembly tool. The fragments used for QUARK vary from 1 to 20 residues and the simulation of the assembly is performed under the guidance from a knowledge-based atom level force field by Monte Carlo replica-exchange simulation.

Many other approaches, including Scrape, PROFESY, FRAGFOLD, etc. are also based on fragment assembly. The main distinction between these approaches and the template-based approaches is that they are not based on any global structural blueprint and they also do not utilize homology or structural similarities between the target and the proteins from which the fragments derive. It is more capable of modeling the target of new folds for template-free methods. However, due to the high computational requirement and low force field accuracy, it is still a major challenge for template-free methods for modeling proteins with a length of >150 residues. Prediction of contact map based on a co-evolution approach has recently shown progress to break down such a length limit of ab initio structure folding.

2.5 Evaluation and Validation of Modeled Structure

The final predicted model has to be evaluated to make sure that the structural features of the model are consistent with the physicochemical rules. This involves checking anomalies in φ–ψ angles, bond lengths, close contacts, and so on. Another way of checking the quality of a protein model is to implicitly take these stereochemical properties into account. This is a method that detects errors by compiling statistical profiles of spatial features and interaction energy from experimentally determined structures. By comparing the statistical parameters with the constructed model, the method reveals which regions of a sequence appear to be folded normally and which regions do not. If structural irregularities are found, the region is considered to have errors and has to be further refined.

SAVES server is a set of programs, which offers to check the model accuracy by just uploading the predicted structure. Procheck program is one of them that is able to check general physicochemical parameters such as φ–ψ angles, chirality, bond lengths, bond angles, and so on. The parameters of the model are used to compare with those compiled from well-defined, high-resolution structures. If the program detects unusual features, it highlights the regions that should be checked or refined further. WHATIF is another comprehensive protein analysis server that validates a protein model for chemical correctness. It has many functions, including checking of planarity, collisions with symmetry axes (close contacts), proline puckering, anomalous bond angles, and bond lengths. It also allows the generation of Ramachandran plots as an assessment of the quality of the model.

Atomic non-local environment assessment (ANOLEA) is a web server that uses the statistical evaluation approach. It performs energy calculations for atomic interactions in a protein chain and compares these interaction energy values with those compiled from a database of protein X-ray structures. If the energy terms of certain regions deviate significantly from those of the standard crystal structures, then it suggests that the corresponding region has not been modeled correctly. The threshold for unfavorable residues is normally set at 5.0. Residues with scores above 5.0 are considered regions with errors. Verify3D is another server using the statistical approach. It uses a pre-computed database containing 18 environmental profile based on secondary structures and solvent exposure, compiled from high-resolution protein structures. To assess the quality of a protein model, the secondary structure and solvent exposure propensity of each residue is calculated. If the parameters of a residue fall within one of the profiles, it receives a high score, otherwise a low score. The result is a two-dimensional graph illustrating the folding quality of each residue of the protein structure. The threshold value is normally set at zero. Residues with scores below zero are considered to have a non-favorable environment.

The assessment results can be different using different verification programs. Although the full-length protein chain of this model is declared favorable by ANO LEA, residues in the C-terminus of the protein are considered to be of low quality by Verify3D. Because no single method is clearly superior to any other, a good strategy is to use multiple verification methods and identify the consensus between them. It is also important to keep in mind that the evaluation tests performed by these programs only check the stereochemical correctness, regardless of the accuracy of the model, which may or may not have any biological meaning. Some tools predict the accuracy of the predicted model based on the Ramachandran plot of amino acid residues. It is a two-dimensional scatter plot showing torsion angles of each amino acid residue in a protein structure. The plot delineates preferred or allowed regions of the angles as well as disallowed regions based on known protein structures. This plot helps in the evaluation of the quality of a new protein model.

2.6 Recent Advances in Prediction Approaches

Two types of qualitatively different approaches for structural modeling are available: comparative modeling and de novo methods. Comparative modeling uses structural templates, while de novo methods model the protein without the detail of structural templates. The computational assessment of structural prediction (CASP) categorizes targets into two groups: (1) template-based modeling (TBM) and (2) free modeling (FM) (Moult et al. 2018).

Several parameters have been considered to improve the accuracy of prediction by TBM. PSI-BLAST and the profile-to-profile alignment methods have improved the accuracy of template identification and alignment. A composite structure assembly simulation utilizes the information from multiple templates, which refined the individual templates to be more similar to the native structures (Yang et al. 2015). In recent years, the availability of vast experimental sequence and structural databases has made it easier to get close homology templates for a target sequence. FM approaches are also referred to as ab initio or de novo structure prediction. Fragment assembly is also a type of FM approach (Dukka 2017). Some recent advances in the FM approach have been made, which considers the evolutionary constraints, contact information, and correlated mutation in scoring functions to improve the accuracy of the predicted protein model.

2.7 Applications

Accessibility of protein 3D structures and other structural analysis tools is facilitating the integration of an immense amount of information, which could be useful to further explore the possibilities to strengthen understanding of protein structure and function in the future. The 3D structures of a protein provide a better insight into the binding site and other functionally important regions, which could be utilized for drug designing. The availability of a target structure is the prerequisite condition to proceed for structure-based drug designing, and it also guides the changes in lead molecules. A 3D complex structure of a protein with a ligand provides better information about the residues involved in the interaction. Interaction of a receptor with a small molecule explains the mechanism of the pharmacological activity of a drug, binding affinity, and lead modification.

Computational modeling also explains how the mutation of an amino acid causes loss of function by destroying the native structure of protein required for its normal function. It can also explain the mechanism of drug resistance by depicting structural changes in the mutant target protein, which causes loss of proper binding or interaction of a drug with the target protein. Numerous forces such as hydrophobic interaction, Van der Waals, hydrogen bonding, and electrostatic are involved between the protein–ligand complexes to provide stability. Modeling of intermolecular connections in the protein–ligand complex is a very complex process due to a large number of degrees of freedom and inadequate information related to the impact of water on the binding.

2.8 Conclusion

Molecular modeling has turned into a significant and fundamental approach available to restorative scientific experts in the field of drug designing. Molecular modeling reveals the three-dimensional structures of proteins to unravel its related physicochemical properties. The protein modeling makes efficient use of computer science algorithms, theoretical science principles, and experimental information to uncover the structural and biological properties of a macromolecule. The selection of different tools for protein structure prediction depends on the nature of the problem to be addressed. This chapter has described the basic approaches and the recent advancements in methods of protein structure prediction. Required emphasis has also been given to the type of errors that may emerge and accumulate during protein modeling work. The development of a highly accurate and precise modeling tool should be the prime necessity as the applications of X-ray crystallography or NMR spectroscopy which being time-intensive do not seem to be the practical approaches for determination of the structure of every given protein. The accuracy of protein structure prediction is a crucial step because it is the basis of structure-based drug designing.

Competing Interest The authors declare that there are no competing interests.

References

Adcock SA, McCammon JA (2006) Molecular dynamics: survey of methods for simulating the activity of proteins. Chem Rev 106(5):1589–1615

Alberts B, Johnson A, Lewis J, Walter P, Raff M, Roberts K (2002) Protein function in: molecular biology of the cell, 4th edn. Garland Science, New York

Altschul SF, Gish W, Miller W, Myers EW, Lipman DJ (1990) Basic local alignment search tool. J Mol Biol 215(3):403–410

Bailey R (2018) Learn about the 4 types of protein structure. Thought Co. thoughtco.com/protein-structure-373563

Bates PA, Sternberg MJ (1999) Model building by comparison at CASP3: using expert knowledge and computer automation. Proteins S3:47–44

Baumann G, Frommel C, Sander C (1989) Polarity as a criterion in protein design. Protein Eng 2 (5):329–334

Berg JM, Tymoczko JL, Stryer L (2002) Protein structure and function. Biochemistry 262:159

Blundell TL, Sibanda BL, Sternberg MJ, Thornton JM (1987) Knowledge-based prediction of protein structures and the design of novel molecules. Nature 326(6111):347–352

Bowie JU, Luthy R, Eisenberg D (1991) Method to identify protein sequences that fold into a known three-dimensional structure. Science 253(5016):164–170

Breda A, Valadares NF, Norberto de Souza O et al (2008) Protein structure, modelling and applications. In: Gruber A, Durham AM, Huynh C et al (eds) Bioinformatics in tropical disease research: a practical and case-study approach [internet]. National Center for Biotechnology Information, Bethesda

Brown TL, LeMay HE, Bursten BE, Burdge JR (2003) Chemistry the central science, 14th edn. Pearson, Upper Saddle River

Burley SK, Kurisu G, Markley JL, et al (2017) PDB-Dev: a prototype system for depositing integrative/hybrid structural models. Structure 25(9):1317–1318

Burley SK, Berman HM, Bhikadiya C et al (2019) RCSB Protein Data Bank: biological macromolecular structures enabling research and education in fundamental biology, biomedicine, biotechnology and energy. Nucleic Acids Res 47(D1):464–474

Callaway E (2015) The revolution will not be crystallized: a new method sweeps through structural biology. Nature 525(7568):172–174

Carroni M, Saibil HR (2016) Cryo electron microscopy to determine the structure of macromolecular complexes. Methods 95:78–75

Chen Y, Su C, Ke M et al (2011) Biochemical and structural insights into the mechanisms of SARS coronavirus RNA ribose 2'-O-methylation by nsp16/nsp10 protein complex. PLoS Pathog 7 (10):e1002294

Chothia C, Lesk AM (1986) The relation between the divergence of sequence and structure in proteins. EMBO J 5(4):823–836

Cohen M, Potapov V, Schreiber G (2009) Four distances between pairs of amino acids provide a precise description of their interaction. PLoS Comput Biol 5(8):e1000470

Csaba G, Birzele F, Zimmer R (2009) Systematic comparison of SCOP and CATH: a new gold standard for protein structure analysis. BMC Struct Biol 9:23

Czaplewski C, Rodziewicz-Motowidlo S, Liwo A, Ripoll DR, Wawak RJ, Scheraga HA (2000) Molecular simulation study of cooperativity in hydrophobic association. Protein Sci 6:1235–1245

Dayringer HE, Tramontano A, Sprang SR, Fletterick RJ (1986) Interactive program for visualization and modelling of proteins, nucleic acids and small molecules. J Mol Graph 6:82–87

Desmet J, De-Maeyer M, Hazes B, Lasters I (1992) The dead-end elimination theorem and its use in protein side-chain positioning. Nature 356(6369):539–542

Dever TE, Green R (2012) The elongation, termination, and recycling phases of translation in eukaryotes. Cold Spring Harb Perspect Biol 4(7):a013706

Dukka BK (2017) Recent advances in sequence-based protein structure prediction. Brief Bioinform 18(6):1021–1032
Fiser A (2010) Template-based protein structure modeling. Methods Mol Bio (Clifton, N.J.) 673:73–94
Fiser A, Do RK, Sali A (2000) Modeling of loops in protein structures. Protein Sci 9(9):1753–1773
Han KF, Baker D (1995) Recurring local sequence motifs in proteins. J Mol Biol 251(1):176–187
Hansen DF, Kay LE (2011) Determining valine side-chain rotamer conformations in proteins from methyl 13C chemical shifts: application to the 360 kDa half-proteasome. J Am Chem Soc 133 (21):8272–8281
Haywood AM (1997) Transmissible spongiform encephalopathies. N Engl J Med 337 (25):1821–1828
He HT, Gursoy RN, Kupczyk-Subotkowska L, Tian J, Williams T, Siahaan TJ (2006) Synthesis and chemical stability of a disulfide bond in a model cyclic pentapeptide: cyclo(1,4)-Cys-Gly-Phe-Cys-Gly-OH. J Pharm Sci 95(10):2222–2234
Hendlich M, Lackner P, Weitckus S, Floeckner H, Froschauer R, Gottsbacher K, Casar G, Sippl JM (1990) Identification of native protein folds amongst a large number of incorrect models: the calculation of low energy conformations from potentials of mean force. J Mol Biol 216 (1):167–180
Hintze BJ, Lewis SM, Richardson JS, Richardson DC (2016) Molprobity's ultimate rotamer-library distributions for model validation. Proteins 84(9):1177–1189
Hooft RW, Vriend G, Sander C, Abola EE (1996) Errors in protein structures. Nature 381 (6580):272
Hooft RW, Sander C, Vriend G (1997) Objectively judging the quality of a protein structure from a Ramachandran plot. Comput Appl Biosci 13(4):425–430
Horiuchi M, Priola SA, Chabry J, Caughey B (2000) Interactions between heterologous forms of prion protein: binding, inhibition of conversion, and species barriers. Proc Natl Acad Sci U S A 97(11):5836–5841
Hospital A, Goni JR, Orozco M, Gelpí JL (2015) Molecular dynamics simulations: advances and applications. Adv Appl Bioinforma Chem 8:37–47
Ihm Y (2004) A threading approach to protein structure prediction: studies on TNF-like molecules, Rev proteins, and protein kinases. Retrospective Theses and Dissertations 948
Jaroszewski L, Rychlewski L, Zhang B, Godzik A (1998) Fold prediction by a hierarchy of sequence, threading, and modeling methods. Protein Sci 7(6):1431–1440
Johnson RL, Laufer E, Riddle RD, Tabin C (1994) Ectopic expression of sonic hedgehog alters dorsal-ventral patterning of somites. Cell 79(7):1165–1173
Jones DT (1999) GenTHREADER: an efficient and reliable protein fold recognition method for genomic sequences. J Mol Biol 287(4):797–715
Jones DT, Taylor WR, Thornton JM (1992) A new approach to protein fold recognition. Nature 358 (6381):86–89
Jonic S (2016) Cryo-electron microscopy analysis of structurally heterogeneous macromolecular complexes. Comput Struct Biotechnol J 14:385–390
Kendrew JC, Bodo G, Dintzis HM, Parrish RG, Wyckoff H, Phillips DC (1958) A three-dimensional model of the myoglobin molecule obtained by X-ray analysis. Nature 181 (4610):662–666
Krieger E, Koraimann G, Vriend G (2002) Increasing the precision of comparative models with YASARA NOVA—a self-parameterizing force field. Proteins 47(3):393–302
Krieger E, Nabuurs SB, Vriend G (2003) Homology modelling, methods of biochemical analysis. Struct Bioinf 44:509–524
Kumar A, Chordia N (2017) Role of bioinformatics in biotechnology. Res Rev Biosci 12(1):116
Liu H, Elstner M, Kaxiras E, Frauenheim T, Hermans J, Yang W (2001) Quantum mechanics simulation of protein dynamics on long timescale. Proteins 44(4):484–489
Lodish H, Berk A, Zipursky SL et al (2000) Molecular cell biology, 4th edn. W. H. Freeman, New York

Masters CL, Beyreuther K (1998) Alzheimer's disease. Br Med J 316(7129):446–448

Morris AL, MacArthur MW, Hutchinson EG, Thorton JM (1992) Stereochemical quality of protein structure coordinates. Proteins 12(4):345–364

Moult J, Fidelis K, Kryshtafovych A, Schwede T, Tramontano A (2018) Critical assessment of methods of protein structure prediction (CASP)-round XII. Proteins 86(S1):7–15

Neil J, Bulleid Ellgaard L (2011) Multiple ways to make disulfides. Trends Biochem Sci 36 (9):485–492

Novotny J, Rashin AA, Bruccoleri RE (1988) Criteria that discriminate between native proteins and incorrectly folded models. Proteins 4(1):419–430

Orlova EV, Saibil HR (2011) Structural analysis of macromolecular assemblies by electron microscopy. Chem Rev 111(12):7710–7748

Peach RJ, Bajorath J, Brady W, Leytze G, Greene J, Naemura J, Linsley PS (1994) Complementarity determining region 1 (CDR1)- and CDR3-analogous regions in CTLA-4 and CD28 determine the binding to B7-1. J Exp Med 180(6):2049–2058

Pearson WR (1990) Rapid and sensitive sequence comparison with FASTP and FASTA. Methods Enzymol 183:63–98

Peitsch M (1997) Large scale protein modeling and model repository. In: Proceedings. International conference on intelligent systems for molecular biology; ISMB. International conference on intelligent systems for molecular biology, vol 5, p 234

Peitsch M, Schwede T, Guex N (2000) Automated protein modelling—the proteome in 3D. Pharmacogenomics 1(3):257–266

Peterson LX, Kang X, Kihara D (2014) Assessment of protein side-chain conformation prediction methods in different residue environments. Proteins 82(9):1971–1984

Pruisner SB (1996) Molecular biology and pathogenesis of prion diseases. Trends Biochem Sci 21 (12):482–487

Rabl J, Leibundgut M, Ataide SF, Haag A, Ban N (2010) Crystal structure of the eukaryotic 40S ribosomal subunit in complex with initiation factor 1. Science 331(6018):730–736

Rankin NJ, Preiss D, Welsh P, Burgess KEV, Nelson SM, Lawlor DA, Sattar N (2014) The emergence of proton nuclear magnetic resonance metabolomics in the cardiovascular arena as viewed from a clinical perspective. Atherosclerosis 237(1):287–200

Rodriguez R, Chinea G et al (1998) Homology modeling, model, and software evaluation: three related resources. Bioinformatics 14(6):523–528

Sali A, Blundell TL (1993) Comparative protein modelling by satisfaction of spatial restraints. J Mol Biol 234(3):779–715

Sanchez R, Sali A (1997) Advances in comparative protein-structure modeling. Curr Opin Struct Biol 7(2):206–214

Scouras AD, Daggett V (2010) The Dynameomics rotamer library: amino acid side chain conformations and dynamics from comprehensive molecular dynamics simulations in water. Protein Sci 20(2):341–352

Serafini A (1989) Linus Pauling: a man and his science, 1st edn. Paragon House, New York

Serdyuk IN, Zaccai NR, Zaccai J (2007) Methods in molecular biophysics: structure, dynamics, function. Cambridge University Press, New York

Shi J, Blundell TL, Mizuguchi K (2001) FUGUE: sequence-structure homology recognition using environment-specific substitution tables and structure-dependent gap penalties. J Mol Biol 310 (1):243–257

Shortle D, Simons KT, Baker D (1998) Clustering of low-energy conformations near the native structures of small proteins. Proc Natl Acad Sci U S A 95(19):11158–11162

Simons KT, Bonneau R, Ruczinski I, Baker D (1999) Ab initio protein structure prediction of CASP III targets using ROSETTA. Proteins 37(S3):171–176

Sippl MJ (1993) Recognition of errors in three-dimensional structures of proteins. Proteins 17 (4):355–362

Sliwoski G, Kothiwale S, Meiler J, Lowe EW (2014) Computational methods in drug discovery. Pharmacol Rev 66(1):334–395

Snyder DA, Chen Y, Denissova NG et al (2005) Comparisons of NMR spectral quality and success in crystallization demonstrate that NMR and X-ray crystallography are complementary methods for small protein structure determination. J Am Chem Soc 127(47):16505–16511

Tappura K (2001) Influence of rotational energy barriers to the conformational search of protein loops in molecular dynamics and ranking the conformations. Proteins 44(3):167–179

Taylor WR (1986) Identification of protein sequence homology by consensus template alignment. J Mol Biol 188:233–258

Thompson JD, Higgins DG, Gibson TJ (1994) CLUSTAL W: improving the sensitivity of progressive multiple sequence alignment through sequence weighting, position-specific gap penalties and weight matrix choice. Nucleic Acids Res 22(22):4673–4680

Voet D, Vote JJ (1990) Biochemistry. Wiley, New York

Wang HW, Wang JW (2017) How cryo-electron microscopy and X-ray crystallography complement each other. Protein Sci 26(1):32–39

Wohlgemuth I, Brenner S, Beringer M, Rodnina MV (2008) Modulation of the rate of peptidyl transfer on the ribosome by the nature of substrates. J Biol Chem 283(47):32229–32235

Wooley JC, Lin HS (2005) National Research Council (US) committee on frontiers at the interface of computing and biology; catalyzing inquiry at the interface of computing and biology. National Academies Press, Washington

Yang J, Yan R, Roy A, Xu D, Poisson J, Zhang Y (2015) The I-TASSER suite: protein structure and function prediction. Nat Methods 12:7–8

Yang J, Zhang W, He B, Walker SE, Zhang H, Govindarajoo B, Virtanen J, Xue Z, Shen HB, Zhang Y (2016) Template-based protein structure prediction in CASP11 and retrospect of I-TASSER in the last decade. Proteins 84(S1):233–246

Cavity/Binding Site Prediction Approaches and Their Applications

3

Himanshu Avashthi, Ambuj Srivastava, and Dev Bukhsh Singh

Abstract

The binding site of a protein governs its function by allowing binding of small and macromolecules such as nucleic acids, proteins, and other molecules. These binding molecules, also known as ligands, generally form non-covalent bonds and have transient interactions and dissociate after performing a function. The binding sites are unique and have shape complementarity to its ligands to maintain the specificity and affinity. For example, molecules such as hormones, activators, inhibitors, neuro-transmitters, and toxins have specificity in their binding sites. A ligand-binding site entails vast information about its biological function, such as the geometry, physicochemical properties, and electrostatic charge, which in turn allows binding for the highly specific ligand. Various experimental methods such as X-ray crystallography, mass spectrometry, nuclear magnetic resonance, and isothermal titration calorimetry are used to determine the binding site of proteins. For drug discovery, it is inevitable to use high throughput screening of binding sites of proteins, and computational methods give an efficient and cost-effective way of analyzing the same. Several algorithms, tools, and software are available to detect protein cavities computa-

H. Avashthi
Department of Computational Biology and Bioinformatics, Sam Higginbottom University of Agriculture, Technology and Sciences, Prayagraj, Uttar Pradesh, India

College of Biotechnology, Uttar Pradesh Pandit Deen Dayal Upadhyaya Pashu Chikitsa Vigyan Vishwavidyalaya Evam Go-Anusandhan Sansthan, Mathura, Uttar Pradesh, India

A. Srivastava
Department of Biotechnology, Indian Institute of Technology, Madras, Chennai, Tamil Nadu, India

D. B. Singh (✉)
Department of Biotechnology, Institute of Biosciences and Biotechnology, Chhatrapati Shahu Ji Maharaj University, Kanpur, Uttar Pradesh, India

D. B. Singh (ed.), *Computer-Aided Drug Design*,
https://doi.org/10.1007/978-981-15-6815-2_3

tionally. The study of binding sites is relevant to various fields of research, including computer-aided drug design, agrochemical design, cancer mechanisms, drug formulation, and physiological regulation.

Keywords

Ligand · Binding site · Cavity · Protein–ligand interaction · Receptor · Binding affinity

3.1 Introduction

Knowledge of the ligand-binding site gives important information about the nature and function of a protein. It is important for performing molecular docking and also for computational drug design and screening (Bradford and Westhead 2004). The information about binding sites improves the prediction of protein–ligand and protein–protein interactions. These interactions can be predicted through a docking approach. Ligands usually bind at specific sites of target proteins, and these sites are known as pockets/cavities (Schmitt et al. 2002).

In general, enzymes and hormones are protein molecules and have a particular shape, which speeds up biochemical reactions within the body and ultimately behaving as a catalyst (Gropper and Smith 2012). The activity of an enzyme depends upon several variables, such as temperature, pH, and concentration. Enzymatic reactions are carried out with the binding of the substrate to the active site of the enzyme (Kirby 1996). The active site is the specific region of an enzyme where the substrate binds and causes changes in the reaction that lead to the formation of the product (Klibanov 2001). The active site has a unique geometric shape which is complementary to the geometric shape of another molecule called substrate (Lehn 1988). The action of an enzyme with a substrate is based on lock and key and induced fit theory. Lock and key theory was postulated in *1894* by Emil Fischer. In this theory, the enzyme acts as a lock, and the substrate acts as a key. Only the exact sized substrate (key) fits into the hole of the key (active site) of the lock (enzyme). Substrate molecule with smaller, larger, or incorrect shape and size does not fit into the active site of the lock (Koshland 1995). Only the exact shape key opens a particular lock, as illustrated in Fig. 3.1.

On the other hand, the induced fit model was proposed by Daniel Koshland in 1958 (Koshland 1995). In the induced fit model, initially, active site and substrate have no matches for each other. The final shape of the enzyme is determined after binding of a ligand with its active site. After binding, many structural rearrangement and conformational changes take place in the enzyme structure.

In the era of industrialization, peoples are getting susceptible to various infections and diseases and excessive use of antibiotics making micro-organisms resistant to these molecules (Walker 1996). To combat and keep pace with these organisms, we have to find new ways or new drugs to limit their propagation and survival. However, a drug takes 15–20 years to come up to society with a very complex

Fig. 3.1 Mechanism of lock and key for the formation of enzyme–substrate complex

process of screening and clinical trials (Bleicher et al. 2003). The computational techniques are known to help in pacing the process of drug discovery. The branch of bioinformatics, which helps in designing the drug using a computational approach, is known as computer-aided drug design (CADD) (Schneider and Fechner 2005). The first step of drug design is target identification and selection. When we confirmed that the target is protein, then we identify the binding site/cavity in that protein (Bleicher et al. 2003).

Experimental methods such as X-ray crystallography, NMR, electron microscopy, and binding essays help determine the binding sites; however, these techniques are time-taking and cost-intensive. The growth of experimentally determined structure in protein data bank (PDB) allows using the knowledge to develop tools for the binding site prediction. The coordinate information of proteins available in PDB gives us information about the binding site and its surrounding residues. These information can be exploited further to understand the binding sites of unknown proteins. There are many tools available to identify these binding sites in a receptor molecule. The binding sites can also be predicted by estimating the site prone to make hydrogen bonds and electrostatic interactions (Singh and Dwivedi 2016). The hydrophilic and charged residues generally help in making these interactions and are also important for post-translational modifications including phosphorylation. Phosphorylation mainly happens in residues carrying hydroxyl group, i.e. serine, threonine, and tyrosine. These phosphorylation sites along with charged patches are often involved in binding, hence are also used for the prediction.

After recognizing the cavity site, we take ligand and receptor molecules through a process that we know by the name of docking. In the docking process, the ligand molecule forms a non-covalent bond with some amino acids of the protein molecule. Remember that these cavity sites are specific to each ligand. Every single ligand cannot bind on every cavity site. Specificity and affinity are two such features that determine the strength of the chemical bond and the nature of ligand that will bind (Ladbury 1996). To find the cure for a disease, we must first identify those target molecules and identify the cavity/binding site present in the target. So all together, recognizing the cavity or binding site in the protein molecule is a complex step. Since, as long as we do not know about the cavity present in the target, we cannot design any drugs. Thus everything depends on the target's cavity site. So, in this chapter, we will talk about what are the cavity or binding sites, their roles, methods, and approaches to identify them, etc.

3.2 Target Molecule

Drug target identification is one of the important steps in drug designing. A ligand or drug binds to a specific site on the 3D structure of a protein or target to generate the therapeutic response. In computer-aided drug designing, the 3D structure of a drug target is retrieved from the PDB database or can be modeled using sequence information, if 3D structural details not available in PDB. Protein, nucleic acid, carbohydrate, or lipids can serve as a drug target but in most cases, drug targets are protein. Exact and accurate prediction of the binding site in a drug target is very essential and important to guide the process of drug discovery. A receptor is a bio-molecule (DNA or protein) that receives chemical signals from ligand molecules. A ligand molecule binds to a protein or receptor to produce a physiological response. Receptors are found on the surface of target cells, which interact with ligands. Sometimes, binding of a ligand to a receptor does not generate an appropriate physiological response, due to only the wrong selection of ligand molecule (Keiser et al. 2009). A receptor is a flexible molecule, and as a result, some structural and conformational changes occur due to the binding of a substrate on its active site. Computational tools mostly consider the flexibility of ligand or substrate interacting with the binding site of a target molecule but do not consider the receptor flexibility. There are limited docking tools that give a certain extent of flexibility to the receptor active site along with ligand.

3.3 Binding Site and Active Site

The active site is the specific region of a target enzyme formed by the composition and 3D arrangement of certain amino acids, which is occupied by a specific substrate to catalyze a chemical reaction. The active sites are present in binding regions of a substrate so they can also be called binding sites. These binding sites mostly have a pocket to provide the base to a ligand molecule to bind, and these pockets are also known as cavity sites. Note that an active site is always a binding site while all binding sites do not perform catalysis, hence cannot always be an active site. For active site prediction, the size of different cavities in a protein is measured, and then in most cases, cavity with the largest area and volume is considered to be used as a binding site for a ligand or substrate. There are many approaches for validation of binding site, which utilizes the information of known cavity composition in the functionally and evolutionary related proteins. It is a structural part of the protein that determines whether the protein is functional or non-functional. The active site consists of a few numbers of residues that form temporary bonds with the substrate-binding site. It is also referred to as the binding site. The site at which the catalysis reaction takes place is known as the catalytic site (Mattos and Ringe 1996; Tiwari et al. 2016).

3.4 Ligand Molecule

It is a small chemical molecule, ion that forms a complex with a bio-molecule (DNA or protein) to serve a biological purpose. In protein–ligand interaction, ligand produces a signal by binding to a site on a receptor protein. Typically binding results change the conformation of the target molecule (Hansch and Klein 1986). There are many natural ligands present in the human body. Plants are also able to synthesize many natural small chemical molecules, which show interaction with different natural targets or receptors. Detailed physicochemical and biological property of many ligand molecules is available in chemical compound databases, which can be utilized for drug designing. Accurate and precise information about the ligand-binding site is required to understand the mechanism of binding and other dynamic perturbations in the target when a ligand binds to it.

3.5 Binding Affinity

The strength of the binding interaction between a single bio-molecule (protein or DNA) to its ligand or binding partner is known as binding affinity. The ligand can be any drug or inhibitor molecule (Aqvist et al. 1994). In CADD, the binding affinity of a ligand with a molecular drug target is predicted using docking tools that utilize energy scoring function for calculation of binding energies of complex and binding affinity. Different docking tools use a different algorithm, scoring function, and parameters for calculation of binding energy of protein–ligand complex, as a result, binding energy calculated using different docking tools is not the same. Binding affinity can be used as a parameter to screen the potential compounds, which can be used for further structural optimization or in vitro testing. In a real system, binding interaction of a ligand not only depends on the target protein but also on the other factors and environments such as temperature, solvent, pH, ions, and presence of cofactor or other molecules.

3.6 Chemical Specificity

The ability of a protein's binding site to bind a specific ligand depends upon the complementarity of both molecules. That is why few numbers of ligands can able to bind with a single protein. It depicts the binding strength between a protein and ligand (Chargaff 1950). Specificity plays a very important role in the recognition and binding of a ligand with a target molecule. A small structural change or mutation in the binding site of a protein can result in the loss of binding because, in mutant protein, the ligand is not able to recognize the 3D shape or spatial arrangement of amino acids required for its binding. In CADD, if a target protein has undergone mutation, then a new potential drug molecule is designed, keeping in mind the new shape and spatial arrangement of amino acids in the binding site of a mutant or resistant protein.

3.7 Binding Site and Molecular Interactions

The binding site provides a base for the interaction of two molecules and describes the ability of the receptor to form bonds with other substances. Based on the bound molecule, the binding site can be protein–protein, protein–nucleic acid, protein–carbohydrate, protein–lipid, and protein–small molecule binding sites. The binding of the drug molecule to plasma proteins (albumin, lipoproteins, and globulins) is a major determinant of drug distribution (Macalino et al. 2015).

3.7.1 Protein–Drug Interactions

DNA and protein are molecules that show interaction with other small molecules such as substrate, drug, or other ligands. Proteins are an important molecule in all living cells and play an essential role in various cellular processes in the form of enzyme, hormone, and receptor. Each protein performs a specific function that is governed by its 3D structure (Alberts et al. 1998). Protein–ligand interactions are vital for all biological processes that occur in living organisms. The function of a protein depends upon the specific sites that are designed to bind with a specific ligand molecule. Ligand-binding interactions can alter the protein conformations and its function. To perform its function properly, binding of a protein with other molecules should be very specific. A drug is a small organic molecule, which binds to the receptor and forms a protein–drug complex and controls the function of biological receptors. Binding can be of two types: intracellular or extracellular (Silverman and Holladay 2014). Based on the drug binding mechanism, drug binding may be of reversible or an irreversible type.

3.7.1.1 Reversible Binding

In reversible binding, usually, the drug binds the proteins with weaker chemical bonds such as hydrogen bonds, hydrophobic bonds, ionic bonds, and van der Waals interactions. The binding of drugs to plasma protein is a reversible process.

3.7.1.2 Irreversible Binding

In the case of irreversible binding, a drug or inhibitor permanently binds with the binding site of the drug target. Irreversible binding of drugs rarely takes place. As a result of covalent bonding or strong force of interaction between drug and target protein, the event of carcinogenicity or cellular toxicity takes place.

3.7.1.3 Factors Affecting Protein–Drug Binding

(a) Drug-related factors: It includes physicochemical characteristics of the drug, concentration of drug in the body, and affinity of a drug.
(b) Protein/tissue-related factors: It includes physicochemical characteristics of protein or drug and concentration of protein or drug.
(c) Drug interactions: It contains allosteric changes in a protein molecule, competition between drugs to occupy the binding site, and competition between drug and biological components.
(d) Patient-related factors: It includes the age of the patient, inter-subject variabilities such as due to genetics, environmental factors, and disease states. Altogether, more protein binding disturbs the absorption and also decreases the distribution and metabolism of drugs (Nayal and Honig 2006).

3.7.1.4 Role of Water Molecules

In the last 10–15 years, the significance of water molecules in drug design and protein structures has become of extensive interest. Traditionally, water molecules play two crucial roles in ligand binding (de Beer et al. 2010). Water molecules stabilize a protein–ligand complex by contributing hydrogen bond interaction between a ligand and a protein. The second role is that water can be displaced by ligands on binding with the target protein. The role of the water molecule in binding interaction of a ligand with the active site of the target protein can be studied using the molecular dynamics simulation. Slight changes in water-based hydrogen bonding networks affect ligand–protein interaction energies and show the effect of solvation or water molecule on the binding. Water molecules also determine the binding or rejection of ligand to the binding site of protein (Sousa et al. 2006). Water molecules can mediate to direct interactions or may cause an effect of electrostatic screening.

3.7.2 Drug–Nucleic Acid Interactions

Nucleic acids are the carrier of genetic information, and hence they are an important molecule for disease prevention. Nucleic acids are targeted for various diseases, including various types of cancer (Sheng et al. 2013). DNA, the carrier of genetic information in humans, is mutated in various diseases, which often result in gene expression alteration. The structures of DNA can be used for designing small molecules to regain the gene expression pattern. The small molecules which bind to DNA can be categorized into two major classes: (1) covalent binder and (2) non-covalent binder. The non-covalent binders can further be classified into major groove binders, minor groove binders, and intercalators (Boer et al. 2009). On the other hand, RNA itself regulates various activities from catalysis to gene expression, which makes RNA a suitable target for binding. Since the structure of RNA is highly variable, designing a small molecule against them is a challenging task. The advancement in technology and growth in the structures of protein–nucleic acid complexes help the computational tools to predict the binding of small molecules with nucleic acids.

3.7.3 Protein–Protein Interactions

Among the following, the protein–protein binding site is most studied and is involved in almost all essential function including replication, transcription, and translation. Protein–protein complexes can be categorized based on the type of subunits, the strength of interactions, and the time of interactions (Jones and Thornton 1996). Based on subunit types, protein–protein complexes can be hetero-oligomer or homo-oligomer. Hetero-oligomers are having different chains, whereas homo-oligomers have the same chains. Based on the strength of interactions, the protein–protein complexes are categorized into weak and strong binding complexes. Strong binding complexes have better shape and chemical complementarity than weak binding complexes, which could be because of various reasons such as the presence of charged residues at the interfaces, presence of interaction coordinating metal ions, and evolutionary selection of the shape of the protein molecule. Based on the time-duration of interactions, the protein–protein complexes can be categorized as transient and permanent complexes (Steed et al. 2007). Transient complex interacts for a short period, whereas once permanent complexes are formed, they cannot be separated into their monomeric forms. Transient complexes are often involved in signaling and regulation and show cascading effect while permanent complexes are important for catalysis, transport, structure protein formation, etc (Singh and Tripathi 2020).

An accurate analysis of protein–protein interaction using docking approaches is a complex problem due to flexibility and conformational space related issues of macromolecules (Gabb et al. 1997). In protein–protein docking, the complexity of considering the flexibility of two macromolecules is challenging. To establish a relationship between two proteins, a few number of search algorithms are available (Table 3.1).

3.7.4 Interaction of Protein with Nucleic Acid, Lipid, and Carbohydrate

Protein–nucleic acid interactions are crucial for various biological processes such as replication, transcription, translation, DNA repair, RNA processing, splicing, and DNA packing (Von Hippel et al. 1984). Based on the flexibility of interacting molecules, the interactions can be categorized into direct and indirect interactions and also known as direct and indirect readouts, respectively. In a direct readout mechanism, there is no flexibility in the molecules, where indirect readout involves conformational change before or during binding.

In proteins, various metal ions are present, which helps in some favorable interaction or regulates the binding by activating or inhibiting proteins. Based on their function, these metal ions or small molecules can be classified into agonists and

Table 3.1 Different types of protein–protein interaction prediction approaches

Search Algorithm	Scoring Parameters	Principle	References
DOT	Global rigid search: Fast Fourier transform (FFT)	Shape complementarity, electrostatics, and VDW	Norel et al. (1994)
GRAMM-X	Global rigid search: FFT	Shape complementarity and Lennard-Jones potential	Tovchigrechko and Vakser (2006)
HADDOCK	Global rigid search	Electrostatic, VDW, and desolvation energy	Dominguez et al. (2003)
HEX	Global rigid search	Shape complementarity	Pagadala et al. (2017)
ICM	Global rigid search: Monte Carlo	Empirical scoring function	Huang et al. (2010)
MolFit	Global rigid search	Shape complementarity	Redington (1992)
PatchDock	Global rigid search	Shape complementarity	Schneidman-Duhovny et al. (2005)
M-ZDOCK	Global rigid search	Shape complementarity	Pierce et al. (2005)
3D-dock suite	Global rigid search: FFT	Shape complementarity and electrostatics	Smith and Sternberg (2003)
3D garden	Global rigid search in ensemble	Shape complementarity and Lennard-Jones potential	Lesk and Sternberg (2008)

antagonists. An agonist molecule is known to an activator, whereas an antagonist molecule inhibits a protein. This activation and inhibition process is chiefly explored in the drug discovery process. Drug discovery is a knowledge-based approach, in which first we understand the molecular processes involved in a disease. Subsequently, we choose our protein target, which is directly involved in the disease, and then we activate or inhibit the target molecule, with the help of a small molecule, depending upon the function of the target protein.

Protein–carbohydrate and protein–lipid interactions are also involved in various important biological processes such as the immune system, digestive system, carbohydrate transport, membranes, and in anabolism and catabolism of carbohydrates and fats (Vyas 1991). Despite having a lot of functional applications, these interactions are not much explored because of the problems in getting the structure of these proteins solved by crystallographic and spectroscopy techniques.

3.8 Binding Site Prediction

Based on the role of protein, the binding site can be categorized into active and regulatory sites. The active site of a protein binds to the ligand molecules and performs the enzymatic activity, whereas the regulatory site binds to the regulator molecule, which either activates or inhibits the process of binding at the active site of the proteins (Bradford and Westhead 2004). Binding sites determine the strength and type of interactions between protein and ligand molecules. To predict protein

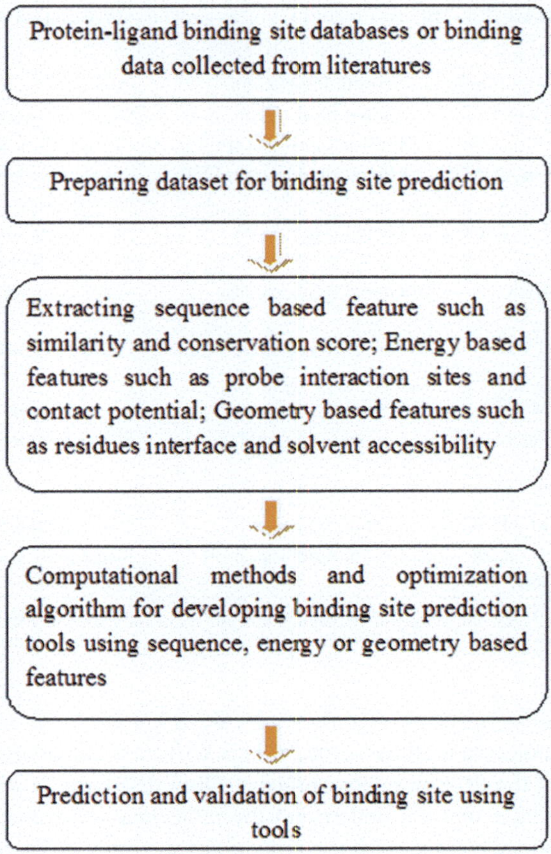

Fig. 3.2 Development of binding site prediction tools

binding sites, mostly machine learning-based approaches are used. The machine learning models are developed using structural, sequence, or evolutionary information (Liang et al. 2006). Protein complexes modeling and docking software are a few examples of using evolutionary and structural information. The evolutionary algorithm of binding site prediction utilizes the information of multiple sequence alignment and assumes that binding site residues are conserved in evolutionarily related proteins, whereas the energy-based methods calculate the interaction potential between the protein and the ligand, and utilizes the 3D structural information of proteins and protein–ligand complex from PDB (Fig. 3.2). Binding site prediction methods are based on evolutionary algorithms, energy-based algorithms, and geometric algorithms.

3.8.1 Evolutionary Algorithms/Sequence-Based Predictions

With the advancement in sequencing technique and a growing number of protein sequences, it is in demand to predict the binding site of proteins using only sequence information. Evolutionary methods work on an assumption that sequence similarity leads to structural similarity. Hence, the binding site in a protein can be predicted by obtaining similar proteins with known binding sites. The multiple sequence alignment, conservation score, and substitution matrix can be used to obtain the sequence similarity.

Sequence-based binding site prediction can find out a ligand-binding motif in non-similar proteins (Ahmad et al. 2004). Different sequence-based approaches were developed, including ConSeq, conservation score, miner, and so on (Hwang et al. 2007). In another sequence-based approach, multiple sequence alignment (MSA) is constructed from homologous sequences of a target protein, and conserved residues among all the sites in the MSA are determined. Sequence-based methods use majorly two kinds of properties for binding site prediction.

3.8.1.1 Single Residue Based Approach

Various physiochemical properties such as hydrophobicity, side-chain pKa, solubility, solvent accessibility, etc. are associated with binding and non-binding residues and are used to develop machine learning-based model. Few models use only one binding partner structure to predict binding sites, whereas others use both partners information.

3.8.1.2 Window Based Approach

In a window based approach, properties of neighboring residue are considered important for the predictions (Capra and Singh 2007). Although just by using the features from the sequence, it is difficult to predict binding residues. Various methods use evolutionary information to increase the accuracy of their prediction model. To associate the evolutionary information, first homologous sequences are obtained by performing alignment of all the known binding proteins with the query sequence, and then the scoring matrix is developed for binding and non-binding residues in homologous sequences.

3.8.2 Energy-Based Algorithms

This method considers the interaction potential between protein and ligand. Here, a simple probe is used to determine the different interaction potential on protein, and favorable binding regions are mapped based on the energy. Energetically favorable probe interaction sites are clustered using their spatial proximity, and the total energy of interaction for probes within each cluster is calculated (Laurie and Jackson 2005). Q-SiteFinder is an energy-based method which determines clusters of energetically favorable methyl probe to locate the binding site. Energetically favorable sites indicate the location on a protein where a ligand could interact and bind. Some scoring function of binding site prediction uses three different probe types to locate the hydrophobic site, hydrogen bond donor, and hydrogen bond acceptor and considers the probe site related to favorable interaction energy (Ruppert et al. 1997).

3.8.3 Geometry-Based Algorithm/Structure-Based Predictions

If the structure of ligand and receptor is available, we can accurately estimate the interface residue by using distance and solvent accessibility based calculations. In the distance-based method, we can choose a cut-off and see if distances between the residues of the receptor and ligand molecules are less than the cut-off value. In the solvent accessibility method, we can calculate the solvent accessibility of complex and separated ligand and receptor molecules and check if there is a change in solvent accessibility in free and complex form, then the residue should involve in binding. When the structure of receptor and ligand is not solved as a complex, we can predict binding site from structure information such as the shape of the protein, hydrophobic patches on the surface, number of charged residues on the surface, compactness of a protein, etc. (Gabb et al. 1997).

3.9 Approaches

Binding site prediction of protein–protein, protein–nucleic acids, and protein–small molecule is targeted by several research groups, and significant accuracy is achieved (Laurie and Jackson 2006). Various methods are developed for prediction of the binding site, which can be broadly categorized as follows.

3.9.1 Statistical Approach

Statistical methods used various features to distinguish between binding and non-binding residue using a statistical approach. This approach is fast and may achieve significant accuracy. However, it needs detailed knowledge about the factor responsible for the occurrence of binding and non-binding sites.

3.9.2 Machine Learning

Machine learning approaches are highly accurate and train themselves for the prediction if the well-labeled dataset is provided. Here various features can be combined for the prediction method development.

3.9.3 Meta-Predictors

Finally, meta-predictors use multiple predictors for predicting the binding and non-binding residues and derive consensus results and try to associate confidence with the prediction. Since meta-predictors are the combination of various approaches, it works well for various prediction problems.

3.10 Prediction Tools and Servers

Several computational tools for pocket and cavity prediction have been developed, which are based on evolutionary algorithms, energy-based algorithms, and geometric algorithms. Some other approaches have also been developed, which combines the two or more methods to improve the accuracy of cavity or pocket prediction (Zhang et al. 2011). Binding site prediction tools use different parameters and methods for prediction and have different accuracy. Some docking tools also provide the facility of the cavity and binding site prediction, before proceeding for docking. In the recent decade, various ligand-binding site prediction and analysis tools have been developed.

The CASTp is a cavity/pocket prediction tool that is based on 3D α-shapes methods. This method generates two α-shape envelopes in a protein, and space between these two envelopes measures the size of the pocket (Liang et al. 1998). A pocket is a concave protein surface region accessible to the outer solvent, whereas a cavity is an inner void inside the protein surface that is not accessible to the outer solvent (Fig. 3.3). For cavity prediction, the probe radius is set to 1.4 Å, i.e. the radius of the water molecule.

To understand protein interfaces and interactions, docking, the accurate prediction of pockets and cavities in a protein is essential. PoCavEDT is a geometric technique used to predict binding pockets and cavities in proteins using Euclidean distance transform (Daberdaku 2019). In this approach, probe spheres are used to identify pocket regions between two solvent-excluded surfaces, and the probe size used depends on the size of the binding ligand. This is a simple geometrical method used to predict the ligand-binding site. The prediction accuracy of this method was evaluated by applying it to a set of protein–ligand complexes and their corresponding unbound protein structures. The comparison of ProBiS, a binding site detection tool, was made with some other tools such as DaliLite, MolLoc, and MultiBind by calculating the RMSD between similar binding site residues of a protein by structural superimposition (Konc et al. 2015). After comparison, the lowest average RMSD value was obtained for ProBiS. The tools and servers used for the prediction of the binding site are listed in Table 3.2.

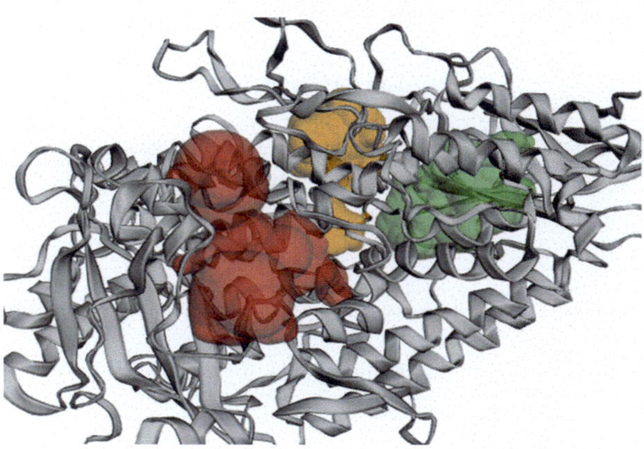

Fig. 3.3 Showing pockets in three different colors, predicted by CASTp. Red color shows pocket 1, orange color pocket 2, and green color pocket 3

Table 3.2 Tools and servers used for the binding sites prediction

Tool/Servers	Description	References
CASTp	Used to locate and measure concave surface regions such as cavity or pockets on 3D structures of proteins	Tian et al. (2018)
LigASite	Collection of biologically relevant binding sites in protein structures	Dessailly et al. (2008)
PDBeMotif	Analysis of binding sites in a protein and conserved structural features within the same species or across different species	Inhester et al. (2017)
fPOP	fPOP refers to footprinting pockets of proteins. It is a collection of spatial patterns of protein binding sites identified by shape analysis	Tseng et al. (2010)
metaPocket	It is a type of meta-predictor used to predict ligand-binding sites or pockets on the protein surface	Huang (2009)
PocketQuery	Web service for interactively exploring protein–protein interactions	Koes and Camacho (2012)
IBIS	It observes experimentally determined biological structures such as protein–protein, protein–nucleic acids, protein–small molecule, and protein–ion interactions	Shoemaker et al. (2012)
KBDOCK	Identify spatially clusters protein binding sites for knowledge-based protein docking	Ghoorah et al. (2014)
Pocketome	Conformational ensembles of all druggable binding sites	An et al. (2005)
sc-PDB	It identifies binding sites suitable for the docking	Meslamani et al. (2011)
The FunFOLD	This tool accurately predicts ligand-binding residues from protein sequences	Roche et al. (2011)
ProBiS	It detects the structurally similar protein binding sites through local structural alignment	Konc and Janezic (2010)
DEPTH	It calculates the depth of a residue from the protein surface	Tan et al. (2013)
FINDSITE	It is a threading-based binding site prediction tool	Skolnick and Brylinski (2009)
PocketDepth	It is a geometry-based method and uses depth based clustering to predict the ligand-binding sites	Kalidas and Chandra (2008)
GHECOM 1.0	This is a program for finding multi-scale pockets on the protein surface	Payandeh et al. (2018)
Pocket-finder	It is based on the Ligsite algorithm	Hetényi and van der Spoel (2011)
Screen2	It is a tool for identifying protein cavities	Xie and Hwang (2015)
ConCavity	This tool identifies functional sites of proteins	
MultiBind and MAPPIS	It is a web server for multiple alignments of protein 3D binding sites and their interactions	Shulman-Peleg et al. (2008)
MolAxis	It reads files in the standard PDB format.	Yaffe et al. (2008)
Fpocket	It is a very fast and open source protein pocket detection tool	Le Guilloux et al. (2009)
SuMo	It is a tool for finding ligand-binding sites	Jambon et al. (2005)

(continued)

Table 3.2 (continued)

Tool/Servers	Description	References
CAVER	It is a tool for the accurate and fast prediction of tunnels and channels in protein or nucleic acid. Tunnels are void buried in a protein core, whereas channels are exposed to the surrounding solvent	Petrek et al. (2006)
SiteHound	This program identifies protein regions that are likely to interact with ligands	Hernandez et al. (2009)
SURFNET	This program generates surfaces and void regions	Laskowski (1995)
MSPocket	It is a tool for the detection and graphical representation of protein surface pockets	Zhu and Pisabarro (2010)
Phosfinder	It is a method for the prediction of phosphate-binding sites in the 3D structure of a protein	Parca et al. (2011)
VOIDOO	It is a program for cavity prediction in macromolecular structures	Kleywegt and Jones (1994)
PocketPicker	It is a tool for the prediction and evaluation of surface binding pockets	Weisel et al. (2007)
McVol	This tool is based on the Monte Carlo algorithm. It can recognize internal cavities as well as surface clefts	Till and Ullmann (2010)

Various other web servers and databases related to binding pockets are available to analyze the shape, size, structural properties, and descriptors in protein–protein, protein–RNA, protein–DNA, and protein–ligand complexes. For example, positively or negatively charged patches can be studied to analyze the binding of charged molecules such as protein and nucleic acids. Similarly split pocket is a tool to identify the charged patch and pockets including the residues present, accessible surface area, pocket volume, and residues present at the pocket mouth (i.e. residues present on the edges of the pocket). To analyze membrane proteins, ChExVis is a channel visualization platform where users can calculate channels and active site information such as pores and transmembrane pores, channel length, and residues location in the proteins (Masood et al. 2015). In addition, PockDrug is a pocket drugability prediction, i.e. finding the probability of a pocket present in a protein to bind to a drug molecule (Hussein et al. 2015). These servers are used in understanding the drugability of various compounds such as diarylamine derived from anthranilic acid and are tested for blocking the ZIKA virus infection by binding with RNA polymerase. Moreover, the drugable sites of essential phosphatase proteins of *Aspergillus fumigatus* were also identified using PockDrug for the prevention of fungal infections.

3.11 Validation of Binding Site

Binding residues predictions are validated by comparing the predicted output with the experimentally labeled dataset. A residue is true positive (TP), if it is predicted as binding when it is actually binding, true negative (TN) if predicted as non-binding when it is a non-binding, false negative (FN) if predicted as non-binding but it is binding, and false positive (FP), if predicted as binding but actually is non-binding (Wang et al. 2002). Using these variables, we make several measures to further cross-validate the residues. Prediction of binding sites can be validated by sensitivity, specificity, and accuracy measure which is given by

$$\text{Sensitivity} = \text{TP}/\text{TP} + \text{FN}$$
$$\text{Specificity} = \text{TN}/\text{TN} + \text{FP}$$
$$\text{Accuracy} = \text{TP} + \text{TN}/(\text{TP} + \text{TN} + \text{FP} + \text{FN})$$

Often a balance between sensitivity and specificity is assumed to be a good predictor; therefore, balanced accuracy is also calculated by estimating the average between specificity and sensitivity.

$$\text{Balanced accuracy} = (\text{sensitivity} + \text{specificity})/2$$
$$\text{Precision} = \text{TP}/(\text{TP} + \text{FP})$$
$$\text{Recall} = \text{TP}/(\text{TP} + \text{FN})$$

Finally, the area under the receiver operator characteristic (ROC) curve is also used to define if there is a fine balance between true positive rate and false positive rate for a prediction model. Precision and recall calculation is often used in the validation of prediction models. There is always a trade-off between precision and recall, and a model having a balance between the two is considered as a good model.

Geometry-based methods provide an accurate prediction of the binding site in a protein. This approach can be further improved by combining the other information, such as the evolutionary conservation of residues in the known binding site. Geometry-based methods return the pockets or cavities based on the size, area, or volume, but the largest cavity is not always associated with the binding site. Later on, a support vector machine-based model has been developed, and their prediction results were found better than the methods based on pure geometry or evolutionary conservation (Wang et al. 2013).

3.12 Role of the Binding Site in Drug Designing

The binding site plays a key role in computer-aided drug design. To predict binding sites, we should know the 3D structure of the drug target (Henrich et al. 2010). So the identification and selection of drug targets is a very crucial and critical step. In this process, we identify a drug target that is present in our body in the form of protein and nucleic acid (Hopkins et al. 2006). A protein may be used as a drug target if it is intrinsically associated with a particular disease mechanism. For structure-based drug design, a disease-related and functional protein can be used as a drug target, and it should have a binding site for a small molecule. Binding site details of the drug targets should be available to proceed for the drug designing. In most of the cases, we select protein molecule as a drug target, but sometimes we prefer nucleic acids. Specific recognition of DNA sequence is achieved by the small molecule, via the combination of hydrogen bond donor/acceptor site available at the minor groove or major groove. These drugs are called minor groove binder, major groove binder, and an intercalator.

Protein–ligand or protein–small molecule interactions are crucial for almost all the biological processes. The interactions of small molecules with proteins are useful for providing stability, which helps in catalyzing reactions (Burgoyne and Jackson 2006). Various small molecules are natural inhibitors that provide a feedback loop to pathways, which help in maintaining a balance. Exploiting this process, we design small molecules to increase or decrease the activity of a protein for balancing the disturbance in a pathway.

3.13 Recent Advances and Future Perspective

Protein–protein interactions are widely studied, and as a result, several prediction servers are available to predict binding site, affinity, and thermodynamics features of such interactions. The absence of data in protein–nucleic acid interactions leads to less availability of servers to predict affinity and other thermodynamics properties of these complexes. Also, it should be noted that we mainly talk about well-structured regions of proteins while discussing protein interactions. On the other hand, intrinsically disordered regions of proteins that lack well-defined 3D structures in solution are also observed to be important for protein–protein and protein–nucleic acid interactions. Intrinsically disordered regions are also found to be present in hub proteins, which are in the center of the biological networks. Various databases such as DisProt, D2P2, DisBind, and MobiDB provide information about intrinsically disordered proteins. Few servers such as ANCHOR, DIBS, and IDPpi are available for protein–protein interactions in disordered regions, whereas, for nucleic acid binding, very few servers are available such as DisoRDPbind. The availability of structures and experimentally validated disordered regions will provide more insights into these interactions and will also give scope for the development of servers and databases.

3.14 Conclusion

Protein interactions are crucial to every living organism, and a single amino acid change in the interaction sites can lead to devastating diseases. Hence, interaction sites should be studied carefully, and computational methods give a reliable and cost-effective approach to predict such interactions. Here we have presented different tools and servers to identify the ligand-binding sites, pocket volume, protein cavity, phosphorylation sites, etc. Although each method has its advantages, the growth of the data in every field allows us to improve the performance of these methods. Finally, these methods are developed for different purposes with different datasets and approaches; hence, they should be chosen carefully for a study.

Competing Interest The authors declare that they have no competing interests.

References

Ahmad S, Gromiha MM, Sarai A (2004) Analysis and prediction of DNA-binding proteins and their binding residues based on composition, sequence and structural information. Bioinformatics 20 (4):477–486

Alberts B, Bray D, Johnson A, Lewis N, Raff M, Roberts K, Walter P (1998) Essential cell biology: an introduction to the molecular biology of the cell. Garland Publishing, New York

An J, Totrov M, Abagyan R (2005) Pocketome via comprehensive identification and classification of ligand binding envelopes. Mol Cell Proteomics 4(6):752–761

Aqvist J, Medina C, Samuelsson JE (1994) A new method for predicting binding affinity in computer-aided drug design. Protein Eng Des Sel 7(3):385–391

Bleicher KH, Böhm HJ, Müller K, Alanine AI (2003) Hit and lead generation: beyond high-throughput screening. Nat Rev Drug Discov 2(5):369–378

Boer DR, Canals A, Coll M (2009) DNA-binding drugs caught in action: the latest 3D pictures of drug-DNA complexes. Dalton Trans 3:399–414

Bradford JR, Westhead DR (2004) Improved prediction of protein–protein binding sites using a support vector machines approach. Bioinformatics 21(8):1487–1494

Burgoyne NJ, Jackson RM (2006) Predicting protein interaction sites: binding hot-spots in protein–protein and protein–ligand interfaces. Bioinformatics 22(11):1335–1342

Capra JA, Singh M (2007) Predicting functionally important residues from sequence conservation. Bioinformatics 23(15):1875–1882

Chargaff E (1950) Chemical specificity of nucleic acids and mechanism of their enzymatic degradation. Experientia 6(6):201–209

Daberdaku S (2019) Identification of protein pockets and cavities by Euclidean Distance Transform. Peer J Preprints 7:e27314v2

de Beer SB, Vermeulen NP, Oostenbrink C (2010) The role of water molecules in computational drug design. Curr Top Med Chem 10(1):55–66

Dessailly BH, Lensink MF, Orengo CA, Wodak SJ (2008) LigASite—a database of biologically relevant binding sites in proteins with known apo-structures. Nucleic Acids Res 36(Database): D667–D673

Dominguez C, Boelens R, Bonvin AM (2003) HADDOCK: a protein-protein docking approach based on biochemical or biophysical information. J Am Chem Soc 125(7):1731–1737

Gabb HA, Jackson RM, Sternberg MJ (1997) Modelling protein docking using shape complementarity, electrostatics and biochemical information. J Mol Biol 272(1):106–120

Ghoorah AW, Devignes MD, Smaïl-Tabbone M, Ritchie DW (2014) KBDOCK 2013: a spatial classification of 3D protein domain family interactions. Nucleic Acids Res 42(Database):D389–D395

Gropper SS, Smith JL (2012) Advanced nutrition and human metabolism, 6th edn. Wadsworth Publishing, Belmont

Hansch C, Klein TE (1986) Molecular graphics and QSAR in the study of enzyme-ligand interactions. On the definition of bioreceptors. Acc Chem Res 19(12):392–400

Henrich S, Salo-Ahen OM, Huang B, Rippmann FF, Cruciani G, Wade RC (2010) Computational approaches to identifying and characterizing protein binding sites for ligand design. J Mol Recognit 23(2):209–219

Hernandez M, Ghersi D, Sanchez R (2009) SITEHOUND-web: a server for ligand binding site identification in protein structures. Nucleic Acids Res 37(Web Server):W413–W416

Hetényi C, van der Spoel D (2011) Toward prediction of functional protein pockets using blind docking and pocket search algorithms. Protein Sci 20(5):880–893
Hopkins AL, Mason JS, Overington JP (2006) Can we rationally design promiscuous drugs? Curr Opin Struct Biol 16(1):127–136
Huang B (2009) MetaPocket: a meta approach to improve protein ligand binding site prediction. OMICS J Integr Biol 13(4):325–330
Huang SY, Grinter SZ, Zou X (2010) Scoring functions and their evaluation methods for protein-ligand docking: recent advances and future directions. Phys Chem Chem Phys 12 (40):12899–12908
Hussein HA, Borrel A, Geneix C, Petitjean M, Regad L, Camproux AC (2015) PockDrug-Server: a new web server for predicting pocket druggability on holo and apo proteins. Nucleic Acids Res 43(W1):W436–W442
Hwang S, Gou Z, Kuznetsov IB (2007) DP-Bind: a web server for sequence-based prediction of DNA-binding residues in DNA-binding proteins. Bioinformatics 23(5):634–636
Inhester T, Bietz S, Hilbig M, Schmidt R, Rarey M (2017) Index-based searching of interaction patterns in large collections of protein-ligand interfaces. J Chem Inf Model 57(2):148–158
Jambon M, Andrieu O, Combet C, Deléage G, Delfaud F, Geourjon C (2005) The SuMo server: 3D search for protein functional sites. Bioinformatics 21(20):3929–3930
Jones S, Thornton JM (1996) Principles of protein-protein interactions. Proc Natl Acad Sci U S A 93(1):13–20
Kalidas Y, Chandra N (2008) PocketDepth: a new depth based algorithm for identification of ligand binding sites in proteins. J Struct Biol 161(1):31–42
Keiser MJ, Setola V, Irwin JJ et al (2009) Predicting new molecular targets for known drugs. Nature 462(7270):175–181
Kirby AJ (1996) Enzyme mechanisms, models, and mimics. Angew Chem Int Ed 35(7):706–724
Kleywegt GJ, Jones TA (1994) Detection, delineation, measurement and display of cavities in macromolecular structures. Acta Crystallogr D Biol Crystallogr 50(Pt 2):178–185
Klibanov AM (2001) Improving enzymes by using them in organic solvents. Nature 409(6817):241
Koes DR, Camacho CJ (2012) PocketQuery: protein-protein interaction inhibitor starting points from protein-protein interaction structure. Nucleic Acids Res 40(Web Server):W387–W392
Konc J, Janezic D (2010) ProBiS algorithm for detection of structurally similar protein binding sites by local structural alignment. Bioinformatics 26(9):1160–1168
Konc J, Lešnik S, Janežič D (2015) Modeling enzyme-ligand binding in drug discovery. J Cheminf 7(1):48
Koshland D Jr (1995) The key–lock theory and the induced fit theory. Angew Chem Int Ed 33 (23–24):2375–2378
Ladbury JE (1996) Just add water! The effect of water on the specificity of protein-ligand binding sites and its potential application to drug design. Chem Biol 3(12):973–980
Laskowski RA (1995) SURFNET: a program for visualizing molecular surfaces, cavities, and intermolecular interactions. J Mol Graph 13(5):323–330
Laurie AT, Jackson RM (2005) Q-SiteFinder: an energy-based method for the prediction of protein-ligand binding sites. Bioinformatics 21(9):1908–1916
Laurie AT, Jackson RM (2006) Methods for the prediction of protein-ligand binding sites for structure-based drug design and virtual ligand screening. Curr Protein Pept Sci 7(5):395–406
Le Guilloux V, Schmidtke P, Tuffery P (2009) Fpocket: an open source platform for ligand pocket detection. BMC Bioinf 10(1):168
Lehn JM (1988) Supramolecular chemistry-scope and perspectives molecules, supermolecules, and molecular devices (Nobel Lecture). Angew Chem Int Ed 27(1):89–112
Lesk VI, Sternberg MJ (2008) 3D-Garden: a system for modelling protein–protein complexes based on conformational refinement of ensembles generated with the marching cubes algorithm. Bioinformatics 24(9):1137–1144
Liang J, Edelsbrunner H, Woodward C (1998) Anatomy of protein pockets and cavities: measurement of binding site geometry and implications for ligand design. Protein Sci 7(9):1884–1897

Liang S, Zhang C, Liu S, Zhou Y (2006) Protein binding site prediction using an empirical scoring function. Nucleic Acids Res 34(13):3698–3707

Macalino SJ, Gosu V, Hong S, Choi S (2015) Role of computer-aided drug design in modern drug discovery. Arch Pharm Res 8(9):1686–1701

Masood TB, Sandhya S, Chandra N, Natarajan V (2015) CHEXVIS: a tool for molecular channel extraction and visualization. BMC Bioinf 16:119

Mattos C, Ringe D (1996) Locating and characterizing binding sites on proteins. Nat Biotechnol 14 (5):595–599

Meslamani J, Rognan D, Kellenberger E (2011) sc-PDB: a database for identifying variations and multiplicity of 'druggable' binding sites in proteins. Bioinformatics 27(9):1324–1326

Nayal M, Honig B (2006) On the nature of cavities on protein surfaces: application to the identification of drug-binding sites. Proteins 63(4):892–906

Norel R, Lin SL, Wolfson HJ, Nussinov R (1994) Shape complementarity at protein-protein interfaces. Biopolymers 34(7):933–940

Pagadala NS, Syed K, Tuszynski J (2017) Software for molecular docking: a review. Biophys Rev 9 (2):91–102

Parca L, Mangone I, Gherardini PF, Ausiello G, Helmer-Citterich M (2011) Phosfinder: a web server for the identification of phosphate-binding sites on protein structures. Nucleic Acids Res 39(suppl 2):W278–W282

Payandeh Z, Rajabibazl M, Mortazavi Y, Rahimpour A, Taromchi AH (2018) Ofatumumab monoclonal antibody affinity maturation through *in silico* modeling. Iran Biomed J 22 (3):180–192

Petrek M, Otyepka M, Banás P, Kosinová P, Koca J, Damborský J (2006) CAVER: a new tool to explore routes from protein clefts, pockets and cavities. BMC Bioinf 7:316

Pierce B, Tong W, Weng Z (2005) M-ZDOCK: a grid-based approach for Cn symmetric multimer docking. Bioinformatics 21(8):1472–1478

Redington PK (1992) Molfit: a computer program for molecular superposition. Comput Chem 16 (3):217–222

Roche DB, Tetchner SJ, McGuffin LJ (2011) FunFOLD: an improved automated method for the prediction of ligand binding residues using 3D models of proteins. BMC Bioinf 12:160

Ruppert J, Welch W, Jain AN (1997) Automatic identification and representation of protein binding sites for molecular docking. Protein Sci 6(3):524–533

Schmitt S, Kuhn D, Klebe G (2002) A new method to detect related function among proteins independent of sequence and fold homology. J Mol Biol 323(2):387–406

Schneider G, Fechner U (2005) Computer-based de novo design of drug-like molecules. Nat Rev Drug Discov 4(8):649–663

Schneidman-Duhovny D, Inbar Y, Nussinov R, Wolfson HJ (2005) PatchDock and SymmDock: servers for rigid and symmetric docking. Nucleic Acids Res 33(Web Server):W363–W367

Sheng J, Gan J, Huang Z (2013) Structure-based DNA-targeting strategies with small molecule ligands for drug discovery. Med Res Rev 33(5):1119–1173

Shoemaker BA, Zhang D, Tyagi M et al (2012) IBIS (Inferred Biomolecular Interaction Server) reports, predicts and integrates multiple types of conserved interactions for proteins. Nucleic Acids Res 40(Database):D834–D840

Shulman-Peleg A, Shatsky M, Nussinov R, Wolfson HJ (2008) MultiBind and MAPPIS: webservers for multiple alignment of protein 3D-binding sites and their interactions. Nucleic Acids Res 36(Web Server):W260–W264

Silverman RB, Holladay MW (2014) The organic chemistry of drug design and drug action. Academic Press, Amsterdam

Singh DB, Dwivedi S (2016) Structural insight into binding mode of inhibitor with SAHH of Plasmodium and human: interaction of curcumin with anti-malarial drug targets. J Chem Biol 9 (4):107–120

Singh DB, Tripathi T (2020) Frontiers in protein structure, function, and dynamics. Springer, Singapore

Skolnick J, Brylinski M (2009) FINDSITE: a combined evolution/structure-based approach to protein function prediction. Brief Bioinform 10(4):378–391
Smith GR, Sternberg MJ (2003) Evaluation of the 3D-Dock protein docking suite in rounds 1 and 2 of the CAPRI blind trial. Proteins: Struct Funct Bioinf 52(1):74–79
Sousa SF, Fernandes PA, Ramos MJ (2006) Protein–ligand docking: current status and future challenges. Proteins: Struct Funct Bioinf 65(1):15–26
Steed JW, Turner DR, Wallace K (2007) Core concepts in supramolecular chemistry and nanochemistry. Wiley, Hoboken
Tan KP, Nguyen TB, Patel S, Varadarajan R, Madhusudhan MS (2013) Depth: a web server to compute depth, cavity sizes, detect potential small-molecule ligand-binding cavities and predict the pKa of ionizable residues in proteins. Nucleic Acids Res 41(Web Server):W314–W321
Tian W, Chen C, Lei X, Zhao J, Liang J (2018) CASTp 3.0: computed atlas of surface topography of proteins. Nucleic Acids Res 46(W1):W363–W367
Till MS, Ullmann GM (2010) McVol - a program for calculating protein volumes and identifying cavities by a Monte Carlo algorithm. J Mol Model 16(3):419–429
Tiwari A, Avashthi H, Jha R et al (2016) Insights using the molecular model of Lipoxygenase from Finger millet (Eleusine coracana (L.)). Bioinformation 12(3):156–164
Tovchigrechko A, Vakser IA (2006) GRAMM-X public web server for protein-protein docking. Nucleic Acids Res 34(Web Server):W310–W314
Tseng YY, Chen ZJ, Li WH (2010) fPOP: footprinting functional pockets of proteins by comparative spatial patterns. Nucleic Acids Res 38(Database):D288–D295
von Hippel PH, Bear DG, Morgan WD, McSwiggen JA (1984) Protein-nucleic acid interactions in transcription: a molecular analysis. Annu Rev Biochem 53:389–446
Vyas NK (1991) Atomic features of protein-carbohydrate interactions. Curr Opin Struct Biol 1 (5):732–740
Walker CB (1996) The acquisition of antibiotic resistance in the periodontal microflora. Periodontol 10:79–88
Wang R, Lai L, Wang S (2002) Further development and validation of empirical scoring functions for structure-based binding affinity prediction. J Comput Aided Mol Des 16(1):11–26
Wang K, Gao J, Shen S, Tuszynski JA, Ruan J, Hu G (2013) An accurate method for prediction of protein-ligand binding site on protein surface using SVM and statistical depth function. Biomed Res Int 2013:409658
Weisel M, Proschak E, Schneider G (2007) PocketPicker: analysis of ligand binding-sites with shape descriptors. Chem Cent J 1:7
Xie ZR, Hwang MJ (2015) Methods for predicting protein-ligand binding sites. Methods Mol Biol 1215:383–398
Yaffe E, Fishelovitch D, Wolfson HJ, Halperin D, Nussinov R (2008) MolAxis: a server for identification of channels in macromolecules. Nucleic Acids Res 36(Web Server):W210–W215
Zhang Z, Li Y, Lin B, Schroeder M, Huang B (2011) Identification of cavities on protein surface using multiple computational approaches for drug binding site prediction. Bioinformatics 27 (15):2083–2088
Zhu H, Pisabarro MT (2010) MSPocket: an orientation-independent algorithm for the detection of ligand binding pockets. Bioinformatics 27(3):351–358

Role of ADMET Tools in Current Scenario: Application and Limitations

4

Rajesh Kumar Kesharwani, Virendra Kumar Vishwakarma, Raj K. Keservani, Prabhakar Singh, Nidhi Katiyar, and Sandeep Tripathi

Abstract

High rates of drug failure cases are a challenge for the pharmaceutical industry to improve preclinical testing. For the ADMET prediction, selection of suitable experimental data and its use in the form of physiological parameters is a challenging task. Nowadays, ADMET prediction is performed at an early stage of drug designing to remove the pharmacokinetic (PK) property of poor compounds. Various ADMET prediction models have been developed using computational algorithms. Experimentally validated ADMET datasets have been analyzed, and related classification features and descriptors were used for the development of in silico models. The current chapter describes the role of ADMET analysis in drug designing, approaches used for model development, existing tools for ADMET prediction, and limitation of predictive models.

Keywords

ADMET · Toxicity database · Structure · Drug design · Pharmacokinetics

R. K. Kesharwani (✉) · V. K. Vishwakarma · S. Tripathi
Department of Advanced Science & Technology, Nims Univerity Rajasthan, Jaipur, India

R. K. Keservani
Faculty of B. Pharmacy, CSM Group of Institutions, Prayagraj, India

P. Singh
Department of Biochemistry, VBS Purvanchal University, Jaunpur, India

N. Katiyar
Dr. APJ Abdul Kalam Technical University (AKTU), Lucknow, India

D. B. Singh (ed.), *Computer-Aided Drug Design*,
https://doi.org/10.1007/978-981-15-6815-2_4

4.1 Introduction

There is a high-risk investment in current pharmaceutical research and development due to increasing cost and risk of failure in the drug discovery process and development (Khanna 2012). The pharmaceutical industry has always been a major concern about balancing the risk and productivity of research and development (Paul et al. 2010). With the advancement in multidisciplinary approaches for any drugs, there are chances of improvement with the application of computational algorithms and data analysis methods (Kesharwani et al. 2019). Every day, a huge amount of new drug-like compounds generated, but they do not become part of usable medicines. The threshold feature like absorption, distribution, metabolism, elimination/excretion (ADMET) of any drug is more important for its successful use. Computer-aided drug discovery methods are very helpful in guiding the development of the new drug candidates and also screening some potential compounds based on binding interaction, selectivity, ADMET, etc. (Kesharwani and Misra 2011; Singh and Dwivedi 2019). Various computational methods and approaches are being used in drug development, and it is expected that the cost of drug development may be reduced by up to 50% through the use of computational approach (Tan et al. 2010).

The drug discovery process can be divided into two important classes: (1) strategies for lead discovery and its optimization for the development of a potential drug against the selective targets and (2) strategies for the prediction of compounds with druggability, which can be used for evaluating the therapeutic utility of the leads. Most of the drugs have reliable values that are related to assimilation, appropriation, digestion, and discharge properties and different toxicities (T) or unfavorable reactions. The traditional methods used to evaluate ADMET properties are time-consuming, costly, and create management problems for large batch chemicals.

4.1.1 ADMET Prediction

In like manner, top to bottom ADMET investigation will not be performed until the point that a set number of inhibitor compounds have been distinguished. In many cases, it has been seen that some set of compounds rejected during in vitro study because of poor druggability and its ADMET property is not within the threshold value. Natural compounds and their metabolites may also have toxic effects, and the labels regarding the safe and secure use of these compounds should be issued. During the drug discovery process, some potential compounds are screened at a different level based on certain criteria and filters. Virtual screening is the most efficient throughput screening system to choose some compounds and is based on the binding affinity of compounds to the target structure. One objective of ADMET screening ought to be the improvement of a database corresponding to synthetic structures and organic chemicals.

It is not possible to perform ADMET related in vitro or in vivo studies for a large set of compounds. Therefore, the development of in silico ADMET prediction model

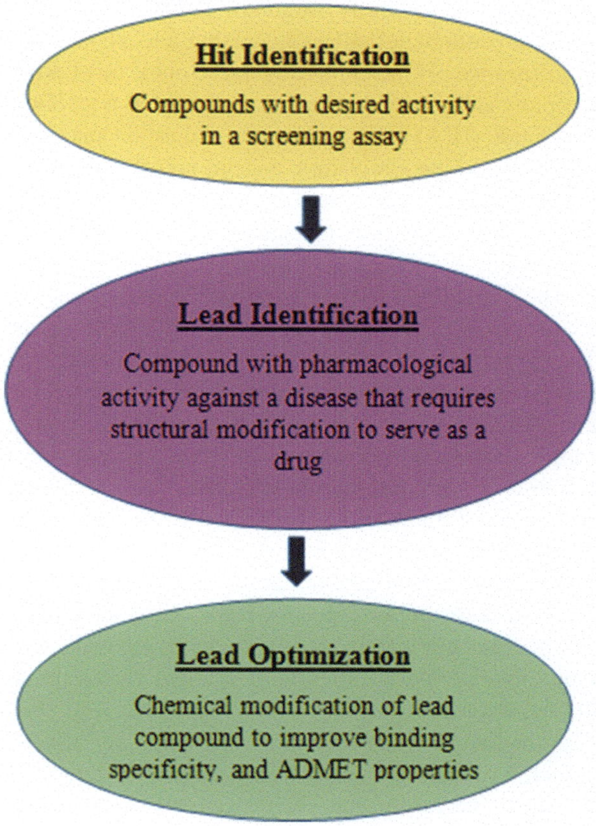

Fig. 4.1 Role of ADMET prediction in optimizing the activity of a lead compound to qualify as a drug candidate

is a good strategy to evaluate the PK of many compounds, and it can guide the necessary structural changes in the lead compounds (Fig. 4.1). The development of efficient and accurate in silico ADMET models will allow the parallel improvement of compound feasibility and druggability (de la Nuez and Rodríguez 2008). In the most recent decade, countless toxicity expectation models have been accounted for, and a few surveys concerning the advancement of these models have been conducted (Cheng et al. 2013). To tackle these issues, a few arrangements have been accounted for concerning how to grow more compelling models and where these models can be utilized (Huang et al. 2013). The accuracy of ADMET models can be improved by integrating the other important parameters or by considering more clinical data and results (Singh 2014).

4.1.2 ADMET Parameters and their Role

Absorption of any drug is a complicated procedure which is affected by various components, including not only the natural properties of the substance (atomic size, aqueous solubility (logSaq), ionization constant (pKa), and octanol/water segment coefficient (logP) values), but also physiological conditions inside the life form (nearby pH, absorptive surface zone), function of catalysts, transporters, and receptors along with the gastrointestinal (GI) tract (George 1981). Human intestinal absorption (HIA) is typically estimated as the amount of the drug that absorbed via intestine into the circulatory system, and this is most challenging for the in silico model developer. The amount of compound which remains after retention and first-pass hepatic digestion is characterized as the oral bioavailability (F) of that compound.

The blood/brain (BB) barrier coefficient typically communicated as logBB and characterized as the proportion between the substance present in the blood and brain. The entry of compounds over the blood/brain barrier (BBB) is an essential determinant of neurotoxicity and depends predominantly on uninvolved dissemination over the BBB layer (Chen et al. 2009). The dynamic transport likewise might be vital. For supplements and endogenous chemical compounds, e.g. monocarboxylic acids, amino acids, amines, thyroid hormones, hexoses, purine bases, and nucleosides, a few transport frameworks controlling the passage of the separate compound classes into the brain have been distinguished.

Digestion is one of the fundamental variables impacting the destiny and danger of synthetic compounds. Digestion incorporates an arrangement of substance responses (set of metabolic pathways) inside the living being, which convert xenobiotic with more polarity and effortlessly discharged via excretion in less lethal forms. Commonly, digestion is divided into two stages—stage I and stage II. Stage I, it is described as the fictionalization stage, majorly affects lipophilic atoms, rendering them more polar and all the more promptly excretable. In stage II, usually suggested as a detoxification phase, such functionalized moieties are in this way conjugated with profoundly polar particles. The two stages are catalyzed by particular chemicals that are either present in the cytosol (cytosolic or dissolvable compounds) or membrane bound protein (microsomal proteins) or the superfamily of cytochrome P450 (CYP450).

Family cytochrome P450 are more than 70 groups of proteins, catalyze the oxidative stage I metabolic responses of different compounds (Werck-Reichhart and Feyereisen 2000). Stage II digestion is represented by different chemicals following up on various sorts of particles. The most noteworthy among them is glutathione S-transferase (GST), N-acetyltransferase (NAT), methyltransferase (MT), sulfotransferase (SULT), and UDP-glucuronosyltransferase (UGT). Other than stage I and stage II digestion, the liver causes particular pre-fundamental (first-pass) impacts, particularly following the oral admission. What's more, stage III digestion alludes to the discharge of cellular metabolites with efflux transporters. Discharge is the way toward wiping out waste metabolic items, the significant course of which is renal (urinary) discharge using the kidneys. The major non-metabolic courses of freedom (CLtot) incorporate bile and urinary end of unaltered compounds.

Drug lethality is a significantly essential drug property. The potential for lethality remains the most variable property of a drug. Toxicity is a level when a chemical compound is harmful to the human or animal. If short-term or one-time exposure of any chemical is harmful to the human or animal is called acute toxicity. Some drugs are in its preliminary preclinical phase and if it harms human health, it will be back to its initial phase of drug development.

Toxicity is the most important property for any type of chemotherapy. Since it could be species-specific, organ-specific, and could include different host factors and dose values. Nonetheless, as hepatotoxicity is a noteworthy sign of drug lethality, compounds produced after the metabolism of drugs can also be toxic. The identification and cataloging of toxic entities or substructure can guide a person to avoid the addition of these toxic entities on a pharmacophore (Singh 2018).

4.2 Importance of ADMET

Since 1990, various studies were focused on physicochemical properties defining the molecules drug-likeness to find the importance of lipophilicity, size, and H-bonding nature of drug molecules and their bioavailability in the subject system (Navia and Chaturvedi 1996; Lipinski et al. 2001). From these investigations, a few dependable guidelines have been drawn to help scientific experts in the structure and choice of atoms that ought to have an improved probability of getting to be effective oral drugs (Muchmore et al. 2010). The objective of the biological analysis is to survey the general ADME attributes of the new substance compounds. Bioanalytical techniques can be used to determine the PK of a compound/drug (Jamalapuram et al. 2012). The role and facilities provided by ADMET predictor tools are shown in Fig. 4.2.

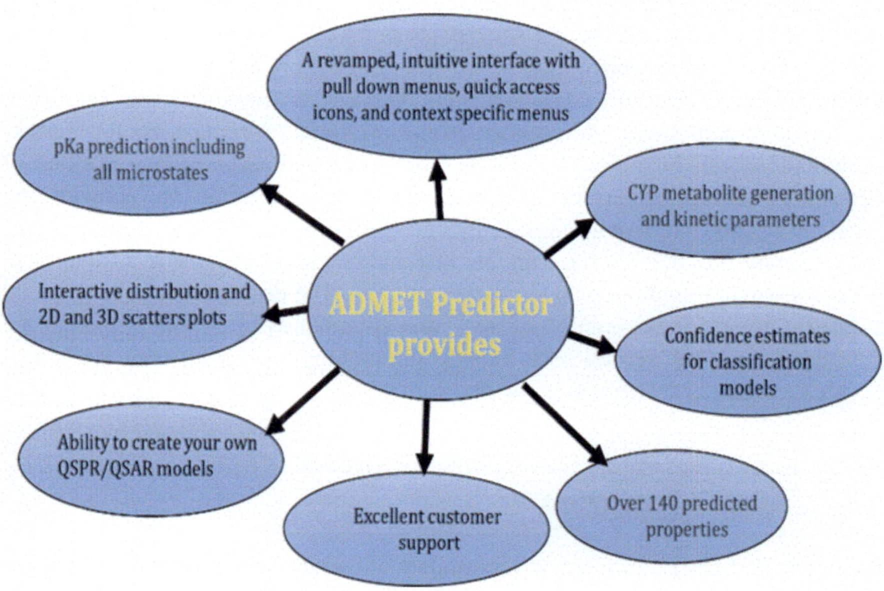

Fig. 4.2 Facilities available at ADMET predictor

4.3 The Evolving Science of ADMET

Since 1950, scientific experts relied on in vivo testing for drug effects on the human body. Bioavailability, tissue appropriation, PK, digestion, and poisonous quality are surveyed in one rat, and one non-rat animal category. Biodistribution is evaluated utilizing radioactively marked chemicals later being developed because it is costly (Oldendorf 1970). Pharmacodynamic (PD) adequacy of test compounds is regularly surveyed through in vitro models, e.g. receptor official, trailed by affirmation through in vivo viability models in mice or rodents. In vivo PK study is utilized for lead improvement to drug trial digestion and retention. Understanding the PK–PD relationship is essential in building up a comprehension of the system of activity and metabolic destiny of compound. In any case, there are huge contrasts in assimilation and digestion among species. In vivo models are commonly used for an investigational new drug application, yet these have shortcomings. The toxicity testing model should be designed for components of human models to avoid the risk of adverse reactions.

4.4 Blood–Brain Barrier Models

A very new drug able to work in the CNS may demonstrate extraordinary remedial guarantee because of their high intensity at the target site. A lot of BBB data available in literature and databases could be connected by analysts to create in silico models of brain infiltration. Different datasets related to drugs were analyzed and many models have been proposed for assurance of logBB for compounds. Thus, there is need of large and more diverse datasets for exact estimations of logBB.

There are various in silico models providing logBB prediction around 0.35–0.45 log units that could be utilized for screening purposes. By inspecting the wide range of valuable sub-atomic descriptors, some essential speculations for further examinations can be made. It is acceptable to recognize two classifications of descriptors. The principal descriptors of size (i.e. molar refraction, network and topological lists, sub-atomic mass, surface zone) while the second descriptors of the extremity (i.e. polar surface region, incomplete charges, elements of hydrogen bond corrosive or hydrogen bond base gatherings). The descriptors of size are vital indicators for the transportation of non-polar compounds in the brain, though the descriptors from the second class express the highlights of polar particles, which decide their liking to segment in the blood.

4.5 ADMET Prediction

Prediction of ADMET property is very important because about 60% of drugs fail in the clinical trials due to poor ADME. Nowadays, ADMET prediction is done at an early stage of drug designing to remove the compounds with the poor PK property.

Several new approaches, such as toxicogenomics, data-integration, and decision making systems could be used for in silico ADMET prediction (Cheng et al. 2013).

BBB is a highly selective barrier which can be used to decide the compounds or drug that can pass the BBB and reach their target. CYP450 enzymes play an important role in the metabolism and detoxification of drugs and other toxic chemicals. A drug should not be rapidly metabolized by CYP450, and it should not inhibit CYP450. If a drug inhibits CYP450, then the level of another co-administered drug will be raised, which may have toxic effects (Brown et al. 2008). The in silico models can predict the interactions between CYP450 and a drug. The human hERG gene codes for a potassium ion channel, which plays an important role in the activity of the heart. hERG blockage can cause arrhythmia and death. Therefore, a drug must qualify the hERG test in in silico models. P-glycoprotein (Pgp) extracts many foreign substances from the cell and decides the PK properties of drugs. Pgp may efflux the drug from the cell surface, and reduce the concentration of drug to target. Therefore, computational models can be used to predict a drug is Pgp substrates or Pgp inhibitors. Mutagenicity or mutation causing the capability of a drug or chemical should also be predicted to ensure its role (Tripathi et al. 2015).

4.6 Strategies for the Designing of ADMET Model

4.6.1 Selection of Experimental Data

Prediction of an accurate ADMET model is a challenging task with respect to time, speed, and accuracy. Before the start of model prediction, the selection of dataset and prediction parameters play a key role to derive desired models with more accuracy and reproducibility. For any predictive model, the initial step is to get accurate experimental data.

4.6.2 Calculation of Physicochemical Parameters or Descriptor Values

Traditionally physiochemical features of chemical compounds or descriptors are mainly 3 types, one-dimensional, two-dimensional, and three-dimensional. Due to the advancement of technology, a huge number of descriptors are available in various tools, e.g. 1800 descriptors described by Todeschini and Consonni (2008). The selection of suitable descriptors is a key factor in ADMET model development after getting suitable experimental data. The representation of the descriptor in the form of mathematical equations using a statistical or machine learning approach is used to identify the molecular properties of chemical compounds. Many scientists have discovered some new class of descriptor as physicochemical (measured or calculated), geometrical, constitutional (including group contributions), topological, electro-topological, quantum chemical and molecular fingerprints, together with some others. Interestingly, a molecule can be represented in the form of a molecular

Table 4.1 Tool/server for the calculation of descriptors

Tool	Description	References
E-Dragon	It is used for the calculation of 1600 molecular descriptors	Tetko et al. (2005)
MOLE db—Molecular descriptors data base	It is a free web online database contains 1124 molecular descriptors	Ballabio et al. (2009)
EPISUITE	This is an estimation program interface suite for the calculation of property and environmental fate estimation programs	Epa (2012)
Online Chemical Modeling Environment (OCHEM)	It is a web-based platform which calculates descriptor and used for designing of in silico model	Sushko et al. (2011)
The Chemistry Development Kit (CDK)	It is an open source database with the chemical processing features	Steinbeck et al. (2006)
ChemDraw	It is a chemical drawing tool having descriptor calculation features	Mills (2006)

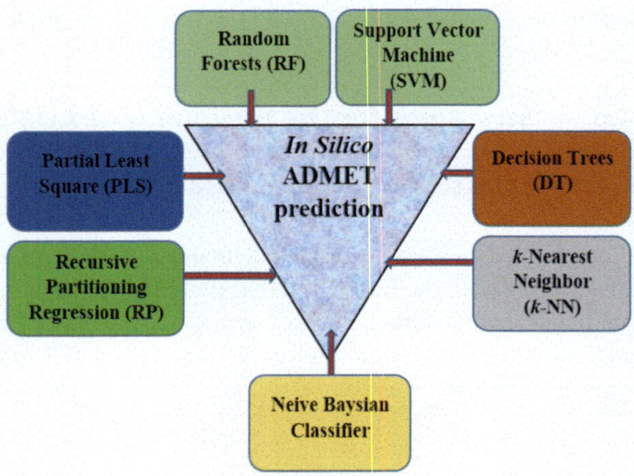

Fig. 4.3 Representation of ADMET prediction methods

fingerprint or terms of computer language binary string. The descriptor representation in the form of binary string for any structure and substructure is given in predefined SMART format files. Currently, many free and proprietary tools/servers are available to calculate the descriptor values for any chemical compounds, e.g. Open Babel (O'Boyle et al. 2011), ChemDraw (Li et al. 2004), and ACD ChemSktech (Spessard 1998) (Table 4.1).

4.6.3 ADMET Prediction Methods and Tools

The development of the ADMET model prediction tool is designed with the help of a statistical and machine learning approach. In the present book chapter we have discussed seven most usable methods (Fig. 4.3).

4.6.3.1 Recursive Partitioning Regression

The recursive partitioning regression analysis approach is a statistical method used in multivariable analysis (Breiman et al. 1984). It is a partitioning method to split the study data or population and sub-population recursively in certain class until and unless the process terminates due to stopping criteria is reached (Cook and Goldman 1984). Recursive partitioning regression techniques are nonparametric and it does not depend on the variable of the predictor. It is widely used for data analysis and in silico model development since the 1980s (Chen et al. 2011). With the generation of a large amount of biological data, e.g. microarray data, DNA sequencing, etc. it gains more popularity due to its capability to analyze multivariate data exploration.

4.6.3.2 Partial Least Square (PLS) Regression

It is a statistical method similar to principal components analysis (Abdi 2010). The relation between the two matrices can be easily derived using the PLS method. It is generally used to find the direction (multidimensional) in one axis space, e.g. x-axis space which explains the maximum variance (multidimensional) in the y-axis space (Bookstein 1994). It has been founded that PLS regression generates good results when the matrix of predictors consists of several variables than observations, and when there is multicollinearity among x values. The development of PLS methods is initially for the social science study, but it gains more importance in other fields too including in silico ADMET model development, biological data analysis, bioinformatics, neurobiology, etc. (Boulesteix and Strimmer 2007; Nikolić et al. 2013).

4.6.3.3 Random Forests (RF)

Random forest is an ensemble learning method used for data analysis and data classification (Breiman et al. 1984). Here, data are in the form of a regression tree, and the selections of features are based on random selection. A Berkeley (California, USA) based statistician Leo Breiman first introduced the RF algorithm in 2001 (Altmann et al. 2010). In these methods, aggregation of a large number of decision trees using ensemble learning techniques results in the reduction of variance compared to single decision trees (Dong et al. 2018).

4.6.3.4 Decision Trees

Decision trees based on sorting techniques are powerful nonparametric supervised learning *algorithms* for the data classification (Kamiński et al. 2018). The decision trees algorithm is used for classification and regression analysis. This algorithm can simply represent and classify data. Currently, many scientists are using this method in their studies (Quinlan 1987). It is highly useful in many fields like computer science, neuroscience, bioinformatics, etc. Among the data mining methods, decision trees continue to be mainly applied in ADMET prediction, especially in the form of ensemble-based random that predicts the value of a target variable based on several input variables (descriptors) (Dong et al. 2018). Compared with other machine learning methods, decision trees have several advantages such as easy to understand and interpret, requiring little preparation, and low computational cost.

4.6.3.5 Naive Bayes Classifiers

Naive Bayes classifiers are a machine learning method based on Bayes' theorem (Shi et al. 2015). It is not a single algorithm, but a family of algorithms where all of them share a common principle, i.e. every pair of features being classified is independent of each other (Fang et al. 2013). It is a probabilistic classifier and has been frequently used since the 1960s. Due to its appropriate prepossessing ability, it is a competitive and interesting algorithm for the scientist irrespective of the presence of many advanced algorithms like the support vector machine (Lian et al. 2016). It has a broad range of applications in many streams including computational biology, drug designing, and ADMET prediction (Klon et al. 2006).

4.6.3.6 k-Nearest Neighbour (k-NN)

k-NN is a supervised learning method and used for pattern identification (Altman 1992). It is widely useful for real-life problems. vNN Web Server is based on the k-NN algorithm and available free of cost for public use (Schyman et al. 2017). It is nonparametric and is not based on the assumption of the distribution of data (Jaskowiak and Campello 2011). The complexity of k-NN is simpler than that of SVM. In k-NN, an object is classified by majority votes of its neighbors, with the object being assigned to the class most common among its k nearest neighbors (Coomans and Massart 1982).

4.6.3.7 Support Vector Machine (SVM)

SVM is a supervised learning algorithm used for data classification and regression analysis (Zhang 2001). For a few decades, it is used frequently due to its good prediction of the model using training and test sets. The biggest challenge in ADMET prediction using dataset is classification and regression (Fradkin and Muchnik 2006; Ivanciuc 2007). The classification and regression models developed by SVM are widely used in ADMET prediction, and have several limitations such as SVM training needs high computational cost for the large sample classification problem. The model built by SVM is like a black box, which lacks interpretation of biological mechanisms for medicinal chemists or biologists. SVM based classification models were originally developed for binary problems (Steinwart and Christmann 2008). The sampling unbalanced problem of classification often reduces the application of SVM (James et al. 2013; Schölkopf et al. 2002).

4.7 ADMET Tools

In silico ADMET prediction is an important component of drug discovery (Singh et al. 2013). Estimation of ADMET in the early phases of drug discovery is important for screening the drug candidate and optimizing the lead. Quantitative structure-property relationships (QSPR), quantitative structure-activity relationships (QSAR) are used to predict drug toxicity. The carcinogenicity, mutagenicity, and liver toxicity should be evaluated during drug designing. Emphasis has been given to develop in silico tool because it reduces animal testing, reduces cost and time, and is reliable.

vNN is a tool for assessments of the PK and toxic properties of a drug. vNN has 15 ADMET models, which predict the important properties of drug compounds such as cytotoxicity, mutagenicity, cardiotoxicity, drug–drug interactions, microsomal stability, and drug-induced liver injury (Schyman et al. 2017). Another tool, ADMETopt is used for ADMET screening for lead optimization. This tool uses the information of 50,000 unique scaffolds extracted chemical compounds deposited in ChEMBL and Enamine database (Yang et al. 2018). It predicts 7 physicochemical and 8 biological properties. Recently, admetSAR 2.0 has been released for ADMET prediction, which is based on 47 models for drug discovery, while an earlier version had only 27 models (Yang et al. 2019). This tool has been developed by the optimization of existing models by considering large and precise training data. Another module ADMETopt has been added to this tool, which can be used for lead optimization based on predicted ADMET properties. Several tools are available for ADMET prediction (Table 4.2).

Many ADMET models have been developed, but still, it is not easy to accurately predict so many ADMET properties. ADMET-score, a scoring function has been generated using 18 ADMET properties to evaluate the drug-likeness of a compound (Guan et al. 2019). The weight of each ADMET property was based on the accuracy rate of the model, the importance of the endpoint in PK, and the usefulness index. The performance of this scoring function was evaluated using FDA-approved drugs, ChEMBL compounds, and withdrawn drugs.

Many studies also indicate that there is no linear correlation between the ADMET score and the quantitative estimate of drug-likeness. admetSAR is used in chemical and pharmaceutical fields for predicting ADMET related outcomes. In admetSAR 2.0, the predictive model is more optimized as it has been developed by considering a large amount of training data (Yang et al. 2019). Now, this tool includes 47 models for assessment of drug discovery or environmental risk. Also, another module ADMETopt has been added for lead optimization based on ADMET properties.

The pharmacokinetics knowledge base (PKKB) has been developed to provide detailed information about ADMET properties. PKKB includes 10,000 experimental ADMET data of 1685 drugs (Cao et al. 2012). It provides information about octanol/water partition coefficient, solubility, the dissociation constant, intestinal absorption, Caco-2 permeability, bioavailability, plasma protein binding, blood-plasma partitioning ratio, the volume of distribution, metabolism, half-life, excretion, urinary excretion, clearance, toxicity, half lethal dose.

QikProp tool is developed by Schrödinger, which is used for the prediction of ADMET related properties such as logPs, logS, BBB, CNS activity, Caco-2, and MDCK cell permeability, log KHSA for human serum albumin and log IC50 for HERG K + channel blockage. QikProp predictions are based on the full 3D molecular structure and provide accurate predictions similar to analogs of well-known drugs (Kesharwani et al. 2015). QikProp is also used for screening/filtering out candidates with unsuitable ADME properties. ProTox-II tool is used for the prediction of acute toxicity, hepatotoxicity, carcinogenicity, mutagenicity,

Table 4.2 Computational web server/tools for ADMET prediction

Tools	Description	References
DSSTox	It is a public database about distributed structure-searchable toxicity	Richard and Williams (2002)
CPDB	It is the carcinogenic potency database, which is unique and widely used the international resource	Nehlin et al. (2018)
Pre-ADMET ADMET prediction	Analysis of binding and permeability across different cellular conformation like MDCK cell, Caco-2 cell and blood–brain barrier, human intestinal absorption, and skin permeability	Rashid (2020)
Pre-ADMET toxicity prediction	This online tool predicts the probability of carcinogenicity as well as toxic potency	Rashid (2020)
Molinspiration	This is used in the mathematical measurement of molecular properties and the likeness of drugs	Qi and Ding (2018)
ChemTree	It is used in prediction of ADMETox properties	Is et al. (2018)
VolSurf	Using energy grid map this software generates 2D molecular characteristic using 3 D molecular interaction	Filipponi et al. (2001)
MetaSite	Using the 3D structure of xenobiotic molecules this software calculates the precise location about their metabolic site	Ajitha et al. (2018)
GRID	Calculate and determine the suitable binding site of a molecule on the known structure	Castellano et al. (2010)
MoKa	It is used in the calculation of pKa values	Milletti et al. (2010)
Shop	Uses in scaffold hopping procedure in discovering the drugs	Tjoe-Nij et al. (2018)
Tsar 3.2	Structure based activity calculation for the identification of new drugs	Li et al. (2009)
Metabase	It is a low-cost radio analytical LIMs based on excel in ADME/PK studies	Bolser et al. (2012)
ADME/toxicity property calculator	It is used for the screening of toxic chemicals through ADMET analysis	Livingstone (2003)
TOPKAT	Used for prediction of toxicology	Beard et al. (2019)
Metabolism	It reflects various metabolic pathways in different species	Johnson et al. (2016)
ADMET	It helps to eliminate unfavorable compounds for ADMET	Ferreira and Andricopulo (2019)
QikProp	Used for the prediction of ADMET related properties	Ioakimidis et al. (2008) and Manidhar et al. (2012)

cytotoxicity, immunotoxicity, adverse outcomes pathways, and toxicity targets (Banerjee et al. 2018). This tool is based on molecular similarity, pharmacophores, fragment propensities, and machine learning models. ProTox-II considers the data from both in vitro assays and in vivo studies. This is a free web server for toxicity prediction. It requires a 2D chemical structure as an input and predicts the toxicity based on 33 models.

4.8 Challenges in Present Scenario and Future Prospective

In the last 15 years, an incredible advancement has been performed in the field of ADMET profiling. This advancement has diminished the rate of drug failure in clinical preliminaries for ADME reasons. The important boundary presently is the poisonous quality segment of ADMET. The prediction for human toxicology must be made strides. The drug development method and ADMET optimization need more open databases and the sharing of experimental prohibitive datasets available in pharmaceutical companies. Accuracy of ADMET prediction relies upon the availability of comparable data related to the model. A high rate of drug failure cases is a challenge for the pharmaceutical industry to improve preclinical testing. In vivo and in vitro ADMET evaluations are time-consuming, costly, and laborious. In silico ADMET prediction tools have been developed to estimate the parameters related to these properties. The accuracy of these prediction tools can also be improved for the reliability of predictions. The accuracy of ADMET tools can be improved by carrying out more PK studies for available drugs, and then including structural information of the drug and related ADMET data into existing models.

4.9 Conclusions

With the advancement of technology, research grants are increasing every year around the world, and very few drugs are available for the safe use of humans. The pharmaceutical companies are more focused to address PK optimization of any drug-like molecules. ADMET prediction is an important step in the drug designing process, which is responsible for the PK optimization for any drugs. As part of an integrated strategy of various approaches used in drug development, the in silico ADMET model can help in better prediction of drug responses in humans and thus can improve the clinical success rates.

Competing Interest The authors declare that there are no competing interests.

References

Abdi H (2010) Partial least squares regression and projection on latent structure regression (PLS regression). WIREs Comput Stat 2(1):97–106

Ajitha M, Sundar K, Arul Mugilan S, Arumugam S (2018) Development of metal-active site and zinc cluster tool to predict active site pockets. Proteins 86(3):322–331

Altman NS (1992) An introduction to kernel and nearest-neighbor nonparametric regression. Am Stat 46(3):175–185

Altmann A, Toloşi L, Sander O, Lengauer T (2010) Permutation importance: a corrected feature importance measure. Bioinformatics 26(10):1340–1347

Ballabio D, Manganaro A, Consonni V, Mauri A, Todeschini R (2009) Introduction to MOLE DB-on-line molecular descriptors database. MATCH Commun Math Comput Chem 62:199–207

Banerjee P, Eckert AO, Schrey AK, Preissner R (2018) ProTox-II: a webserver for the prediction of toxicity of chemicals. Nucleic Acids Res 46(W1):W257–W263

Beard DJ, Davies LJ, Cook JA et al (2019) The clinical and cost-effectiveness of total versus partial knee replacement in patients with medial compartment osteoarthritis (TOPKAT): 5-year outcomes of a randomised controlled trial. Lancet 394(10200):746–756

Bolser DM, Chibon PY, Palopoli N et al (2012) MetaBase—the wiki-database of biological databases. Nucleic Acids Res 40(D1):D1250–D1254

Bookstein FL (1994) Partial least squares: a dose-response model for measurement in the behavioral and brain sciences. Psycoloquy 5:23

Boulesteix AL, Strimmer K (2007) Partial least squares: a versatile tool for the analysis of high-dimensional genomic data. Brief Bioinform 8(1):32–44

Breiman L, Friedman J, Stone CJ, Olshen RA (1984) Classification and regression trees. CRC Press, Boca Raton

Brown CM, Reisfeld B, Mayeno AN (2008) Cytochromes P450: a structure-based summary of biotransformations using representative substrates. Drug Metab Rev 40(1):1–100

Cao D, Wang J, Zhou R, Li Y, Yu H, Hou T (2012) ADMET evaluation in drug discovery. 11. PharmacoKinetics Knowledge Base (PKKB): a comprehensive database of pharmacokinetic and toxic properties for drugs. J Chem Inf Model 52(5):1132–1137

Castellano M, Mastronardi G, Bellotti R, Tarricone GA (2010) Bioinformatics knowledge discovery in text application for grid computing. BMC Bioinf 10(Suppl 6):S23

Chen Y, Zhu QJ, Pan J, Yang Y, Wu XP (2009) A prediction model for blood–brain barrier permeation and analysis on its parameter biologically. Comput Methods Prog Biomed 95 (3):280–287

Chen L, Li Y, Zhao Q, Peng H, Hou T (2011) ADME evaluation in drug discovery. 10. Predictions of P-glycoprotein inhibitors using recursive partitioning and naive Bayesian classification techniques. Mol Pharm 8(3):889–900

Cheng F, Li W, Liu G, Tang Y (2013) *In silico* ADMET prediction: recent advances, current challenges and future trends. Curr Top Med Chem 13(11):1273–1289

Cook EF, Goldman L (1984) Empiric comparison of multivariate analytic techniques: advantages and disadvantages of recursive partitioning analysis. J Chronic Dis 37(9):721–731

Coomans D, Massart DL (1982) Alternative k-nearest neighbour rules in supervised pattern recognition: part 1. k-Nearestneighbour classification by using alternative voting rules. Anal Chim Acta 136:15–27

de la Nuez A, Rodríguez R (2008) Current methodology for the assessment of ADME-Tox properties on drug candidate molecules. Biotecnol Apl 25(2):97–110

Dong J, Wang NN, Yao ZJ, Zhang L, Cheng Y, Ouyang D, Lu AP, Cao DS (2018) ADMETlab: a platform for systematic ADMET evaluation based on a comprehensively collected ADMET database. J Chem Inf 10(1):29

Epa U (2012) Estimation programs interface suite™ for Microsoft® windows, v 4.11. United States Environmental Protection Agency, Washington, DC, USA

Fang J, Yang R, Gao L, Zhou D, Yang S, Liu AL, Du GH (2013) Predictions of BuChE inhibitors using support vector machine and naive Bayesian classification techniques in drug discovery. J Chem Inf Model 53(11):3009–3020

Ferreira LLG, Andricopulo AD (2019) ADMET modeling approaches in drug discovery. Drug Discov Today 24(5):1157–1165

Filipponi E, Cruciani G, Tabarrini O, Cecchetti V, Fravolini A (2001) QSAR study and VolSurf characterization of anti-HIV quinolone library. J Comput Aided Mol Des 15(3):203–217

Fradkin D, Muchnik I (2006) Support vector machines for classification. DIMACS Ser Discrete Math Theoret Comput Sci 70:13–20

George CF (1981) Drug metabolism by the gastrointestinal mucosa. Clin Pharmacokinet 6 (4):259–724

Guan L, Yang H, Cai Y, Sun L, Di P, Li W, Liu G, Tang Y (2019) ADMET-score–a comprehensive scoring function for evaluation of chemical drug-likeness. Med Chem Commun 10(1):148–157

Huang SM, Abernethy DR, Wang Y, Zhao P, Zineh I (2013) The utility of modeling and simulation in drug development and regulatory review. J Pharm Sci 102(9):2912–2923

Ioakimidis L, Thoukydidis L, Mirza A, Naeem S, Reynisson J (2008) Benchmarking the reliability of QikProp. Correlation between experimental and predicted values. QSAR Comb Sci 27 (4):445–456

Is YS, Durdagi S, Aksoydan B, Yurtsever M (2018) Proposing novel MAO-B hit inhibitors using multidimensional molecular modeling approaches and application of binary QSAR models for prediction of their therapeutic activity, pharmacokinetic and toxicity properties. ACS Chem Neurosci 9(7):1768–1782

Ivanciuc O (2007) Applications of support vector machines in chemistry. Rev Comput Chem 23:291

Jamalapuram S, Vuppala PK, Mesangeau C, McCurdy CR, Avery BA (2012) Determination of a highly selective mixed-affinity sigma receptor ligand, in rat plasma by ultra-performance liquid chromatography mass spectrometry and its application to a pharmacokinetic study. J Chromatogr B 891:1–6

James G, Witten D, Hastie T, Tibshirani R (2013) An introduction to statistical learning. Springer, New York

Jaskowiak PA, Campello RJ (2011) Comparing correlation coefficients as dissimilarity measures for cancer classification in gene expression data. In: Proceedings of the Brazilian symposium on bioinformatics 1-8, Brasília

Johnson CH, Ivanisevic J, Siuzdak G (2016) Metabolomics: beyond biomarkers and towards mechanisms. Nat Rev Mol Cell Biol 17(7):451–459

Kamiński B, Jakubczyk M, Szufel PA (2018) A framework for sensitivity analysis of decision trees. Cen Eur J Oper Res 26(1):135–159

Kesharwani RK, Misra K (2011) Prediction of binding site for curcuminoids at human topoisomerase II α protein; an in silico approach. Curr Sci 101:1060–1065

Kesharwani RK, Srivastava V, Singh P, Rizvi SI, Adeppa K, Misra K (2015) A novel approach for overcoming drug resistance in breast cancer chemotherapy by targeting new synthetic curcumin analogues against aldehyde dehydrogenase 1 (ALDH1A1) and glycogen synthase kinase-3 β (GSK-3β). Appl Biochem Biotechnol 176(7):1996–2017

Kesharwani RK, Misra K, Singh DB (2019) Perspectives and challenges of tropical medicinal herbs and modern drug discovery in the current scenario. Asian Pac J Trop Med 12(1):1–7

Khanna I (2012) Drug discovery in pharmaceutical industry: productivity challenges and trends. Drug Discov Today 17(19–20):1088–1102

Klon AE, Lowrie JF, Diller DJ (2006) Improved naive Bayesian modeling of numerical data for absorption, distribution, metabolism and excretion (ADME) property prediction. J Chem Inf Model 46(5):1945–1956

Li Z, Wan H, Shi Y, Ouyang P (2004) Personal experience with four kinds of chemical structure drawing software: review on ChemDraw, ChemWindow, ISIS/Draw, and ChemSketch. J Chem Inf Comput Sci 44(5):1886–1890

Li H, Sun J, Sui X et al (2009) First-principle, structure-based prediction of hepatic metabolic clearance values in human. Eur J Med Chem 44(4):1600–1606

Lian W, Fang J, Li C, Pang X, Liu AL, Du GH (2016) Discovery of Influenza A virus neuraminidase inhibitors using support vector machine and Naïve Bayesian models. Mol Divers 20 (2):439–451

Lipinski CA, Lombardo F, Dominy BW, Feeney PJ (2001) Experimental and computational approaches to estimate solubility and permeability in drug discovery and development settings. Adv Drug Deliv Rev 23(1–3):3–25

Livingstone DJ (2003) Theoretical property predictions. Curr Top Med Chem 3(10):1171–1192

Manidhar DM, Kesharwani RK, Reddy NB et al (2012) Designing, synthesis, and characterization of some novel coumarin derivatives as probable anticancer drugs. Med Chem Res 22:4146–4157

Milletti F, Storchi L, Goracci L et al (2010) Extending pKa prediction accuracy: high-throughput pKa measurements to understand pKa modulation of new chemical series. Eur J Med Chem 45 (9):4270–4279

Mills N (2006) ChemDraw Ultra 10.0 CambridgeSoft, 100 CambridgePark Drive, Cambridge, MA 02140. www.Cambridgesoft.com. Accessed February 2020

Muchmore SW, Edmunds JJ, Stewart KD, Hajduk PJ (2010) Cheminformatic tools for medicinal chemists. J Med Chem 53(13):4830–4841

Navia MA, Chaturvedi PR (1996) Design principles for orally bioavailable drugs. Drug Discov Today 1(5):179–189

Nehlin C, Carlsson K, Öster C (2018) Patients' experiences of using a cellular photo digital breathalyzer for treatment purposes. J Addict Med 12(2):107–112

Nikolić K, Filipić S, Smolinski A, Kaliszan R, Agbaba D (2013) Partial least square and hierarchical clustering in ADMET modeling: prediction of blood-brain barrier permeation of alpha-adrenergic and imidazoline receptor ligands. J Pharm Pharm Sci 16(4):622–647

O'Boyle NM, Banck M, James CA, Morley C, Vandermeersch T, Hutchison GR (2011) Open babel: an open chemical toolbox. J Chem Inf 3(1):33

Oldendorf WH (1970) Measurement of brain uptake of radiolabeled substances using a tritiated water internal standard. Brain Res 24(2):372–376

Paul SM, Mytelka DS, Dunwiddie CT, Persinger CC, Munos BH, Lindborg SR, Schacht AL (2010) How to improve R&D productivity: the pharmaceutical industry's grand challenge. Nat Rev Drug Discov 9(3):203–214

Qi L, Ding Y (2018) TNK2 as a key drug target for the treatment of metastatic colorectal cancer. Int J Biol Macromol 119:48–52

Quinlan JR (1987) Simplifying decision trees. Int J Man Mach Stud 27(3):221–234

Rashid M (2020) Design, synthesis and ADMET prediction of bis-benzimidazole as anticancer agent. Bioorg Chem 96:103576

Richard AM, Williams CR (2002) Distributed structure-searchable toxicity (DSSTox) public database network: a proposal. Mutat Res 499(1):27–52

Schölkopf B, Smola AJ, Bach F (2002) Learning with kernels: support vector machines, regularization, optimization, and beyond. MIT Press, Cambridge

Schyman P, Liu R, Desai V, Wallqvist A (2017) vNN web server for ADMET predictions. Front Pharmacol 4(8):889

Shi H, Tian S, Li Y, Li D, Yu H, Zhen X, Hou T (2015) Absorption, distribution, metabolism, excretion, and toxicity evaluation in drug discovery. 14. Prediction of human pregnane X receptor activators by using Naive Bayesian classification technique. Chem Res Toxicol 28 (1):116–125

Singh DB (2014) Success, limitation and future of computer aided drug designing. Transl Med (Sunnyvale) 4:e127. https://doi.org/10.4172/2161-1025.1000e127

Singh DB (2018) Natural lead compounds and strategies for optimization. In: Ul-Haq Z, Wilson AK (eds) Frontiers in computational chemistry. Bentham Science, Sharjah, pp 1–47

Singh DB, Dwivedi S (2019) Computational screening and ADMET-based study for targeting Plasmodium S-adenosyl-L-homocysteine hydrolase: top scoring inhibitors. Netw Model Anal Health Inform Bioinf 8:4

Singh DB, Gupta MK, Kesharwani RK, Misra K (2013) Comparative docking and ADMET study of some curcumin derivatives and herbal congeners targeting β-amyloid. Netw Model Anal Health Inform Bioinf 2(1):13–27

Spessard GO (1998) ACD Labs/LogP dB 3.5 and ChemSketch 3.5. J Chem Inf Comput Sci 38 (6):1250–1253

Steinbeck C, Hoppe C, Kuhn S, Floris M, Guha R, Willighagen EL (2006) Recent developments of the chemistry development kit (CDK)-an open-source java library for chemo-and bioinformatics. Curr Pharm Des 12(17):2111–2120

Steinwart I, Christmann A (2008) Support vector machines. Springer, New York

Sushko I, Novotarskyi S, Körner R, Pandey AK, Rupp M, Teetz W, Brandmaier S, Abdelaziz A, Prokopenko VV, Tanchuk VY, Todeschini R (2011) Online chemical modeling environment (OCHEM): web platform for data storage, model development and publishing of chemical information. J Comput Aided Mol Des 25(6):533–554

Tan JJ, Cong XJ, Hu LM, Wang CX, Jia L, Liang XJ (2010) Therapeutic strategies underpinning the development of novel techniques for the treatment of HIV infection. Drug Discov Today 15 (5–6):186–197

Tetko IV, Gasteiger J, Todeschini R, Mauri A, Livingstone D, Ertl P, Palyulin VA, Radchenko EV, Zefirov NS, Makarenko AS, Tanchuk VY (2005) Virtual computational chemistry laboratory–design and description. J Comput Aided Mol Des 19(6):453–463

Tjoe-Nij E, Rochin C, Berne N, Sassi A, Leplay A (2018) Chemical risk assessment screening tool of a global chemical company. Saf Health Work 9(1):84–94

Todeschini R, Consonni V (2008) Handbook of molecular descriptors. Wiley, Weinheim

Tripathi A, Singh DV, Kesharwani RK, Misra K (2015) P-glycoprotein: a critical comparison of models depicting mechanism of drug efflux and role of modulators. Proc Natl Acad Sci India Sect B 85(2):359–375

Werck-Reichhart D, Feyereisen R (2000) Cytochromes P450: a success story. Genome Biol 6:3003–3001

Yang H, Sun L, Wang Z, Li W, Liu G, Tang T (2018) ADMETopt: a web server for ADMET optimization in drug design via scaffold hopping. J Chem Inf Model 58(10):2051–2056

Yang H, Lou C, Sun L, Li J, Cai Y, Wang Z, Li W, Liu G, Tang Y (2019) admetSAR 2.0: web-service for prediction and optimization of chemical ADMET properties. Bioinformatics 35 (6):1067–1079

Zhang T (2001) An introduction to support vector machines and other kernel-based learning methods. AI Mag 22(2):103

Database Resources for Drug Discovery

5

Anil Kumar and Praffulla Kumar Arya

Abstract

Drug discovery aims to find such molecules that bind and modulate the function of a molecular target involved in a certain disease. A drug molecule must have certain geometry and physicochemical properties for high binding affinity against a given molecular target. Database searches considering required parameters for biological activity can find molecules suitable for further studies to achieve the desired activity. Chemical databases, as well as molecular target databases, are the backbone of drug discovery, which catalyze the development of computational methods to reduce the time and cost and to build a hypothesis for discovery and design of new drug molecules.

A huge amount of chemical information is available in public domain databases for use by researchers. However, due to limitations of curated resources and software, and dissimilarities in database applications, the exact molecule equivalence among databases is not possible. Advances in methodology to find molecules with a similar structure are now possible due to the hyperlinking of similar molecules in various databases. This chapter discusses the chemical and molecular target databases in detail, which play a vital role in drug discovery in recent times.

Keywords

Drug discovery · Chemical databases · Molecular targets · Disease targets · Biological activity · Metabolic pathways

A. Kumar (✉) · P. K. Arya
Department of Bioinformatics, Central University of South Bihar, Gaya, India
e-mail: kumaranil@cub.ac.in

D. B. Singh (ed.), *Computer-Aided Drug Design*,
https://doi.org/10.1007/978-981-15-6815-2_5

5.1 Introduction

In the drug discovery process, novel therapeutic molecules are identified, which involves several scientific disciplines, including chemistry, biology, and pharmacology. Historically, serendipitous experiments or identification of the active ingredient from traditional remedies led to the discovery of several drugs. Later, libraries of natural products and chemically synthesized small molecules were screened against cell lines to find molecules with a promising therapeutic potential in a process called classical pharmacology. Completion of the human genome project allowed cloning followed by synthesis of enough quantities of purified proteins, which were utilized for high-throughput screening of chemical libraries against biological targets. Hits from these screening results were then tested in cells and animals, respectively, for efficacy.

The modern drug discovery process includes the identification of potential hits and their optimization to improve the affinity, selectivity, efficacy, and oral bioavailability. Once a molecule with all of these required properties has been identified, the process of drug development followed by their clinical trials begins. One or more steps of the drug discovery process involve computer-aided drug design. Thus, the modern drug discovery process involves a huge amount of investments by the pharmaceutical industry. Even after rapid advances in technology and a deep understanding of biological systems, drug discovery is still a lengthy and difficult process with few therapeutic discoveries (Mohs and Greig 2017). The research and development cost for each new therapeutic molecule or drug was around US$1.8 billion in 2010 (Paul et al. 2010). The end product of the drug discovery process is a patent on the potential therapeutic molecule that needs a very expensive phase I, II, and III clinical trials. The commercial success of newly discovered drugs involves a typical interaction between industry, academia, investors, patent laws, and marketing, which also require maintaining secrecy with communication (Warren 2011).

In silico analysis accelerates the identification of drug targets followed by a screening of drug candidates and their refinement. It also facilitates the prediction of potential side effects and drug resistance. High-throughput data such as genomic, transcriptomic, proteomic, and ribosome profiling have made an important contribution to drug discovery and drug repurposing. Development of homology model and protein structure simulation coupled with databases of small molecules and metabolites have provided the way for more informative virtual screening for drug discovery. In this chapter, we discuss databases for therapeutic targets, chemical and drug molecule, metabolic pathways, disease and physiology, and peptide information, as depicted in Fig. 5.1.

5.2 Therapeutic Target Information

Biological macromolecules such as proteins and nucleic acids are potential therapeutic targets. Pharmaceutical agents bind to a therapeutic target to show their effect. Increased understanding of genetic, structural, and functional information of

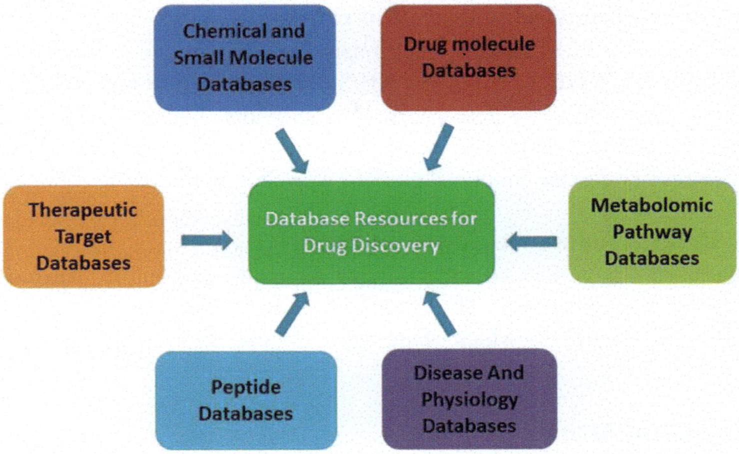

Fig. 5.1 Database resources for drug discovery

Table 5.1 Databases for therapeutic target information

S. No.	Database name	Description
1	Universal protein resource (UniProt)	UniProt consortium maintains the UniProt KnowledgeBase, UniProt reference clusters, and UniProt archive
2	Protein data Bank (PDB)	3D database of X-ray or NMR determined biomolecules
3	Molecular Modeling database (MMDB)	Collection of experimentally determined 3D bio-molecular structures
4	Therapeutic target database (TTD)	Collection of information of known therapeutic protein and nucleic acid and corresponding drugs targets
5	Herbal ingredients targets database (HIT)	Curated database on protein targets and precursors for FDA-approved drugs
6	SuperTarget	Collection of information for drug–target interactions

disease-related genes and proteins raised strong interest in the search of new therapeutic targets and also promoted the study of the underlying mechanism of their binding agents. Algorithms and parameters of drug designing approaches have been refined and tested using more datasets. Therefore, a free database can provide more detailed information about the target. Some important databases providing information about these therapeutic targets are listed in Table 5.1.

5.2.1 Universal Protein Resource (UniProt)

The Universal Protein Resource (UniProt) consortium is formed by the Swiss Institute of Bioinformatics (SIB), the European Bioinformatics Institute (EBI), and the Protein Information Resource (PIR) (UniProt Consortium 2015). It provides information on protein sequences and their functions.

5.2.1.1 UniProtKnowledgeBase (UniProtKB)

It is a database for protein sequences. It consists of two sections, UniProtKB/Swiss-Prot and UniProtKB/TrEMBL. UniProtKB/Swiss-Prot contains manually annotated entries that are curated by researchers and provides hyperlinks to more than 100 related databases with access to additional tools. However, UniProtKB/TrEMBL stores computer-annotated entries (Apweiler et al. 2004).

5.2.1.2 UniProt Reference Clusters (UniRef)

It is clusters of the protein sequences based on sequence identity that have been stored in different databases. The examples of these databases are UniRef100, UniRef90, and UniRef50, which share 100, 90, and 50% identity, respectively (Suzek et al. 2007).

5.2.1.3 UniProt Archive (UniParc)

UniParc database provides data about all protein sequences, which also includes obsolete data, which is excluded from UniProtKB (Leinonen et al. 2004). The UniProt databases are freely accessible online, and their data are available for download in several formats from the FTP server (ftp://ftp.uniprot.org/pub). New releases are included at every 2 weeks.

5.2.2 Protein Data Bank (PDB)

PDB is a freely accessible resource for the three-dimensional (3D) structures of biological macromolecules such as proteins and nucleic acids (Berman et al. 2000). The data submitted to PDB is obtained by 3D-structure elucidation techniques such as NMR spectroscopy, X-ray crystallography, or cryo-electron microscopy. 3D-structure of proteins alone and their complexes with other molecules are available with different resolutions at PDB, which can be used as a target for drug designing. PDB is available on the internet through the websites of its member organizations (PDBe, PDBj, and RCSB) (Berman et al. 2000; Mir et al. 2018; Kinjo et al. 2018). Its archive is maintained by an organization called Worldwide Protein Data Bank (wwPDB) (Berman et al. 2007). The PDB is an important resource for researchers working in the areas of structural biology and drug discovery. Funding agencies and major scientific journals require researchers to submit their structural data to PDB, which is used by several other databases.

5.2.3 Molecular Modeling Database

The Molecular Modeling Database (MMDB) is a freely available repository of experimentally determined 3D structures of biological macromolecules maintained by the National Center for Biotechnology Information (NCBI), USA. MMDB is integrated with NCBI's Entrez search and retrieval system and presents the contents of PDB. MMDB provides detailed and pre-computed structural alignments obtained by the Vector Alignment Search Tool (VAST). It also provides tools for 3D structure visualization and structure/sequence alignment with molecular graphics tool Cn3D (Madej et al. 2012).

5.2.4 Therapeutic Target Database

Therapeutic Target Database (TTD) is a comprehensive repository maintained by Bioinformatics and Drug Design Group (BIDD) at the National University of Singapore and Innovative Drug Research and Bioinformatics Group (IDRB) in Zhejiang University, China (Li et al. 2018). TTD includes information on protein and nucleic acid targets, related disease, pathway-related information, and available drugs against each of these targets (Hopkins and Groom 2002; Overington et al. 2006; Zheng et al. 2006). This database is well referenced to related databases storing information of target function, sequence, 3D structure, enzyme nomenclature, drug structure, ligand properties, therapeutic class, and status of clinical development (Li et al. 2018). Multi-target agents have been explored to increase the therapeutic activity, improved safety profiles, and less resistance by modulating the activity of a primary target (Larder et al. 1995; Keith et al. 2005; Smalley et al. 2006). These agents are available for download from the multi-target agent's page. TTD 2018 update includes 3101 targets, which include 445 successful targets, 1121 clinical trial targets, and 1535 research targets. The total number of drugs included in the database is 34,019, which include 2544 approved drugs, 8103 clinical trial drugs, and 18,923 investigative drugs (Li et al. 2018).

5.2.5 Herbal Ingredients Targets (HIT) Database

HIT is a curated database of protein targets and precursors for FDA-approved drugs. Currently, it contains 1301 known protein targets curated from 3250 literature reports, which include around 586 active molecules from more than 1300 herbs. IC_{50} and K_i values are also collected from the literature reports (Ye et al. 2011).

5.2.6 SuperTarget

SuperTarget is a web-based resource dedicated to drug–target interactions. It provides important information on drug molecules, such as side effects of drug, metabolism, pathways, and gene ontology for target proteins. Most of the interactions have binding affinities that are cross-linked to the related published reports. The user interface enables the user to make complex queries and provides tools for drug screening (Hecker et al. 2012).

5.3 Chemical Information

Chemical databases are considered as a powerful tool in drug design and discovery. Possible requirements based searches in the database can find molecules with desired biological activity that might be an appropriate candidate for further analysis. Some important resource databases for chemical information are summarized in Table 5.2.

Table 5.2 Chemical databases

S. No.	Database name	Description
1	PubChem	A freely available database of chemical molecules and their biological response
2	ZINC	Curated collection of chemical compounds which are commercially available
3	ChEMBL	A manually curated repository of bioactive compounds with drug-like properties
4	Chemical entities of biological interest (ChEBI)	Explicitly referenced database and ontology of molecular entities focused on small molecules
5	NCI database	Collection of more than 250,000 small molecule structures
6	ChemDB	A public database of small molecules built using catalogs of more than 150 vendors and other public sources
7	ChemSpider	Collection of more than 63 million unique chemical entities
8	BindingDB	A publicly available database of small molecules which currently contains about 1.2 million binding data for 5500 proteins and over 0.52 million drug-like molecules
9	PDBbind	Collection of binding affinity data of bio-molecular complexes
10	Toxin and toxin-target database (T3DB)	Database of common substances toxic to humans
11	BIAdb	Collection of 846 benzylisoquinoline alkaloids (BIAs)
12	SuperNatural II	A freely available database of more than 325,508 natural compounds
13	NPACT	Collection of plant-derived anti-cancerous compounds
14	Dictionary of natural products online	Collection of natural products
15	Ligand expo	Collection of chemical/structural information of small molecules from the PDB entries
16	SuperLigands	Collection of ligand structures derived from the PDB entries
17	Toxicology data network (TOXNET)	Collections of toxicology databases freely available online

5.3.1 PubChem

PubChem is a public database that provides information on chemical compounds and their therapeutic roles (Kaiser 2005). It stores small molecules and substances with less than 1000 atoms and 1000 bonds from more than 80 database vendors, which can be download via file transfer protocol (FTP). It consists of three primary databases. PubChem has:

- Compounds: 96.5 million entries for compounds containing pure and characterized chemical compounds.
- Substances: 247.4 million entries for substances containing extracts, complexes, mixtures, and uncharacterized substances.
- BioAssay: Several million bioactivity values from 1.25 million high-throughput screening programs.

This database provides different searching criteria and has a molecule editor facility with SMILES and InChI. Here, each hit provides the information on synonyms of the compound, IUPAC name, chemical formula, 2D/3D chemical structure, molecular weight, logP, H-bond donor and acceptor, SMILES, therapeutic role, and hyperlinks to other related resources.

5.3.2 Zinc

The ZINC database is a vast collection of chemical compounds, which can be searched using different query parameters. These compounds can be downloaded from the ZINC, and used for virtual screening against a target of the disease. It is used by researchers in pharmaceutical and biotech companies as well as in research universities. ZINC database contains over 35 million purchasable molecules available for virtual screening. Data is freely available for download in several file formats, including SMILES, mol2, 3D SDF, and DOCK flexibase format. Searching, browsing, and molecular drawing interface facility are available on the ZINC database (Irwin and Shoichet 2005).

5.3.3 ChEMBL

ChEMBL is a repository of bioactive compounds with drug-like properties managed by the EBI. It is accessible through a user-friendly web interface. Bioactivity data of compounds against drug targets can also be downloaded by FTP (Mok and Brenk 2011; Gaulton et al. 2012). Data can be used to develop compound screening libraries for lead identification in the drug discovery process (Brenk et al. 2008). In total, this database provides more than 1.6 million compounds with 14 million biological activity values from approximately 1.2 million assays. These assays are mapped to approximately 11,000 targets, including 9052 proteins, out of which 4255

Table 5.3 ChEMBL tools and resources

S. No.	Database name	Description
1	UniChem	Provides the cross-referencing between identifiers from different chemical databases
2	SureChEMBL	A database for patent information
3	Malaria data	Compounds, targets, assays, and data for the malaria-related study
4	ChEBML-NTD	Primary screening and medicinal chemistry data of tropical diseases
5	ADME SARfari	Tool for prediction and comparison of (absorption, distribution, metabolism, and excretion) ADME targets
6	Kinase SARfari	A chemogenomics workbench for kinases incorporating and linking kinase sequence
7	GPCR SARfari	Chemogenomics workbench for G protein-coupled receptor (GPCR)

are from humans (Gaulton et al. 2017). The ChEMBL group also provides several tools and resources for data mining purposes, which are summarized in Table 5.3.

ChEMBL tools and resources include Kinase SARfari and GPCR SARfari, both of which are integrated chemogenomics workbench focused on kinases and GPCR, respectively. ChEMBL-NTD contains data related to endemic tropical diseases of Asia, Africa, and America (Bender 2010; Bellis et al. 2011). Medicines for Malaria Venture (MMV) stores the information of compounds from the malaria box screening set. It also includes malaria data stored in ChEMBL-NTD. ADME SARfari is used for predicting and comparing cross-species ADME targets (Davies et al. 2015). There is need to explore and validate the traditional knowledge of medicinal herbs and their bioactives using modern proteomics, genomics, and metabolomics approaches (Singh et al. 2019).

5.3.4 Chemical Entities of Biological Interest (ChEBI)

ChEBI is an open access referenced repository for molecular entities and their ontology based on small chemical compounds. This database is a part of the Open Biomedical Ontologies (OBO) effort, which does not include nucleic acids and peptides (Degtyarenko et al. 2008; de Matos et al. 2010).

5.3.5 NCI Database

The NCI database has more than 250,000 small molecule structures. Its graphical user interface has been developed using the chemistry information toolkit CACTVS to perform rapid searches by numerous criteria. It includes all structures, anticancer and anti-HIV screening data supplemented by predicted data such as logP, and biological activities. This database can be searched by using Boolean searches and flexible substructure searches. The user can perform 3D pharmacophore-based queries. 2D and 3D visualization and numerous output format options are available (Ihlenfeldt et al. 2002).

5.3.6 ChemDB

ChemDB is a public database of small molecules built using the digital catalogs of more than 150 vendors and other public resources. It is a database of more than five million chemicals annotated with physicochemical properties such as three-dimensional structure, melting temperature, and solubility. It is periodically updated and supports multiple molecular formats. This database includes chemical reaction capabilities as well as unique search capabilities (Chen et al. 2005). Text-based efficient searches can be performed based on more than 65 million annotations. It utilizes fuzzy text matching algorithms to produce better results (Chen et al. 2007).

5.3.7 ChemSpider

ChemSpider is a database of chemical structures, which stores more than 67 million entities collected from different resources (Pence and Williams 2010). These sources include databases of curated literature, vendor catalogs, molecular properties, toxicity, and analytical data. ChemSpider is maintained by the Royal Society of Chemistry. Its objective is to store all chemical structures and also to provide the hyperlink to the related information.

5.3.8 BindingDB

BindingDB is a publically available database of small molecules, which currently contains about 1.2 million binding data for 5500 proteins and over 0.52 million drug-like molecules. It facilitates several search options such as query by protein target name, journal citations, chemical similarity, and substructure. It also provides data download tools which help the user to download the data by target or query results (Liu et al. 2007). This database provides binding affinity data based on protein–ligand (chemical/drug-like molecule) complexes (Chen et al. 2001). It includes data extracted from the scientific reports, PubChemBioAssays, and ChEMBL entries for established targets (Chen et al. 2001b). The purpose of BindingDB is to help researchers from various disciplines such as computational chemistry, medicinal chemistry, chemical biology, and drug discovery (Chen et al. 2002).

5.3.9 PDBbind

It is a collection of binding affinity data derived from the bio-molecular complexes available in PDB (Wang et al. 2004; Wang et al. 2005). It stores valuable information on protein–ligand complexes, which is useful for understanding various interactions occurring in biological systems. The 2017 release of PDBbind provides binding data of 17,900 bio-molecular complexes, which include 14,761 protein–ligand, 121 nucleic acid–ligand, 837 protein–nucleic acid, and 2181protein–protein complexes. All binding data are curated from over 32,000 original references. Free registrations are provided to the users to access all the functionalities of the database, which also include PDBbind content download (Liu et al. 2015).

5.3.10 Toxin and Toxin-Target Database (T3DB)

T3DB is a collection of common substances, which are toxic to humans. Currently, this database contains approximately 3700 toxic compounds or poisons along with their synonyms. Common pollutants, food toxins, pesticides, household, industrial, and cigarette toxins are aggregated in the database. Each toxin is linked to respective molecular targets. There are 42,433 toxin–toxin target pairs recorded in T3DB. Each entry (ToxCard) in the database stores detailed information of the toxin, which includes its chemical properties, mechanisms of action, toxicity dose values, molecular and cellular interactions, symptoms and treatment, spectral information, and modulated genes. All the information is curated from thousands of scientific reports, books, and related databases (Wishart et al. 2015). Its objective is to develop a better understanding of the mode of action of toxins and identifying their molecular targets. T3DB can be queried based on keyword, sequence, chemical structure, and spectral searches. It is linked to the related databases such as DrugBank and Human Metabolome Database (HMDB) (Lim et al. 2010).

5.3.11 BIAdb

BIAdb is a collection of benzylisoquinoline alkaloids (BIAs) that stores information of around 846 unique BIAs. Many BIA's possess therapeutic properties and can be considered as potential lead molecules. Hence BIAdb can be useful to the researchers working on natural alkaloids as potential therapeutic agents. These are produced by a variety of organisms, such as bacteria, fungi, plants, and animals. These are known to have pharmacological properties and have been traditionally used to treat several diseased conditions. Cocaine as a local anesthetic and stimulant; caffeine and nicotine as a stimulant; morphine as an analgesic; and quinine as an antimalarial drug are good examples of BIAs (Singla et al. 2010).

5.3.12 Super Natural II

Super Natural II is a freely available database of natural compounds (NCs). It stores more than 325,508 compounds with information on 2D structures, physicochemical properties, and toxicity. The structural diversity of natural products provides an opportunity for research and innovations in drug discovery, nutrition, and agrochemical research. Most of the current drugs and beauty products are derived from natural products (Banerjee et al. 2015).

5.3.13 Naturally Occurring Plant-Based Anti-Cancer Compound-Activity-Target Database (NPACT)

NPACT is a collection of anti-cancer compounds derived from plant sources. It contains more than 1574 entries of compounds. Each entry/record provides information on a compound's structure, inhibitory values such as $IC_{50}/ED_{50}/EC_{50}/GI_{50}$, physical and topological properties, drug-likeness, cancer types, target information, references, and vendors of compounds. It provides various options for browsing, searching for users. An online similarity search can also be performed. Further, each record is hyperlinked to related databases so that the users can refer to existing data if interested (Mangal et al. 2013).

5.3.14 Dictionary of Natural Products Online

The Dictionary of Natural Products Online includes all compounds contained in the dictionary of natural products (http://dnp.chemnetbase.com). It is derived from a Dictionary of Organic Compounds (DOC), a repository of natural product information since its inception in 1930. Its information has been aggregated by a team at Chapman and Hall, UK. Its online version provides information on all known natural products. Similar compounds are organized into a single entry simplifying the relationships of those closely related compounds. Each compound is indexed by its structural/biogenetic type. There is extensive coverage of natural products of unknown structure, which is being enhanced by various retrospective searches.

5.3.15 Ligand Expo

Ligand Expo is an updated version of Ligand Depot. It aggregates the chemical as well as structural information of small molecules from the entries of PDB. It provides various tools to search for chemical components in the PDB dictionary. Structural entries with a specific small molecule can also be identified. It also provides a sketch tool that can be used to build new chemical definitions (Feng et al. 2004).

5.3.16 SuperLigands

SuperLigands is a repository that stores small molecule structures present as ligands in the PDB database. It aggregates information about drug-likeness and binding properties. The structural similarity of these compounds can be estimated by calculating Tanimoto coefficients and by 3D superposition. 2D similarity search for compounds based on fingerprints can be also performed. This database could be a useful resource for prediction and analysis in the field of biological research (Michalsky et al. 2005).

5.3.17 Toxicology Data Network

The Toxicology Data Network (TOXNET) is one of the world's largest collections of toxicology databases freely available online (TOXNET, 2020). It provides effective access to the online group of databases developed by the National Library of Medicine (NLM). These databases are the resource for information on toxicology, environmental health, hazardous chemicals, and toxic releases. Some of the popular databases of TOXNET are listed in Table 5.4.

5.4 Drug Molecule Information

The internet is becoming the first port for all kinds of information searches. Drug-related resources that are currently available online provide researchers a convenient path to the information. The diversity of the information that is accessible today online is growing at an exponential rate, and freely available resources are making a very significant contribution in terms of the benefits to research as well as to the society. With the recent explosion in biological and chemical information, our knowledge about drugs and their molecular targets and their mechanism of action cannot be compiled in a few encyclopedic books. There is a huge amount of data from many sources available through different public databases, which were accumulated over the past half-century. Some important drug resource databases are listed in Table 5.5.

5.4.1 DrugBank

The DrugBank is a freely available online comprehensive database of drugs and their target information. DrugBank associates detailed chemical and pharmacological data of drugs with corresponding drug target data such as sequence, structure, and pathway information. The information in the drug bank is very well referenced to the published scientific reports and other related databases, which make it more useful to the researchers as well as to the pharmaceutical industry. Nearly all drugs listed in Wikipedia have links to DrugBank. It supports the drug discovery and

Table 5.4 TOXNET databases (https://toxnet.nlm.nih.gov/)

S. No.	Database name	Description
1	HSDB	Toxicology data of more than 5000 hazardous chemicals
2	TOXLINE	Collection of literature references on biochemical, pharmacological, and toxicological effects of drug molecules
3	ChemIDplus	Repository of more than 400,000 chemicals
4	LactMed	Drugs and other chemical molecules to which breastfeeding mothers may be exposed
5	Dart	References to developmental and reproductive toxicology reports
6	TOXMAP	A tool for exploring environmental health data
7	Toxics release inventory (TRI)	TRI stores data of annual environmental releases of toxic chemicals
8	Comparative Toxicogenomics database (CTD)	CTD is a collection of data describing relationships of chemicals, genes, and human diseases
9	Household products database	Health effects of chemicals in household products
10	Haz-map	Database of occupational diseases and their symptoms
11	Integrated risk information system (IRIS)	Provides dose-response assessment for hazardous chemicals
12	International toxicity estimates for risk (ITER)	Provides risk information for more than 600 chemicals
13	Resources on alternatives to the use of live vertebrates in biomedical research and testing (ALTBIB)	Use of live vertebrates in biomedical research
14	Chemical carcinogenesis research information system (CCRIS)	Provides carcinogenicity and mutagenicity test results for more than 8000 chemicals
15	Carcinogenic potency database (CPDB)	Stores results of 6540 chronic, long-term animal cancer tests
16	GENE-TOX	It stores genetic toxicology test data for more than 3000 chemicals

Table 5.5 Drug molecule databases

S. No.	Database name	Description
1	DrugBank	A freely accessible comprehensive database of drugs and drug targets information
2	SuperDRUG2	SuperDRUG version 2.0 is a unique resource for more than 4600 approved/marketed drugs
3	PharmGKB	Pharmacogenomics related resource, managed at Stanford University since its inception

repurposing of many existing drugs to treat rare and newly identified diseases (Wishart et al. 2006; Wishart et al. 2018a). Each entry comprises more than 200 data fields, out of which 50% are dedicated to drug/chemical information and the rest to drug target information.

5.4.2 SuperDRUG2

SuperDRUG Version 2.0 is a notable resource for approved and marketed drugs. Currently, it provides more than 4600 drugs/pharmaceutical agents (Siramshetty et al. 2018). Each entry for the drug is annotated with 2D and 3D chemical structures, dosage, biological targets, physicochemical properties, side effects, and other necessary details. A database search can be performed with different methods. It is provided with a 2D chemical structure search and a 3D superposition feature that superposes a drug with known ligand molecules found in the protein–ligand complexes. Simulation of "physiologically-based" pharmacokinetics of drugs can also be performed. Potential drug–drug interactions can be identified by the interaction check feature, which also provides alternative recommendations for elderly patients. SuperDRUG2 is freely available for academic users. It needs a free browser plugin "Chime" for visualization (Goede et al. 2005).

5.4.3 PharmGKB

The Pharmacogenomics Knowledgebase (PharmGKB) provides information on the impact of human genetic variation on drug responses. It is funded by the National Institutes of Health (NIH) and managed at Stanford University since its inception. It is a partner of the NIH Pharmacogenomics Research Network (PGRN). It includes data from clinical as well as basic pharmacokinetics and pharmacogenomics research in cardiovascular, cancer, pulmonary and metabolic pathways domains. Its goal is to explore the role of genetic variation among individuals in contributing to differences in drug reactions. Currently, it provides information on 645 drugs with variant annotations, 132 pathways, 100 dosing guidelines, 509 annotated drug labels (Mcdonagh et al. 2011; Whirl-Carrillo et al. 2012).

5.4.4 Search Tool for Interactions of Chemicals (STITCH)

Small molecules play a very crucial role in biological systems by their interaction with target biomolecules (Sharan et al. 2007). Interaction network is even more important for drug development as diseases occur due to several changes in the same pathway. The interaction network leads to a better understanding of a drug's impact in a biological system (Hopkins 2008; Barabási et al. 2011). To provide access to this data, which is important for computational steps in drug discovery, STITCH aggregates information on interactions from pathways, crystal structures, binding studies, and drug–target relationships. A network of chemical relations in associated binding proteins can also be explored. Each proposed interaction is well referenced to the original data sources (Kuhn et al. 2008). Its chemical space has also grown more than 430,000 compounds (Szklarczyk et al. 2015). It is available online through a newly redesigned web interface and via an extensive application program interface (API) (Szklarczyk et al. 2016).

5.5 Metabolomic Pathway Information

The role of metabolomics in drug discovery is undeniable. The metabolic profile is a footprint of phenotype and biochemical activity following any irregularities in the system. Small molecules found in a biological system, as well as drugs and their metabolic by-products, are called metabolites. Metabolomics can suggest interesting molecular targets for drug discovery and provide information about possible novel therapeutic agents. Metabolomics, as well as compound databases, is growing exponentially with the incorporation of more molecular and spectral information. More numbers of biological systems are being represented by metabolic network models (Gupta et al. 2012). A combination of experimental as well as computational tools with high-throughput screening experiments can provide new promising lead molecules. Important metabolic databases are listed in Table 5.6.

5.5.1 Kyoto Encyclopedia of Genes and Genomes (KEGG)

KEGG is a collection of resources for genomes, biological pathways, diseases, drugs, and chemical substances (Kanehisa and Goto 2000). It is utilized for research

Table 5.6 Metabolomic pathway databases

S. No.	Database name	Description
1	KEGG	Databases for genomes, pathways, diseases, and chemical compounds
2	HMDB	Collection of information about small molecule metabolites
3	SMPDB	A comprehensive, repository of more than 30,000 small molecule pathways
4	BiGG	Provides free biochemical, genetic, and genomic knowledgebase
5	MetaboLights database	MetaboLights is a manually curated, freely accessible database of metabolomics experiments and derived information at EMBL
6	BioCyc	BioCyc is a collection of 13,075 (as of April 2018) Pathway/Genome Databases (PGDBs)
7	Reactome	Biological pathways which include metabolic pathways (pathways related to protein trafficking and signaling)
8	WikiPathways	Collection of manually curated biological pathways created using MediaWiki software

Table 5.7 Resources available at KEGG database

S. No.	Database name	Description
A. Systems information		
1	PATHWAY	Contains pathway maps with molecular interaction, and reaction
2	MODULE	KEGG modules are manually defined functional units identified by the M numbers. These are used for annotation of sequenced genomes
3	BRITE	A collection of hierarchical text (htext) files storing functional hierarchies of biological objects (KEGG objects)
B. Genomic information		
1	GENOME	A collection of organisms with complete genome sequences and selected viruses with relevance to diseases
2	GENES	A collection of gene catalogs from NCBI RefSeq and GenBank
3	ORTHOLOGY	A database of molecular functions as functional orthologs
C. Chemical information		
1	COMPOUND	A repository of small molecules, biopolymers, and other chemicals relevant to biological systems
2	GLYCAN	A collection of experimentally determined glycan structures
3	REACTION	A repository of chemical reactions from KEGG metabolic pathway maps enzyme nomenclature
4	ENZYME	Collection of information about enzyme nomenclature (EC number system) based on ExplorEnz database
D. Health information		
1	DISEASE	A collection of disease entries focusing only on the perturbation basis
2	DRUG	Collection of approved drugs in the USA, Europe, and Japan
3	NETWORK	To capture knowledge on diseases and drugs in terms of perturbed molecular networks
4	MEDICUS	An integrated resource of diseases, drugs, and health-related substances
5	ENVIRON	A collection of health-promoting natural products of plants such as crude drugs, essential oils, etc.

in various fields of biological sciences, including data analysis in genomics, metabolomics, systems biology, and translational research in drug discovery and development (Table 5.7). KEGG integrates genetic building blocks and pathways of the biological system together, which includes genes, proteins, small molecules, chemical reactions, and reaction networks.

5.5.2 Human Metabolome Database (HMDB)

HMDB is an online freely accessible database, which provides information on experimentally verified small molecule metabolites. HMDB stores chemical, clinical, and biochemical data. Currently, it provides information on more than 114,100 metabolites and 652 diseases. Additionally, 5779 proteins are linked to these metabolite entries. This database is connected with other resources such as KEGG, PubChem, ChemSpider, ChEBI, PDB, and DrugBank (Wishart et al. 2007; Wishart et al. 2009). The HMDB is considered to be the first metabolomics database dedicated to human metabolomics research. The chemical data includes 135,138 compounds with spectra along with 3897 NMR spectra, 22,247 experimental LC-MS/MS spectra, and 7418 experimental GC-MS spectra (Wishart et al. 2018b).

5.5.3 Small Molecule Pathway Database (SMPDB)

SMPDB is a high-quality and freely available repository storing more than 30,000 small molecule pathways found in humans. Most of these pathways are not available in any other database. It facilitates research in pathway elucidation in humans by providing data of metabolic, physiological disease, signaling, drug metabolism, and drug-action pathways. Each small molecule is hyperlinked. SMPDB is a user-friendly database and facilitates text, sequence, and chemical structure-based searches. It can be queried with lists of metabolite names, drug names, and genes, or protein names. The query can also be made using various identifiers such as UniProt, GenBank, and Affymetrix or Agilent microarray IDs (Frolkis et al. 2010; Jewison et al. 2014).

5.5.4 BiGG

BiGG database is a biochemical genetic and genomic knowledgebase of metabolic models that utilize the Constraint-based Reconstruction and Analysis (COBRA) framework (Schellenberger et al. 2010). It is freely available for academic use. It aggregates various genome-scale metabolic networks into a single resource and follows standard nomenclature so that components can be compared across different organisms. Data is hyperlinked to several related databases for more information on genes, proteins, metabolites, reactions, and references. This database addresses the need for systems biology researchers by providing 75 genome-scale high-quality metabolic models (King et al. 2016).

5.5.5 MetaboLights Database

MetaboLights is a manually curated and freely accessible online database for metabolomics research maintained at the EMBL-EBI (Kale et al. 2016). This database provides experimental data from the metabolomics experiments, which are Metabolomics Standards Initiative (MSI) compliant. It has strong reporting capabilities and also offers user-friendly submission tools. Studies are assigned with a unique identifier for reference. The reference layer of the database combines metabolites with metabolomics experiments. The database also provides a guide to downloading and using experimental data as well as for data submission.

5.5.6 BioCyc

BioCyc is a cluster of 13,075 (as of April 2018) Pathway/Genome Databases (PGDBs). Each database provides the genome and metabolic pathways of a specific organism. These databases are organized into tiers based on the amount of manual review and update. Tier 1 PGDBs are manually curated and frequently updated. Tier 1 PGDBs include HumanCyc, EcoCyc, MetaCyc, and the BioCyc Open Compounds Database (BOCD) (Krieger et al. 2004; Romero et al. 2005). Tier 2 PGDBs are generated computationally and have moderate manual updating. However, Tier 3 PGDBs were computationally generated and receive no manual updates.

This database serves as a reference for the genomes and metabolic pathways of thousands of sequenced organisms. It also compiles protein features and gene ontology information from the UniProt database. A suite of software tools has also been provided with the website for various purposes, such as database searching and visualization, data analysis and comparative genomics, and pathway queries.

5.5.7 Reactome

Reactome is an open-source database for metabolic pathways and other pathways related to protein trafficking and signaling. It provides pathway data on humans and some other organisms. Reactome provides data for proteins, reactions, and pathways for humans (Croft et al. 2011). It includes experimentally established, manually, and electronically deduced reactions in its pathway diagram collection. It also provides tools for the visualization and analysis of biological pathways.

5.5.8 WikiPathways

WikiPathways is an open biological pathways curation platform. It also provides tools for data analysis and visualization. It provides a tool for graphical pathway editing and well hyperlinked to databases covering major gene, protein, and small-molecule systems. Currently, it has over 2300 pathways for 25 species. It contains more than 640 pathways from human covering more than 7500 genes. It also stores pathways with more than 1000 metabolites (Kutmon et al. 2016; Slenter et al. 2018).

5.6 Disease and Physiology Information

Disease and physiology information is useful for physicians, genetics researchers in science and medicine, and other professionals concerned with genetic disorders. Disease and physiology databases provide information on a medical or genetic condition used for the diagnosis of a certain disease. Some of the important databases of this kind are listed in Table 5.8.

5.6.1 Online Mendelian Inheritance in Man (OMIM)

OMIM is a comprehensive database for human genes and genetic disorders. OMIM provides referenced data on known Mendelian disorders and more than 15,000 genes. The entries are updated daily and contain many links to other genetic resources. OMIM has approximately 24,667 entries, out of which 8704 entries represent phenotypes, and the rest 15,963 entries represent genes related to known phenotypes (Amberger et al. 2011).

5.6.2 METAGENE

METAGENE is a knowledge base supporting the diagnosis of inborn errors of metabolism. It provides comprehensive information about 428 metabolic diseases, differential diagnoses, associated laboratory findings, and recent publications. It helps health professionals dealing with rare metabolic disorders in diagnosing or treating patients with these disorders. It is updated regularly. It has a tool for facilitating the treatment of patients with phenylketonuria/hyperphenylalaninemia who may be responsive to tetrahydrobiopterin (BH4) by knowing the genotype. Information is based on published cases in the literature or on patients documented in Rare Metabolic Diseases Database (RAMEDIS) (Trefz et al. 2008).

Table 5.8 Disease and physiology databases

S. No.	Database name	Description
1	OMIM	A comprehensive knowledge base for human genes, genetic phenotypes, and genetic disorders
2	METAGENE	A database which facilitates the diagnosis of inborn errors of metabolism in a practical approach
3	RAMEDIS	A web-based repository for rare metabolic diseases
4	OMMBID	A web-based comprehensive encyclopedia which covers genes and genetic mechanisms involved in human disease states

5.6.3 RAMEDIS

RAMEDIS is a web-based repository for rare metabolic diseases. Information on rare metabolic diseases with all possible details, which include symptoms, laboratory findings, molecular data, and therapy, was collected in close cooperation with clinical partners to develop this database. The database content is simple to compare and to analyze by using standard medical terms and conditions. Using RAMEDIS, doctors, biochemists, and scientists can publish their case studies electronically in a comfortable way. So far, it stores data of 93 genetic metabolic diseases from 818 patients. It is a universal resource that allows researchers/medical practitioners to extract diverse clinical, biochemical, and molecular data (Töpel et al. 2006; Trefz et al. 2008).

5.6.4 Online Metabolic and Molecular Basis of Inherited Disease (OMMBID)

OMMBID is a web-based encyclopedia that covers about genes and genetic mechanisms involved in human disease states. It describes the metabolism, diagnosis, and treatment of metabolic disorders. It also stores detailed pathways information, chemical structures, and physiological data, useful for clinical biochemists.

5.7 Peptide Information

Bioactive peptides are widely distributed in nature, with a variety of biological activities which have attracted researchers/scientists from biological/medical fields and pharmaceutical industry. The information on the structure of bioactive peptide is important for the development of peptide-based therapeutic agents. Many bioactive peptide databases were designed by mining literature information. Some popular peptide databases are listed in Table 5.9.

Table 5.9 Peptide databases

S. No.	Database name	Description
1	PepBank	PepBank is a peptide sequence database with 21,691 individual entries as of September 2018
2	StraPep	StraPep is a structure database for bioactive peptides with a user-friendly browser and search engine
3	Antimicrobial peptide database (APD)	APD is a manually curated database of natural antimicrobial peptides (AMPs) with their biological activity
4	CAMPR3	Collection of antimicrobial peptides with family-specific sequence composition
5	CancerPPD	A manually curated collection of anticancer peptides (ACPs) and anticancer proteins

5.7.1 PepBank

PepBank is a peptide sequence database with 21,691 individual entries. It has a user-friendly interface with the advanced search function, text-based search, BLAST, and Smith--Waterman search facilities. MEDLINE abstracts were mined as a major source of peptide sequence data in this database. Public databases (Artificial Selected Proteins/Peptides Database (ASPD) and UniProt) and full-text articles were the other sources of peptide sequence data in the PepBank. The database can be used to discover peptide-based drugs (Shtatland et al. 2007).

5.7.2 StraPep

StraPep is a structure database for bioactive peptides with a user-friendly browser and search engine. Currently, it provides bioactive peptide structures, which include toxin and venom peptide, cytokine and growth factor, neuropeptide, hormone, and antimicrobial peptide (Wang et al. 2018). Each entry related to cystine knot provides detailed information of a particular peptide, which includes the location of disulfide bonds, experimental structure, secondary structure, classification, and post-translational modification. Several user-friendly tools have been provided for browsing as well as sequence and structure-based searching, which make this database very useful for the researchers (Wang et al. 2018).

5.7.3 Antimicrobial Peptide Database (APD)

APD is a manually curated database dedicated to natural AMPs with a known sequence and biological activity (Wang and Wang 2004; Wang et al. 2009). Currently, it contains a total of 3016 AMPs, which include 2533 antibacterial, 182 antiviral, 109 anti-HIV, 1083 antifungal, 47 antiparasitic, and 217 anticancer peptides. AMPs with antioxidants, anti-biofilm, spermicidal, antimalarial, insecticidal, chemotactic, and wound healing are also available in this database. It can also be searched based on molecule-binding partners, target pathogens, post-translational modifications, and animal models (Wang et al. 2016).

5.7.4 CAMPR3

CAMPR3 database has been created to support antimicrobial peptide family-based research. These peptides have family-specific conserved sequences, which can be mined to design novel AMPs. It contains information on sequence signatures that are captured as patterns and Hidden Markov models (HMMs). It also provides hyperlinks to UniProt, PubMed, and other related databases. Presently, it holds more than 8000 sequences, more than 700 structures, and above 100 family-specific signatures of AMPs. This database also provides web-based tools for sequence alignment, pattern creation, and HMM-based search (Waghu et al. 2016).

5.7.5 CancerPPD

CancerPPD is a manually curated database of ACPs and anticancer proteins. It contains 3491 ACPs and 121 anticancer proteins. Each peptide entry provides information about the origin, nature, anticancer property, N- and C-terminal modifications, etc. It also includes the information of 249 types of cancer cell lines and 16 assays that were used for testing the ACPs. Tertiary structures of peptides were predicted and stored using PEPstr, and secondary structures were assigned using the database of secondary structure assignments (DSSP). Several web-based tools such as keyword-based search, browsing, sequence, and structural similarity search have been integrated to assist the users. CancerPPD could be very useful in finding novel anticancer therapeutics (Tyagi et al. 2015).

5.8 Challenges and Future Perspective

In this digital era, databases are an infrastructural need for the research, which is utilized by the high-performance computing platforms. The huge amount of data is being generated from various research areas such as genomics, transcriptomics, proteomics, metabolomics, and metagenomics. This data requires the creation of new databases and advanced bioinformatics tools for data analysis. The future of database development is bright. However, the annotation of existing data available through these databases is a challenge for the researchers. The integration of a huge number of databases available is also crucial. Problems of nomenclature and standardization are needed to be addressed to resolve this issue. The growth of databases will pave the way for further research in biological, pharmaceutical, and related fields.

5.9 Summary

With the use of high-throughput technologies, research laboratories have started generating huge amounts of data. This data is being stored and managed in various databases. Depending on the information stored, these databases can be categorized into several groups. This chapter provides an overview of databases useful for drug discovery and development. Most of these databases are user-friendly and freely accessible online. It discusses the kind of information stored in these databases and the way it can be retrieved and used by various bioinformatics tools for analysis.

Competing Interest The authors declare that there are no competing interests.

References

Amberger J, Bocchini C, Hamosh A (2011) A new face and new challenges for Online Mendelian Inheritance in Man (OMIM®). Hum Mutat 32(5):564–567

Apweiler R, Bairoch A, Wu CH, Barker WC, Boeckmann B et al (2004) UniProt: the universal protein knowledgebase. Nucleic Acids Res 32(Database):D115–D119

Banerjee P, Erehman J, Gohlke BO, Wilhelm T, Preissner R et al (2015) Super natural II—a database of natural products. Nucleic Acids Res 43(Database):D935–D939

Barabási AL, Gulbahce N, Loscalzo J (2011) Network medicine: a network-based approach to human disease. Nat Rev Genet 12:56–68

Bellis LJ, Akhtar R, Al-Lazikani B, Atkinson F, Bento AP et al (2011) Collation and data-mining of literature bioactivity data for drug discovery. Biochem Soc Trans 39(5):1365–1370

Bender A (2010) Databases: compound bioactivities go public. Nat Chem Biol 6:309

Berman HM, Westbrook J, Feng Z, Gilliland G, Bhat TN et al (2000) The Protein Data Bank. Nucleic Acids Res 28(1):235–242

Berman H, Henrick K, Nakamura H, Markley JL (2007) The worldwide Protein Data Bank (wwPDB): ensuring a single, uniform archive of PDB data. Nucleic Acids Res 35(Database): D301–D303

Brenk R, Schipani A, James D, Krasowski A, Gilbert IH et al (2008) Lessons learnt from assembling screening libraries for drug discovery for neglected diseases. ChemMedChem 3 (3):435–444

Chen X, Lin Y, Gilson MK (2001) The binding database: overview and user's guide. Biopolymers 61(2):127–141

Chen X, Liu M, Gilson MK (2001b) BindingDB: a web-accessible molecular recognition database. Comb Chem High Throughput Screen 4(8):719–725

Chen X, Lin Y, Liu M, Gilson MK (2002) The Binding Database: data management and interface design. Bioinformatics 18(1):130–139

Chen J, Swamidass SJ, Dou Y, Bruand J, Baldi P (2005) ChemDB: a public database of small molecules and related chemoinformatics resources. Bioinformatics 21(22):4133–4139

Chen JH, Linstead E, Swamidass SJ, Wang D, Baldi P (2007) ChemDB update—full-text search and virtual chemical space. Bioinformatics 23(17):2348–2351

Croft D, O'Kelly G, Wu G, Haw R, Gillespie M et al (2011) Reactome: a database of reactions, pathways and biological processes. Nucleic Acids Res 39(Database):D691–D697

Davies M, Dedman N, Hersey A, Papadatos G, Hall MD et al (2015) ADME SARfari: comparative genomics of drug metabolizing systems. Bioinformatics 31(10):1695–1697

de Matos P, Alcántara R, Dekker A, Ennis M, Hastings J et al (2010) Chemical entities of biological interest: an update. Nucleic Acids Res 38(Database):D249–D254

Degtyarenko K, de Matos P, Ennis M, Hastings J, Zbinden M et al (2008) ChEBI: a database and ontology for chemical entities of biological interest. Nucleic Acids Res 36(Database):D344–D350

Feng Z, Chen L, Maddula H, Akcan O, Oughtred R et al (2004) Depot: a data warehouse for ligands bound to macromolecules. Bioinformatics 20(13):2153–2155

Frolkis A, Knox C, Lim E, Jewison T, Law V et al (2010) SMPDB: the small molecule pathway database. Nucleic Acids Res 38(Database):D480–D487

Gaulton A, Bellis LJ, Bento AP, Chambers J, Davies M et al (2012) ChEMBL: a large-scale bioactivity database for drug discovery. Nucleic Acids Res 40(Database):D1100–D1107

Gaulton A, Hersey A, Nowotka M, Bento AP, Chambers J et al (2017) The ChEMBL database in 2017. Nucleic Acids Res 45(D1):D945–D954

Goede A, Dunkel M, Mester N, Frommel C, Preissner R (2005) SuperDrug: a conformational drug database. Bioinformatics 21(9):1751–1753

Gupta MK, Singh DB, Rath SK, Misra K (2012) Metabolic modeling and simulation analysis of thyroid disorder pathway. J Comput Sci Syst Biol 5(2):52–61

Hecker N, Ahmed J, von Eichborn J, Dunkel M, Macha K et al (2012) SuperTarget goes quantitative: update on drug-target interactions. Nucleic Acids Res 40(Database):D1113–D1117

Hopkins AL (2008) Network pharmacology: the next paradigm in drug discovery. Nat Chem Biol 4:682–690

Hopkins AL, Groom CR (2002) The druggable genome. Nat Rev Drug Discov 1(9):727–730

Ihlenfeldt WD, Voigt JH, Bienfait B, Oellien F, Nicklaus MC (2002) Enhanced CACTVS browser of the Open NCI Database. J Chem Inf Comput Sci 42(1):46–57

Irwin JJ, Shoichet BK (2005) ZINC—a free database of commercially available compounds for virtual screening. J Chem Inf Model 45(1):177–182

Jewison T, Su Y, Disfany FM, Liang Y, Knox C et al (2014) SMPDB 2.0: big improvements to the small molecule pathway database. Nucleic Acids Res 42(Database):D478–D484

Kaiser J (2005) Science resources. Chemists want NIH to curtail database. Science 308(5723):774

Kale NS, Haug K, Conesa P, Jayseelan K, Moreno P et al (2016) MetaboLights: an open-access database repository for metabolomics data. Curr Protoc Bioinformatics 53:1–18

Kanehisa M, Goto S (2000) KEGG: Kyoto encyclopedia of genes and genomes. Nucleic Acids Res 28(1):27–30

Keith CT, Borisy AA, Stockwell BR (2005) Multicomponent therapeutics for networked systems. Nat Rev Drug Discov 4(1):71–78

King ZA, Lu JS, Dräger A, Miller PC, Federowicz S et al (2016) BiGG models: a platform for integrating, standardizing, and sharing genome-scale models. Nucleic Acids Res 44(D1):D515–D522

Kinjo AR, Bekker GJ, Wako H, Endo S, Tsuchiya Y et al (2018) New tools and functions in data-out activities at Protein Data Bank Japan (PDBj). Protein Sci 27(1):95–102

Krieger CJ, Zhang P, Mueller LA, Wang A, Paley S et al (2004) MetaCyc: a multiorganism database of metabolic pathways and enzymes. Nucleic Acids Res 32(Database):D438–D442

Kuhn M, von Mering C, Campillos M, Jensen LJ, Bork P (2008) STITCH: interaction networks of chemicals and proteins. Nucleic Acids Res 36(Database):D684–D688

Kutmon M, Riutta A, Nunes N, Hanspers K, Willighagen EL et al (2016) WikiPathways: capturing the full diversity of pathway knowledge. Nucleic Acids Res 44(D1):D488–D494

Larder BA, Kemp SD, Harrigan PR (1995) Potential mechanism for sustained antiretroviral efficacy of AZT-3TC combination therapy. Science 269(5224):696–699

Leinonen R, Diez FG, Binns D, Fleischmann W, Lopez R et al (2004) UniProt archive. Bioinformatics 20(17):3236–3237

Li YH, Yu CY, Li XX, Zhang P, Tang J et al (2018) Therapeutic target database update 2018: enriched resource for facilitating bench-to-clinic research of targeted therapeutics. Nucleic Acids Res 46(D1):D1121–D1127

Lim E, Pon A, Djoumbou Y, Knox C, Shrivastava S, Guo AC et al (2010) T3DB: a comprehensively annotated database of common toxins and their targets. Nucleic Acids Res 38(Database): D781–D786

Liu Z, Li Y, Han L, Li J, Liu J et al (2015) PDB-wide collection of binding data: current status of the PDBbind database. Bioinformatics 31(3):405–412

Liu T, Lin Y, Wen X, Jorissen RN, Gilson MK (2007) BindingDB: a web-accessible database of experimentally determined protein-ligand binding affinities. Nucleic Acids Res 35(Database): D198–D201

Madej T, Addess KJ, Fong JH, Geer LY, Geer RC et al (2012) MMDB: 3D structures and macromolecular interactions. Nucleic Acids Res 40(Database):D461–D464

Mangal M, Sagar P, Singh H, Raghava GP, Agarwal SM (2013) NPACT: naturally occurring plant-based anti-cancer compound-activity-target database. Nucleic Acids Res 41(Database):D1124–D1129

McDonagh EM, Whirl-Carrillo M, Garten Y, Altman RB, Klein TE (2011) From pharmacogenomic knowledge acquisition to clinical applications: the PharmGKB as a clinical pharmacogenomic biomarker resource. Biomark Med 5(6):795–806

Michalsky E, Dunkel M, Goede A, Preissner R (2005) SuperLigands - a database of ligand structures derived from the Protein Data Bank. BMC Bioinf 6:122

Mir S, Alhroub Y, Anyango S, Armstrong DR, Berrisford JM et al (2018) PDBe: towards reusable data delivery infrastructure at protein data bank in Europe. Nucleic Acids Res 46(D1):D486–D492

Mohs RC, Greig NH (2017) Drug discovery and development: role of basic biological research. Alzheimers Dement 3(4):651–657

Mok NY, Brenk R (2011) Mining the ChEMBL database: an efficient chemoinformatics workflow for assembling an ion channel-focused screening library. J Chem Inf Model 51(10):2449–2454

Overington JP, Al-Lazikani B, Hopkins AL (2006) How many drug targets are there? Nat Rev Drug Discov 5(12):993–996

Paul SM, Mytelka DS, Dunwiddie CT, Persinger CC, Munos BH et al (2010) How to improve R&D productivity: the pharmaceutical industry's grand challenge. Nat Rev Drug Discov 9 (3):203–214

Pence HE, Williams A (2010) ChemSpider: an online chemical information resource. J Chem Educ 87:1123–1124

Romero P, Wagg J, Green ML, Kaiser D, Krummenacker M et al (2005) Computational prediction of human metabolic pathways from the complete human genome. Genome Biol 6(1):R2

Schellenberger J, Park JO, Conrad TM, Palsson BO (2010) BiGG: a biochemical genetic and genomic knowledgebase of large scale metabolic reconstructions. BMC Bioinf 1:213

Sharan R, Ulitsky I, Shamir R (2007) Network-based prediction of protein function. Mol Syst Biol 3:88

Shtatland T, Guettler D, Kossodo M, Pivovarov M, Weissleder R (2007) PepBank—a database of peptides based on sequence text mining and public peptide data sources. BMC Bioinf 8:280

Singh S, Singh DB, Singh S et al (2019) Exploring medicinal plant legacy for drug discovery in post-genomic era. Proc Nat Acad Sci India Sect B Biol Sci 89:1141–1151

Singla D, Sharma A, Kaur J, Panwar B, Raghava GP (2010) BIAdb: a curated database of benzylisoquinoline alkaloids. BMC Pharmacol 10:4

Siramshetty VB, Eckert OA, Gohlke BO, Goede A, Chen Q et al (2018) SuperDRUG2: a one stop resource for approved/marketed drugs. Nucleic Acids Res 46(D1):D1137–D1143

Slenter DN, Kutmon M, Hanspers K, Riutta A, Windsor J et al (2018) WikiPathways: a multifaceted pathway database bridging metabolomics to other omics research. Nucleic Acids Res 46 (D1):D661–D667

Smalley KS, Haass NK, Brafford PA, Lioni M, Flaherty KT et al (2006) Multiple signaling pathways must be targeted to overcome drug resistance in cell lines derived from melanoma metastases. Mol Cancer Ther 5(5):1136–1144

Suzek BE, Huang H, McGarvey P, Mazumder R, Wu CH (2007) UniRef: comprehensive and non-redundant UniProt reference clusters. Bioinformatics 23(10):1282–1288

Szklarczyk D, Franceschini A, Wyder S, Forslund K, Heller D et al (2015) STRING v10: protein-protein interaction networks, integrated over the tree of life. Nucleic Acids Res 43(Database): D447–D452

Szklarczyk D, Santos A, von Mering C, Jensen LJ, Bork P et al (2016) STITCH 5: augmenting protein-chemical interaction networks with tissue and affinity data. Nucleic Acids Res 44(D1): D380–D384

Töpel T, Hofestädt R, Scheible D, Trefz F (2006) RAMEDIS: the rare metabolic diseases database. Appl Bioinf 5(2):115–118

TOXNET (2020) National Library of Medicine. https://www.nlm.nih.gov/toxnet/index.html. Accessed 30 Mar 2020

Trefz F, Scheible D, Götz H, Töpel T, Hofestädt R et al (2008) METAGENE and RAMEDIS: databases for metabolic diseases and patients with inborn errors on metabolism. J Inherit Metab Dis 31:289

Tyagi A, Tuknait A, Anand P, Gupta S, Sharma M et al (2015) CancerPPD: a database of anticancer peptides and proteins. Nucleic Acids Res 43(Database):D837–D843

UniProt Consortium (2015) UniProt: a hub for protein information. Nucleic Acids Res 43(Database):D204–D212
Waghu FH, Barai RS, Gurung P, Idicula-Thomas S (2016) CAMPR3: a database on sequences, structures and signatures of antimicrobial peptides. Nucleic Acids Res 44(D1):D1094–D1097
Wang Z, Wang G (2004) APD: the antimicrobial peptide database. Nucleic Acids Res 32:D590–D592
Wang R, Fang X, Lu Y, Wang S (2004) The PDBbind database: collection of binding affinities for protein-ligand complexes with known three-dimensional structures. J Med Chem 47 (12):2977–2980
Wang R, Fang X, Lu Y, Yang CY, Wang S (2005) The PDBbind database: methodologies and updates. J Med Chem 48(12):4111–4119
Wang G, Li X, Wang Z (2009) APD2: the updated antimicrobial peptide database and its application in peptide design. Nucleic Acids Res 37:D933–D937
Wang G, Li X, Wang Z (2016) APD3: the antimicrobial peptide database as a tool for research and education. Nucleic Acids Res 44(D1):D1087–D1093
Wang J, Yin T, Xiao X, He D, Xue Z et al (2018) StraPep: a structure database of bioactive peptides. Database 2018:bay038. https://doi.org/10.1093/database/bay038
Warren J (2011) Drug discovery: lessons from evolution. Br J Clin Pharmacol 71(4):497–503
Whirl-Carrillo M, McDonagh EM, Hebert JM, Gong L, Sangkuhl K et al (2012) Pharmacogenomics knowledge for personalized medicine. Clin Pharmacol Ther 92(4):414–417
Wishart DS, Knox C, Guo AC, Shrivastava S, Hassanali M et al (2006) DrugBank: a comprehensive resource for *in silico* drug discovery and exploration. Nucleic Acids Res 34(Database): D668–D672
Wishart DS, Tzur D, Knox C, Eisner R, Guo AC et al (2007) HMDB: the human metabolome database. Nucleic Acids Res 35(Database):D521–D526
Wishart DS, Knox C, Guo AC, Eisner R, Young N et al (2009) HMDB: a knowledgebase for the human metabolome. Nucleic Acids Res 37(Database):D603–D610
Wishart DS, Arndt D, Pon A, Sajed T, Guo AC et al (2015) T3DB: the toxic exposome database. Nucleic Acids Res 43(Database):D928–D934
Wishart DS, Feunang YD, Guo AC, Lo EJ, Marcu A et al (2018a) DrugBank 5.0: a major update to the DrugBank database for 2018. Nucleic Acids Res 46(D1):D1074–D1082
Wishart DS, Feunang YD, Marcu A, Guo AC, Liang K et al (2018b) HMDB 4.0: the human metabolome database for 2018. Nucleic Acids Res 46(D1):D608–D617
Ye H, Ye L, Kang H, Zhang D, Tao L et al (2011) HIT: linking herbal active ingredients to targets. Nucleic Acids Res 39(Database):D1055–D1059
Zheng CJ, Han LY, Yap CW, Ji ZL, Cao ZW et al (2006) Therapeutic targets: progress of their exploration and investigation of their characteristics. Pharmacol Rev 58(2):259–279

Molecular Docking and Structure-Based Drug Design

6

Shikha Agnihotry, Rajesh Kumar Pathak, Ajeet Srivastav,
Pradeep Kumar Shukla, and Budhayash Gautam

Abstract

Computer-aided drug designing (CADD) relates to drug discovery, also characterized as a cost-effective and active tool that manages or creates theoretical models that would be used by large databases for discovery and virtual screening. Till now, several algorithms have been developed and managed through CADD to study different prospects like protein structure and function prediction, identification of ligands interaction, residues of the active site, and study of protein–ligand interactions, which possibly leads to the discovery of

S. Agnihotry
Department of Computational Biology and Bioinformatics, Jacob Institute of Biotechnology and Bioengineering, Sam Higginbottom University of Agriculture, Technology and Sciences, Prayagraj, India

Department of Gastroenterology, Sanjay Gandhi Postgraduate Institute of Medical Sciences, Lucknow, India

R. K. Pathak
Department of Biotechnology, Govind Ballabh Pant Institute of Engineering & Technology, Pauri Garhwal, Uttarakhand, India

A. Srivastav
Department of Photobiology, CSIR-Indian Institute of Toxicology Research, Lucknow, Uttar Pradesh, India

P. K. Shukla
Department of Biological Sciences, Sam Higginbottom University of Agriculture, Technology and Sciences, Prayagraj, India

B. Gautam (✉)
Department of Computational Biology and Bioinformatics, Jacob Institute of Biotechnology and Bioengineering, Sam Higginbottom University of Agriculture, Technology and Sciences, Prayagraj, India
e-mail: budhayash.gautam@shiats.edu.in

D. B. Singh (ed.), *Computer-Aided Drug Design*,
https://doi.org/10.1007/978-981-15-6815-2_6

newer therapeutic agents or drugs. As per in terms of new medicine discovery, designing and binding of small molecules (ligand) with DNA, RNA, or protein (target) is the key step, defined as docking. Docking actively identifies specific hit from large data libraries through simple rigid or flexible docking approaches with the receptor to maximize hit rates in virtual screening. The calculated scores of free energy of binding (poses) define the active compounds involved in interactions. Different new prospects in docking programs are now being used that more focuses accuracy on molecular interaction energy calculation without stringent parameters. The quantum-chemical methods, implicit solvent models, and new global optimization algorithms are now being used to improve flexibility and mobility of ligands and proteins, respectively. This chapter presents some basic algorithms, molecular docking programs based on rigid and flexible receptor/ligand-based on various machine learning techniques used in CADD and molecular docking.

Keywords

Computer-aided drug designing · Molecular docking · Ligands · Molecular interaction · Genetic algorithm

6.1 Introduction

The molecular biology and genomic sciences have largely contributed research related to drugs, chemistry, pharmacology, and clinical sciences for progressing and the development of medicine. The process of both drug discovery and development is very tedious (Drews 2000). Consequently, the application of computational resources to chemical and biological space is under large-scale research. Molecular docking is a computational method and a simulation procedure to study molecular recognition, molecular interaction, and conformation of a receptor–ligand complex. Protein or nucleic acid is usually the receptor, and the ligand is another protein or a small molecule (Meng et al. 2011). More recently, docking is also applied to predict the binding mode between two macromolecules, for instance, protein–protein docking.

Computer-aided drug discovery is employed to escalate the processes of hit identification, lead selection and optimization, analysis of absorption, distribution, metabolism, excretion, and toxicity (ADMET) profile for a lead compound (Qidwai 2017). Bioinformatics tools, along with genomic sciences, have provided better insight into the genetic basis of multifactorial diseases. These approaches reveal more suitable targets for designing future medicines and increasing therapeutic options (Qidwai 2017). For reproducing the experimental data, molecular docking simulations may be used involving docking validations algorithms in which the in silico protein–ligand conformations are compared with the complex structure of protein–ligand obtained from X-ray crystallography or NMR (Mattick et al. 2014).

As per reports, biological activity data and structure and inhibition data availability has enhanced considerably. Potential drug targets and thousands of protein structures are available in structural databases like Protein Data Bank (PDB) (Ferreira et al. 2015) and Worldwide Protein Data Bank (wwPDB). Structures of binary complexes with their binding affinities are also available in some specific databases such as PDBBIND (Wang et al. 2005), PLD (Puvanendrampillai and Mitchell 2003), AffinDB (Block et al. 2006), and BindDB (Livyatan et al. 2015). The three-dimensional experimental structure and affinity data are important as a source of information for docking algorithms development and validation (Dias et al. 2008). Currently, molecular mechanics is the basis for most docking programs. Molecular mechanics involves the description of a polyatomic system using classical physics. As such, molecular force fields are sets of equations with different parameters for systems description (Young 2004). Mostly force fields based on five terms having physical interpretation: potential energy, torsional terms, bond geometry, electrostatic terms, and Lennard-Jones potential. Examples of prominent force fields are Assisted Model Building and Energy Refinement (AMBER) (Case et al. 2005), Groningen Molecular Simulation (GROMOS) (Scott et al. 1999), Merck molecular force field 94 (MMFF) (Halgren 1996), Chemistry at Harvard Macromolecular Mechanics (CHARMM) (Brooks et al. 1983), and Universal force field (UFF) (Rappé et al. 1992). For the protein–ligand binding, two general methodologies were developed: (1) the rigid body approach that relates to the classic model of Emil Fischer. In this model, the ligand and receptor are regarded as two independent bodies that recognize each other based on shape and volume. (2) the flexible docking approach considers a reciprocal effect of protein–ligand recognition on the conformation of each part (Meng et al. 2011). It is generally advisable to use more than one docking program. Different studies have shown that, overall, taking a consensus from various docking protocols yields a better assessment of protein–ligand interactions and more reliable pose ranking (Hevener et al. 2009).

6.2 Docking Guidelines

6.2.1 Hardware and Software Requirements for Molecular Docking

Molecular docking and docking-based virtual screening of public repositories may escalate quickly requiring more computing resources to finish in a couple of weeks (Vyas et al. 2015). The most notable example and success case for this are molecular dynamics, as many pioneering efforts were made to make these calculations scalable. Noteworthy examples include AMBER, GROMACS, Desmond (Higham 2001), NAMD (Phillips et al. 2005), and CHARMM, all of which have been ported to make use of Compute Unified Device Architecture (CUDA) developed by NVI DIA Corporation. With the use of GPUs, a workstation may process the same amount of information as a CPU-cluster enabling the simulation of large systems or conducting longer simulations. Following the success of these approaches, other methods were optimized for CUDA are the ab initio (GAMESS, Firefly), semiempirical calculations (MOPAC), FEP calculations (Desmond), and similarity searching (FastROCS) (Brooijmans and Kuntz 2003).

6.2.2 Docking Process

The docking process may be divided into three main parts: (1) ligand and macromolecule preparation. This is made based on force fields allowing for surface representation and cavities as potential ligand sites; (2) defining the docking type: rigid or flexible; and (3) setting the search strategy for ligand conformations: systematic or stochastic (Guedes et al. 2014). Different scoring functions and search algorithms applied in molecular docking have been highlighted in Fig. 6.1.

6.2.3 Ligand and Protein Preparation

Protein and ligand selection and preparation is an important process for any calculation. The very first step is to obtain a three-dimensional with a high-resolution structure of the protein. For some proteins, structures with previous reports of docking or structural studies may be used (Elokely and Doerksen 2013). Several parameters assignment for docking includes: a selection of the parameterization method depends on the software; for example, AutoDock and SwissDock use an in-house force field, whereas MOE (25) and LeDock (Zhang and Zhao 2016) use AMBER and CHARMM charges and atom types, respectively. Consequently, the same preparation protocol should be used for all the docking calculations. In ligand preparation the first step often involves its extraction from the protein structure, or PDB (e.g., Public repositories like PubChem (Wang et al. 2009), organic synthesis, or virtual compounds) or involves the construction of such molecule from its simplified molecular input line entry (SMILES) format or sketching the molecule

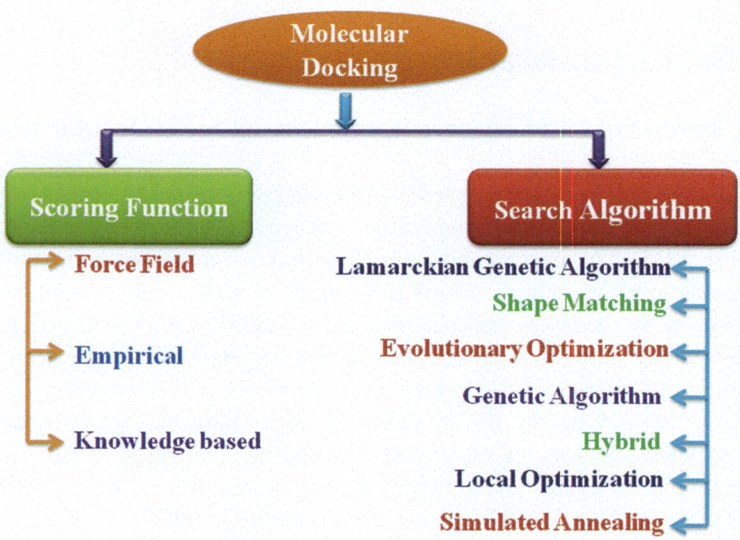

Fig. 6.1 Search algorithms and scoring functions used in molecular docking

performing several optimizations: geometry and/or charge assignment using ab initio or semiempirical methods, energy minimization, or conformational/tautomer search.

Protein structures may also be prepared similarly, and the complex optimization and the use of detailed parameters significantly improve results and modeling of interactions. Following the ligand and protein preparation, the binding site must be selected and delimited. This step can be done manually by specifying the coordinates or automatically using the coordinates of any bound ligand. Additionally, some programs allow for the calculation of cavities or probable binding sites. This is especially useful in cases when the binding site is not known. Also, some programs are capable of blind docking, and the search space involves the entirety of the macromolecule. Binding site mapping is performed through the GRID methodology. In a study, the influence of box size on the identification of hits and virtual screening time is defined (Feinstein and Brylinski 2015). Examples are AutoDock and AutoDock Vina that need the pre-calculation of the GRID for molecular docking (Trott and Olson 2010).

Defining different parameters such as the generation of grid box size, energy range, and exhaustiveness is the fundamental and necessary task before conducting molecular docking. Here, we can generate a grid box size based on amino acid residues found in the binding site cavity for scoring and accurately predicting the binding nature of the ligand. Besides, defining energy range and exhaustiveness especially in the case of AutoDock Vina is helpful in the investigation of the correct conformation with good probability (Trott and Olson 2010; Mamgain et al. 2015).

In the rigid molecule docking problem, we will relate to the molecules as a rigid body, which cannot change its 3D spatial shape during the docking, whereas in flexible docking molecules have the flexibility to rearrange within a defined region in response to the receptor. After ligand binding, flexibility in the receptor can range from small side-chain reorganization to large-scale backbone rearrangements. Rigid docking contributes to the development of several computational concepts that play important roles in flexible docking and design: (a) shape descriptors, (b) local minimization, (c) soft potentials, (d) grid-based energy evaluation, and (e) multiple-copy techniques.

Molecular docking involves several search algorithms, such as fragment-based, genetic algorithm, Monte Carlo, and dynamics algorithms. There are some important basic tools such as FlexX, DOCK, GOLD, and internal coordinate mechanics (ICM), which are specifically used for high throughput docking simulations. Based on the objectives of docking simulations, the different molecular docking procedures involve either a rigid or flexible docking approach. Softening the interaction potential involves the modification to the parameters of interaction that involves the Lennard-Jones potential (Honeycutt and Andersen 1987).

6.2.4 Ligand Conformations Strategies

Docking algorithms play an important role in ligand conformation and placement, which involves systematic or random conformational search and the selection of the optimal solution per the scoring function. Systematic search evaluates conformers individually and also involves a comprehensive sampling of conformations and a combination of structural parameters. Stochastic search used randomly by using two algorithms: Monte Carlo (MC) (Liu and Wang 1999) or genetic algorithms (GA) (Jones et al. 1997). Each develops different conformations based on bond rotations as degrees of freedom. After that, the structures are submitted to the scoring function for pose selection and filtering.

6.2.5 Scoring Functions

Scoring functions are used to discriminate between different solutions evaluating a broad range of properties including, but not limited to, intermolecular interactions, desolvation, electrostatic, and entropic effects. It can be classified as force field-based, empirical, and knowledge-based (Wang et al. 2002). For operational performance, two theoretical aspects of the scoring function are being described. First is the degree to which a scoring function has a global extremum within the ligand pose landscape at the proper location. The second is the degree to which the magnitude of the function at the extremum is accurate. This function predicts the absolute binding affinity between protein and ligand. An ideal scoring function would rank the experimentally determined binding mode most highly. Once the binding mode of ligand is determined, this interaction map can be used to understand the changes required in ligand to achieve better binding affinity. This can also help in depicting the site of all modifications, such as insertion or extension, deletion, and replacement.

The scoring function terms are hydrophobic complementarity, polar complementarity, and entropy (Jain 1996). The four equations related to steric score, polar score, polar repulsion score, and entropy score are used to define the scoring function. Variables used for defining steric score and polar score are based on pair-wise Van der Waals surface distance r between coupled atoms including data regarding the status of atoms (H bond acceptor or donor), charge, and type of element. Both the polar and hydrophobic interactions are distance-dependent and made of sigmoid Gaussian term. Polar repulsion term calculates the penalty of arranging atoms of similar polarity near to each other and is scaled by direction. Entropic term encodes the rotational and translational degrees of freedom lost to the ligand on the binding.

6.2.5.1 Force Field

Force field scoring function uses the parameter contribution for bond stretching, electrostatics, and non-bonding interactions (Vanommeslaeghe and Guvench 2014). These approaches usually involve longer computing times, and the need for distance cut off decreases the accuracy of long-range effects. According to physics principles, quantum mechanics calculation, and experimental data, both are required to generate force field functions and parameters. Solvent treatment in ligand binding is a major point of concern.

6.2.5.2 Empirical Scorings

Empirical scoring functions involve experimental values reproduction and prediction is based on the count of interaction between interacting molecules by observing change in solvent accessible surface area (SASA) value (Durham et al. 2009). It calculates the binding affinity of a structural complex based on the weighted energy set of terms. The empirical scoring functions are much faster in comparison to force field scoring functions due to their simple energy terms. 3D structures and binding affinities data for several protein–ligand complexes are available, which can be used to generate more precise and general empirical scoring function by training with mining the information from known protein–ligand complexes.

6.2.5.3 Knowledge-Based Scoring

Knowledge-based scoring functions are for structures rather than energies; it is based on statistical observations of intermolecular contacts in 3D databases. The structure is constructed using pairwise potentials from known receptor–ligand complexes (Neudert and Klebe 2011). In knowledge-based scoring functions, pairwise potentials are taken from the occurrence frequency of atom pairs in a structured database using the inverse Boltzmann relation.

The major approaches of molecular docking are discussed below.

6.2.6 Ensemble Docking

In the early-stage field of drug discovery, this docking corresponds or relates to the drug target conformations generation in computational structure-based drug discovery that involves important factors like molecular dynamics simulation and also considered as four-dimensional (4D) docking (Amaro et al. 2018). This type of docking is based on multiple docking simulations on different protein conformations and relates to protein flexibility.

6.2.7 Consensus Docking

The consensus method in docking uses data selection from a dynamic benchmark and through multiple docking programs to determine the best program combinations to improve the docking success rate. Consensus scoring improves pose selection, pose enrichment, rank-based, and intersection but directly depends on the scores that are combined, e.g., a strong correlation among them may increase the error rate. Besides, scoring functions are sensitive to specific features of the binding sites. The following sections address crucial topics in molecular docking and some perspectives on the field. In different biological processes, molecular interactions including enzyme–substrate, protein–protein, protein–nucleic acid, drug–nucleic acid, and drug–protein play important roles.

6.3 Different Types of Docking Based on Interactions

The selection of appropriate algorithms, tools, and parameters for docking is an important challenge in molecular docking. In nature, different types of molecular interactions such as protein–ligand (small molecule), protein–peptide like molecule, protein–protein, protein–nucleic acid, or nucleic acids–ligand take place. Different types of docking tools have been developed keeping in mind the nature of interacting molecules, possible forces, and other parameters. In the field of medicinal chemistry, ligand promiscuity is the topic of discussion. Different folding patterns and structural arrangements were deposited in large repositories like PDB, etc. The search for patterns and similarities in binding sites and protein pockets allows the detection of structural changes and behavior. Docking has been classified into many categories based on the nature of the molecules involved in the interaction.

6.3.1 Protein–Ligand Docking

Structure-based design is a very powerful approach to druggable targets. Docking predicts the pose or orientation of a ligand on the binding site of a target molecule or enzyme (Fig. 6.2). For flexible proteins, protein-energy landscape exploration

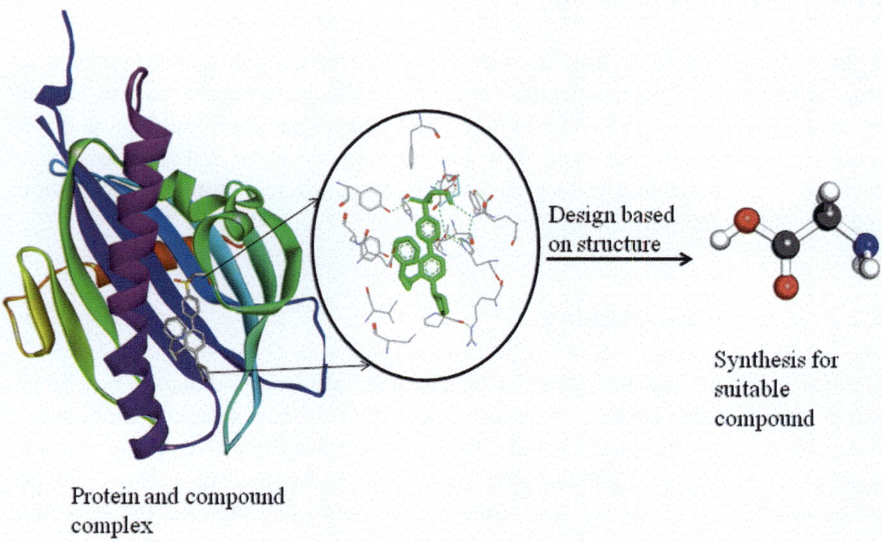

Fig. 6.2 Structure-based drug design: guiding the process of drug designing based on residue composition of target binding site

(PELE) is used for the correct assessment of binding sites and poses (Borrelli et al. 2005). Through machine learning and molecular dynamics using techniques, like self-organizing maps (SOMs) or *k-means* determines the complementarity of protein and ligand conformations (Tamayo et al. 1999). For free energy calculation, MTflex uses Monte Carlo integration and generating rotamers for binding residues based on low-energy values along the free energy surface.

6.3.2 Protein–Peptide like Ligand Docking

The peptide as a sample is highly variable due to high flexibility. Nowadays, peptides are being used in the medicinal areas proving their polypharmacological effects and suitability of protein–protein interaction. It involves calculations that relate to confirmations and poses highlighted in Fig. 6.3. Protein–protein interaction networks can be perturbed by differential gene expression and disease mutations. Molecular modeling approaches play an important role in optimizing the activity of known peptide and also in designing the novel peptide as an inhibitor.

6.3.3 Protein–Protein Docking

In protein–protein docking, protein complexes are determined through sequence alignments, structural comparisons, and multiple protein–protein interactions, within their defined confirmations and docking positions. Protein structure initiative

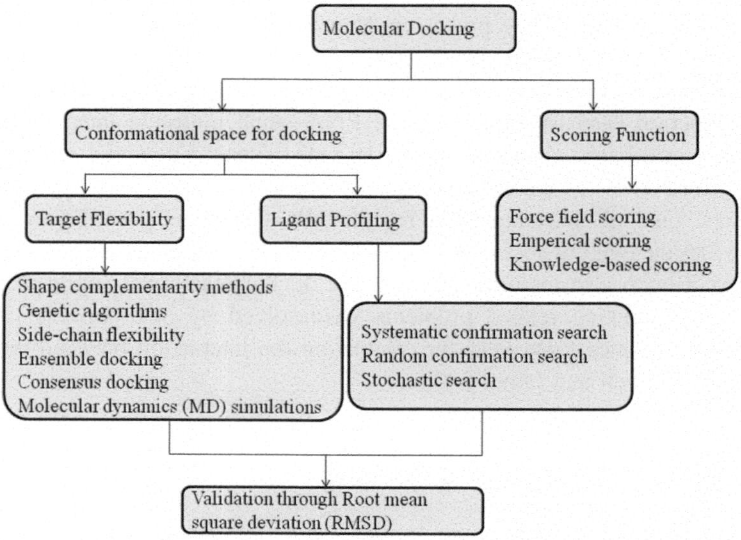

Fig. 6.3 Flowchart describing conformational search space and scoring function algorithm used in the molecular docking

provides significant structural information for the community assessment of structure prediction (CASP) (Moult et al. 1999). For protein–protein docking and macromolecular interactions, critical assessment of protein interactions (CAPRI; http://capri.ebi.ac.uk) acts as a contest-space to challenge different human groups (Janin et al. 2003), software, and servers into correctly predict the conformation of interacting protein–protein pre-chosen targets. Protein–protein docking can be approached as a prediction for the whole complex minimizing each protein by coarse grain models and using local search for the binding sites. Thus, the major challenge for protein–protein docking is the flexibility of the backbone. For this reason, comprehensive computational studies need to be conducted to successfully distinguish realistic complexes from unrealistic predictions.

6.3.4 Protein–Nucleic Acid Docking/Nucleic Acid–Ligand Docking

Proteins and nucleic acids are the two main biological macromolecules which act as a target for many processes/functions. Protein–RNA and protein–DNA interactions are very important for replication, transcription, splicing, translation, and nucleic acids degradation. Abnormalities in protein–nucleic acid interactions are associated with a number of neurological diseases, cancer, and many other metabolism associated issues (Tuszynska et al. 2015). Protein–nucleic acid complexes are being solved by the researchers which may help in understanding different interactions. NPDock is a protein–nucleic acid docking tool and it uses the DARS-RNP and QUASI-RNP statistical potentials for scoring interactions of protein–RNA complexes (Vakser and Aflalo 1994; Tuszynska and Bujnicki 2011).

RNA molecules have recently got an attraction as a drug target due to their importance in biological key processes. However, as of now, the structure-based docking that involves RNA molecules binding with a small molecule (ligands) is not well established that lies under the protein ligand docking category. LigandRNA is a scoring function used for predicting the RNA–small molecule interactions. It is based on a grid-based algorithm, and a knowledge-based potential for scoring is derived from sites of ligand-binding interactions in the known RNA–ligand complexes. LigandRNA takes RNA receptor file and ligand pose file as an input and provides the ranking poses consistent with their score as an output. The modified version of Dock6 also includes RNA–ligand docking facility (Lang et al. 2009). Ligand–RNA docking related problems were solved by incorporating classical molecular mechanics force field for calculating the interaction between the RNA and ligand (Guilbert and James 2008).

6.4 Water Solvation and Docking

Removal of water molecules or solvent has been a constant problem in docking and is not just limited to the inclusion of water molecules during calculations, but to correctly evaluate water contribution to binding and its implications. First efforts to include water molecules during docking calculations suggested that no significant improvement in scoring was obtained. However, some ligands are specially designed to displace water. In such cases, docking simulations are more accurate with the correct treatment of water molecules. Currently, most docking programs can assess the presence of water molecules during calculations (Meng et al. 2011).

6.5 Docking Tools

Covalent inhibitors or modifiers are mostly not prioritized to use as potential drug candidates due to toxicity factors. Notable examples are modulators of acetylcholinesterase. Covalent modifiers may be more selective and effective and more specific for infectious diseases. The overall interest towards rational design and development of covalent inhibitors is expanding. Current programs for covalent docking include AutoDock (Morris et al. 1998), Glide (Halgren et al. 2004), DOCK (Ewing et al. 2001), FlexX (Schellhammer and Rarey 2004), GOLD (Verdonk et al. 2003), etc. A list of some software and their docking algorithms is presented in Table 6.1.

Table 6.1 Software used for molecular docking and their search algorithm

Software name	Search algorithm	References
AUTODOCK4	Lamarckian genetic algorithm	Morris et al. (1998)
DOCK	Shape matching	Ewing et al. (2001)
SWISSDOCK	Evolutionary optimization	Grosdidier et al. (2011)
GOLD	Genetic algorithm	Verdonk et al. (2003)
GLIDE	Hybrid	Halgren et al. (2004)
VINA	Local optimization	Trott and Olson (2010)
RDOCK	Hybrid	Li et al. (2003)
LEDOCK	Simulated annealing	Wang et al. (2016)
HADDOCK	Hybrid	Dominguez et al. (2003)
SURFLEX-DOCK	Shape matching	Jain (2007)
FLEXX	Shape matching	Schellhammer and Rarey (2004)
LIGANDFIT	Shape matching	Venkatachalam et al. (2003)
MTiOpenScreen	Lamarckian genetic algorithm	Labbé et al. (2015)
ZDOCK	Fast Fourier transform algorithm	Pierce et al. (2014)
HEXSERVER	Fourier transform (FFT) algorithm	Macindoe et al. (2010)
UCSF DOCK	Geometric matching algorithm	Allen et al. (2015)
Internal coordinate mechanics (ICM) software	Stochastic global optimization algorithm	Neves et al. (2012)
FRED	Shape-based Gaussian docking function	Mcgann et al. (2003)
MOE-Dock	Hybrid	Corbeil et al. (2012)

6.6 Virtual Screening

In our quest to discover novel drug like molecules, virtual screening emerged as one of the key tools. It plays a tremendous role in the drug discovery program. It is a powerful computational approach used for screening of a set of compounds or chemical compounds database to a target macromolecular structure. It facilitates researchers to select appropriate molecule(s) as a lead from the available chemical compounds database based on lead–target interaction, i.e. binding free energy and interacting amino acid residues through non-covalent bonding (hydrogen and hydrophobic bond) for further validation. Docking is used to perform an interaction study using a single ligand against the binding site residues of the target macromolecular structure at a time while in virtual screening, a set of compounds of the library is used for screening purposes (Pathak et al. 2017, 2018).

6.7 Analysis of Docking Results

Analysis of docked complex structure obtained from molecular docking study is one of the essential tasks for visualization of protein–ligand interaction at the atomic level using molecular modeling tools in 2D or 3D. In this analysis we can identify the number of hydrogen bonds formed among a different functional group of the ligand with amino acid residues present in the binding site of protein along with their bond length, because hydrogen bonding plays a significant role in protein–ligand interaction (Singh and Dwivedi 2016). Besides, we can also analyze hydrophobic and cation–pi interaction. This analysis facilitates researcher to choose the best interacting ligand because in some cases, the binding energy of two or more ligands is the same but the number of interacting amino acid residues are less or more. Therefore, in such a situation, generally we choose ligand having more interaction with the target in terms of interacting amino acid residue numbers. PyMOL and Chimera are widely accepted tool analyses of docking results by selecting different poses of ligand generated during molecular docking and visualization of interacting residues in 3D. Besides, LigPlot is one of the highly cited and recommended tools for analysis of docking results in 2D format (Fig. 6.4) (Mamgain et al. 2015; Pathak et al. 2018; Rana et al. 2019).

6.8 Limitations of Docking Algorithms and Future Scope

Several protein–ligand complexes determined by X-rays are available in databases, which have greatly improved the scoring function of docking. Still, it is challenging to generate an accurate pose by docking. A large number of docking tools have been developed to measure and understand the different types of interaction between the protein–ligand complex. The accuracy of different docking tools has been compared and still exists the chance to refine the scoring function of docking. Different docking tools estimate the different value of the binding energy of the protein–

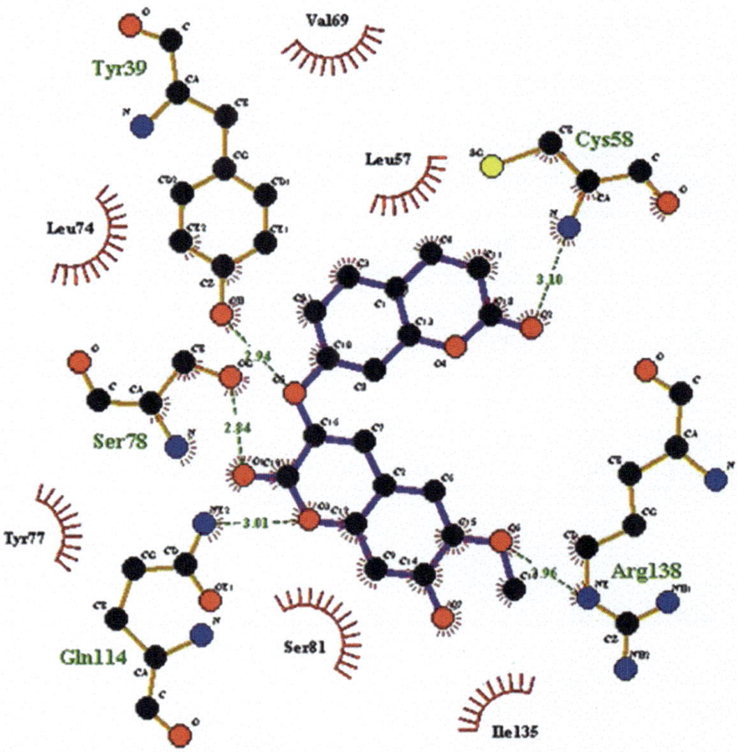

Fig. 6.4 Daphnoretin docked with E6 protein of HPV-16, showing hydrogen bond interaction with TYR39, CYS58, SER78, GLN114, ARG138, CYS58 (green line), whereas amino acid residues participated in interaction through hydrophobic bonding are depicted as arcs (red color), generated by LigPlot

ligand complex (Singh 2014). Energy minimization algorithms also rely on these scoring functions. Therefore, an accurate and precise scoring function is required to achieve the correct binding mode and ranking of a ligand. Besides, a hybrid algorithm is considered more effective to tackle the protein–ligand docking problem because it uses the combination of algorithms (Guan et al. 2018).

6.9 Major Developments in Docking

A small molecule that can bind to the protein responsible for a disease is the main step of the complete process of the new concept discovery. To improve the effectiveness of such design, atomistic computer modeling can work significantly. The accuracy in the calculation of the free energy of binding to the target protein is the most important problem of such modeling. Several important and well-known conventional docking programs are being in consideration. Their search method is based on the problems related to global optimization. To solve this problem, different algorithms are being used, and the heuristic genetic algorithm is distinguished and acknowledged by its elaborate design and popularity among other algorithms. On the bottom, more possible accurate approaches of solvent implicit models are being used often for more clear separation energy calculations. Recently, the new generation programs of docking are developed. They detected the low energy minima spectrum of a ligand–protein complex. These should be more accurate programs because they do not use a pre-calculated grid of ligand–protein interaction potentials. New docking algorithms designed and they work by docking a versatile active ligand into a versatile active protein with many dozen mobile atoms on the bottom of the surrounding energy minimum search (Sulimov et al. 2019). Such docking algorithms improve the accuracy of ligand positioning in the docking. The advancement within the quantum chemistry methods has improved the accuracy of docking. Much advancement has been made in molecular energy calculations, including implicit solvent models and quantum-chemical methods, as well as in ligands flexibility and mobility of atoms of the protein.

6.10 Conclusion

The augmentation of modeling tools and growth in structures determined through X-ray crystallography of the target alone, or in the complex has become important for drug discovery. Structure-based drug design techniques are important and applicable to target-based therapies. The new approaches are being prioritized in the computational database and optimizing compounds having drug-like features. Drug discovery projects should be more focused, and specific so that compounds with druggable activity can be screened easily using docking and other approaches. Along with the availability, studies are being constantly running for providing improvements related to ligand/protein selection, virtual screening, molecular docking, dynamics simulation, and score calculation for drug design and optimization.

Competing Interest The authors declare that there are no competing interests.

References

Allen WJ, Balius TE, Mukherjee S, Brozell SR, Moustakas DT, Lang PT, Case DA, Kuntz ID, Rizzo RC (2015) DOCK 6: impact of new features and current docking performance. J Comput Chem 36(15):1132–1156

Amaro RE, Baudry J, Chodera J, Demir Ö, McCammon JA, Miao Y, Smith JC (2018) Ensemble docking in drug discovery. Biophys J 114(10):2271–2278

Block P, Sotriffer CA, Dramburg I, Klebe G (2006) AffinDB: a freely accessible database of affinities for protein–ligand complexes from the PDB. Nucleic Acids Res 34(suppl1):D522–D526

Borrelli KW, Vitalis A, Alcantara R, Guallar V (2005) PELE: protein energy landscape exploration. A novel Monte Carlo based technique. J Chem Theory Comput 1(6):1304–1311
Brooijmans N, Kuntz ID (2003) Molecular recognition and docking algorithms. Annu Rev Biophys Biomol Struct 32(1):335–373
Brooks BR, Bruccoleri RE, Olafson BD, States DJ, Swaminathan SA, Karplus M (1983) CHAR MM : a program for macromolecular energy, minimization, and dynamics calculations. J Comput Chem 4(2):187–217
Case DA, Cheatham TE, Darden T, Gohlke H, Luo R, Merz KM, Onufriev A, Simmerling C, Wang B, Woods RJ (2005) The Amber biomolecular simulation programs. J Comput Chem 26 (16):1668–1688
Corbeil CR, Williams CI, Labute P (2012) Variability in docking success rates due to dataset preparation. J Comput Aided Mol Des 26(6):775–786
Dias R, de Azevedo J, Walter F (2008) Molecular docking algorithms. Curr Drug Targets 9 (12):1040–1047
Dominguez C, Boelens R, Bonvin AM (2003) HADDOCK: a protein– protein docking approach based on biochemical or biophysical information. J Am Chem Soc 125(7):1731–1737
Drews J (2000) Drug discovery: a historical perspective. Science 287(5460):1960–1964
Durham E, Dorr B, Woetzel N, Staritzbichler R, Meiler J (2009) Solvent accessible surface area approximations for rapid and accurate protein structure prediction. J Chem Inf Model 15 (9):1093–1108
Elokely KM, Doerksen RJ (2013) Docking challenge: protein sampling and molecular docking performance. J Chem Inf Model 53(8):1934–1945
Ewing TJ, Makino S, Skillman AG, Kuntz ID (2001) DOCK 4.0: search strategies for automated molecular docking of flexible molecule databases. J Comput Aided Mol Des 15(5):411–428
Feinstein WP, Brylinski M (2015) Calculating an optimal box size for ligand docking and virtual screening against experimental and predicted binding pockets. J Cheminformatics 7(1):18
Ferreira L, dos Santos R, Oliva G, Andricopulo A (2015) Molecular docking and structure-based drug design strategies. Molecules 20(7):13384–13421
Grosdidier A, Zoete V, Michielin O (2011) SwissDock, a protein-small molecule docking web service based on EADock DSS. Nucleic Acids Res 39(suppl 2):W270–W277
Guan B, Zhang C, Zhao Y (2018) An efficient ABC_DE_based hybrid algorithm for protein–ligand docking. Int J Mol Sci 19(4):1181
Guedes IA, de Magalhães CS, Dardenne LE (2014) Receptor–ligand molecular docking. Biophys Rev 6(1):75–87
Guilbert C, James TL (2008) Docking to RNA via root-mean-square-deviation-driven energy minimization with flexible ligands and flexible targets. J Chem Inf Model 48(6):1257–1268
Halgren TA (1996) Merck molecular force field. III. Molecular geometries and vibrational frequencies for MMFF94. J Comput Chem 17(5–6):553–586
Halgren TA, Murphy RB, Friesner RA, Beard HS, Frye LL, Pollard WT, Banks JL (2004) Glide: a new approach for rapid, accurate docking and scoring. 2. Enrichment factors in database screening. J Med Chem 47(7):1750–1759
Hevener KE, Zhao W, Ball DM, Babaoglu K, Qi J, White SW, Lee RE (2009) Validation of molecular docking programs for virtual screening against dihydropteroate synthase. J Chem Inf Model 49(2):444–460
Higham DJ (2001) An algorithmic introduction to numerical simulation of stochastic differential equations. SIAM Rev 43(3):525–546
Honeycutt JD, Andersen HC (1987) Molecular dynamics study of melting and freezing of small Lennard-Jones clusters. J Phys Chem 91(19):4950–4963
Jain AN (1996) Scoring noncovalent protein-ligand interactions: a continuous differentiable function tuned to compute binding affinities. J Comput Aided Mol Des 10(5):427–440
Jain AN (2007) Surflex-Dock 2.1: robust performance from ligand energetic modeling, ring flexibility, and knowledge-based search. J Comput Aided Mol Des 21(5):281–306

Janin J, Henrick K, Moult J, Ten Eyck L, Sternberg MJ, Vajda S, Vakser I, Wodak SJ (2003) CAPRI: a critical assessment of predicted interactions. Proteins: Struct Funct Bioinf 52(1):2–9

Jones G, Willett P, Glen RC, Leach AR, Taylor R (1997) Development and validation of a genetic algorithm for flexible docking. J Mol Biol 267(3):727–748

Labbé CM, Rey J, Lagorce D, Vavruša M, Becot J, Sperandio O, Villoutreix BO, Tufféry P, Miteva MA (2015) MTiOpenScreen: a web server for structure-based virtual screening. Nucleic Acids Res 43(W1):W448–W454

Lang PT, Brozell SR, Mukherjee S, Pettersen EF, Meng EC, Thomas V, Rizzo RC, Case DA, James TL, Kuntz ID (2009) DOCK 6: combining techniques to model RNA–small molecule complexes. RNA 15(6):1219–1230

Li L, Chen R, Weng Z (2003) RDOCK: refinement of rigid-body protein docking predictions. Proteins: Struct Funct Bioinf 53(3):693–707

Liu M, Wang S (1999) MCDOCK: a Monte Carlo simulation approach to the molecular docking problem. J Comput Aided Mol Des 13(5):435–451

Livyatan I, Aaronson Y, Gokhman D, Ashkenazi R, Meshorer E (2015) BindDB: an integrated database and webtool platform for "reverse-ChIP" epigenomic analysis. Cell Stem Cell 17 (6):647–648

Macindoe G, Mavridis L, Venkatraman V, Devignes MD, Ritchie DW (2010) HexServer: an FFT-based protein docking server powered by graphics processors. Nucleic Acids Res 38 (suppl_2):W445–W449

Mamgain S, Sharma P, Pathak RK, Baunthiyal M (2015) Computer aided screening of natural compounds targeting the E6 protein of HPV using molecular docking. Bioinformation 11 (5):236

Mattick JS, Dziadek MA, Terrill BN, Kaplan W, Spigelman AD, Bowling FG, Dinger ME (2014) The impact of genomics on the future of medicine and health. Med J Aust 201(1):17–20

Mcgann MR, Almond HR, Nicholls A, Grant JA, Brown FK (2003) Gaussian docking functions. Biopolymers 68(1):76–90

Meng XY, Zhang HX, Mezei M, Cui M (2011) Molecular docking: a powerful approach for structure-based drug discovery. Curr Comput Aided Drug Des 7(2):146–157

Morris GM, Goodsell DS, Halliday RS, Huey R, Hart WE, Belew RK, Olson AJ (1998) Automated docking using a Lamarckian genetic algorithm and an empirical binding free energy function. J Comput Chem 19(14):1639–1662

Moult J, Hubbard T, Fidelis K, Pedersen JT (1999) Critical assessment of methods of protein structure prediction (CASP): round III. Proteins: Struct Funct Bioinf 37(S3):2–6

Neudert G, Klebe G (2011) DSX: a knowledge-based scoring function for the assessment of protein–ligand complexes. J Chem Inf Model 51(10):2731–2745

Neves MA, Totrov M, Abagyan R (2012) Docking and scoring with ICM: the benchmarking results and strategies for improvement. J Comput Aided Mol Des 26(6):675–686

Pathak RK, Baunthiyal M, Shukla R, Pandey D, Taj G, Kumar A (2017) *In silico* identification of mimicking molecules as defense inducers triggering jasmonic acid mediated immunity against Alternaria blight disease in brassica species. Front Plant Sci 8:609

Pathak RK, Gupta A, Shukla R, Baunthiyal M (2018) Identification of new drug-like compounds from millets as Xanthine oxidoreductase inhibitors for treatment of Hyperuricemia: a molecular docking and simulation study. Comput Biol Chem 76:32–41

Phillips JC, Braun R, Wang W, Gumbart J, Tajkhorshid E, Villa E, Schulten K (2005) Scalable molecular dynamics with NAMD. J Comput Chem 26(16):1781–1802

Pierce BG, Wiehe K, Hwang H, Kim BH, Vreven T, Weng Z (2014) ZDOCK server: interactive docking prediction of protein-protein complexes and symmetric multimers. Bioinformatics 30 (12):1771–1773

Puvanendrampillai D, Mitchell JB (2003) Protein Ligand Database (PLD): additional understanding of the nature and specificity of protein–ligand complexes. Bioinformatics 19(14):1856–1857

Qidwai T (2017) QSAR modeling, docking and ADMET studies for exploration of potential anti-malarial compounds against *Plasmodium falciparum*. In Silico Pharmacol 5(1):6

Rana G, Pathak RK, Shukla R, Baunthiyal M (2019) *In silico* identification of mimicking molecule (s) triggering von Willebrand factor in human: a molecular drug target for regulating coagulation pathway. J Biomol Struct Dyn 14:1–3

Rappé AK, Casewit CJ, Colwell KS, Goddard WA III, Skiff WM (1992) UFF, a full periodic table force field for molecular mechanics and molecular dynamics simulations. J Am Chem Soc 114 (25):10024–10035

Schellhammer I, Rarey M (2004) FlexX-Scan: fast, structure-based virtual screening. Proteins: Struct Funct Bioinf 57(3):504–517

Scott WR, Hünenberger PH, Tironi IG, Mark AE, Billeter SR, Fennen J, Torda AE, Huber T, Krüger P, van Gunsteren WF (1999) The GROMOS biomolecular simulation program package. J Phys Chem 103(19):3596–3607

Singh DB (2014) Success, limitation and future of computer aided drug designing. Transl Med (Sunnyvale) 4:e127. https://doi.org/10.4172/2161-1025.1000e127

Singh DB, Dwivedi S (2016) Structural insight into binding mode of inhibitor with SAHH of Plasmodium and human: interaction of curcumin with anti-malarial drug targets. J Chem Biol 9 (4):107–120

Sulimov VB, Kutov DC, Sulimov AV (2019) Advances in docking. Curr Med Chem 26 (42):7555–7580

Tamayo P, Slonim D, Mesirov J, Zhu Q, Kitareewan S, Dmitrovsky E, Lander ES, Golub TR (1999) Interpreting patterns of gene expression with self-organizing maps: methods and application to hematopoietic differentiation. Proc Natl Acad Sci U S A 96(6):2907–2912

Trott O, Olson AJ (2010) AutoDock Vina: improving the speed and accuracy of docking with a new scoring function, efficient optimization, and multithreading. J Comput Chem 31(2):455–461

Tuszynska I, Bujnicki JM (2011) DARS-RNP and QUASI-RNP: new statistical potentials for protein-RNA docking. BMC Bioinf 12(1):348

Tuszynska I, Magnus M, Jonak K, Dawson W, Bujnicki JM (2015) NPDock: a web server for protein–nucleic acid docking. Nucleic Acids Res 43(W1):W425–W430

Vakser IA, Aflalo C (1994) Hydrophobic docking: a proposed enhancement to molecular recognition techniques. Proteins: Struct Funct Bioinf 20(4):320–329

Vanommeslaeghe K, Guvench O (2014) Molecular mechanics. Curr Pharm Des 20(20):3281–3292

Venkatachalam CM, Jiang X, Oldfield T, Waldman M (2003) LigandFit: a novel method for the shape-directed rapid docking of ligands to protein active sites. J Mol Graph Model 21 (4):289–307

Verdonk ML, Cole JC, Hartshorn MJ, Murray CW, Taylor RD (2003) Improved protein–ligand docking using GOLD. Proteins: Struct Funct Bioinf 52(4):609–623

Vyas R, Karthikeyan M, Nainaru G, Muthukrishnan M (2015) Pharmacophore and docking based virtual screening of validated *Mycobacterium tuberculosis* targets. Comb Chem High Throughput Screen 18(7):624–637

Wang R, Fang X, Lu Y, Yang CY, Wang S (2005) The PDBbind database: methodologies and updates. J Med Chem 48(12):4111–4119

Wang R, Lai L, Wang S (2002) Further development and validation of empirical scoring functions for structure-based binding affinity prediction. J Comput Aided Mol Des 16(1):11–26

Wang Y, Xiao J, Suzek TO, Zhang J, Wang J, Bryant SH (2009) PubChem: a public information system for analyzing bioactivities of small molecules. Nucleic Acids Res 37(suppl 2):W623–W633

Wang Z, Sun H, Yao X, Li D, Xu L, Li Y, Tian S, Hou T (2016) Comprehensive evaluation of ten docking programs on a diverse set of protein–ligand complexes: the prediction accuracy of sampling power and scoring power. Phys Chem Chem Phys 18(18):12964–12975

Young D (2004) Computational chemistry: a practical guide for applying techniques to real world problems. Wiley, New York

Zhang N, Zhao H (2016) Enriching screening libraries with bioactive fragment space. Bioorg Med Chem Lett 26:3594–3597

Molecular Dynamics Simulation of Protein and Protein–Ligand Complexes

7

Rohit Shukla and Timir Tripathi

Abstract

Biomacromolecules, including proteins and their complexes, adopt multiple conformations that are linked to their biological functions. Though some of the structural heterogeneity can be studied by methods like X-ray crystallography, NMR, or cryo-electron microscopy, these methods fail to explain the detailed conformational transitions and dynamics. The dynamic structural states in proteins are covered in magnitude between 10^{-11} and 10^{-6} m and time-scales from 10^{-12} s to 10^{-5} s. For a comprehensive analysis of the biomolecular dynamics, molecular dynamics (MD) simulation has evolved as the most powerful technique. With the advent of high-end computational power, MD simulations can be performed between μs to the ms time-scale that can accurately describe the dynamics of any system. Various force fields like GROMOS, AMBER, and CH ARMM have been developed for MD simulations. Tools like GROMACS, AMBER, CHARMM-GUI, and NAMD are the most widely used methods for MD simulation that can provide precise information on the motions and flexibility of a protein, which contributes to the interaction dynamics of protein–ligand complexes. MD simulation has several other practical applications in diverse research areas, including molecular docking and drug design, refining protein structure predictions, and studying the unfolding

R. Shukla
Molecular and Structural Biophysics Laboratory, Department of Biochemistry, North-Eastern Hill University, Shillong, India

Department of Biotechnology and Bioinformatics, Jaypee University of Information Technology (JUIT), Solan, India

T. Tripathi (✉)
Molecular and Structural Biophysics Laboratory, Department of Biochemistry, North-Eastern Hill University, Shillong, India

D. B. Singh (ed.), *Computer-Aided Drug Design*,
https://doi.org/10.1007/978-981-15-6815-2_7

pathway of a protein. Combining MD simulation with wet-lab experiments has become an indispensable complement in the investigation of several important and intricate biological processes. Various tools like principal component analysis, cross-correlation analysis, and residues interaction network analysis are additional useful approaches for analyzing MD data. In this chapter, we will discuss MD simulation for a layman understanding and explain how it can be used for protein–ligand characterization as well as for use in diverse biomolecular applications.

Keywords

Molecular dynamics · Principal component analysis · Cross-correlation analysis · Virtual screening · Mutational analysis · Structure-function relation · Protein unfolding · Force fields

7.1 History and Background

Proteins perform a wide range of cellular functions in living organisms, such as catalysis of metabolic reactions, transport, and cell signaling; all those depend on the structure and dynamics of the protein. Several non-covalent and covalent interactions help to stabilize the native conformation of protein that dictates its function (Singh and Tripathi 2020). In practice, the degree of folded nature is generally determined by wet-lab experiments, including fluorescence and circular dichroism studies. All-atom molecular dynamics (MD) simulation has been developed as a new tool to understand the dynamics of protein motions at the atomic level. It provides information about the motion of an individual atom as a function of time, and thus describes the dynamic behavior of a molecule. The advantage of MD simulation is that it provides information about the folding/unfolding mechanism like the final folded structure, the time dependency of these events, and the inter-residue interactions. The pioneers of MD simulation were Alder and Wainright, who introduced this technique in the late 1950s to study the interactions of a hard-sphere (Alder and Wainwright 1957, 1959). In 1964, first simulation using the realistic potential for liquid argon was carried out by Rahman (Rahman 1964). The first realistic system (liquid water) was done in the 1970s to perform simulation (Rahman and Stillinger 1971). However, the first simulation of protein was conducted in 1977 (McCammon et al. 1977). Soon in the 1980s, simulations on protein interacting with small molecules, their thermodynamics (free energy calculations), and rapid calculations of biomolecules were developed (McCammon et al. 1986). In 1998, Duan and Kollman revealed the folding mechanism of the small sub-domain of villin using a μs simulation (Duan and Kollman 1998). In 2013, the Nobel Prize in Chemistry was awarded to Martin Karplus, Michael Levitt, and Arieh Warshel for the development of multi-scale models for complex chemical systems, a technique to simulate the behavior of molecules at various scales from single molecules to proteins.

MD simulation is an emerging field. We entered the keyword "molecular dynamics simulation" to NCBI PubMed database (https://www.ncbi.nlm.nih.gov/pubmed/) resulted in 56,833 articles as of 31/07/2020, suggesting the growing importance of MD simulation. Development of techniques such as potential sampling methods, force field advancement, and high-end computational power is allowing us to perform simulations in a range of μs to ms time-scale. MD simulation can thus be highly useful in the study of biomolecular dynamics. However, the use of MD simulation requires optimal models that can mimic the cellular environment, physical applications that can provide the motions to the model, and large-scale computations. However, recently MD simulations have been extended to cellular scales, and simulations of an entire cell have been performed (Heidari et al. 2016). The development of a more robust algorithm and theory for modeling, docking, scoring, energy-calculations will make the MD simulation more effective. In this chapter, we will discuss the principle, methods, tools, and important applications of MD simulation.

7.2 Introduction

To date, it is not possible to accurately predict detailed biomolecular conformational dynamics in vitro. Techniques like X-ray crystallography, nuclear magnetic resonance (NMR), and recent cryo-electron microscopy (cryo-EM) methods have provided breakthroughs in structural biology. Still, a vast gap exists between the numbers of the available protein sequences and protein structures. There are 177,754,527 protein sequences in the latest release of UniProtKB as of 12/04/2020, while the protein data bank (PDB) has only 1,62,259 protein structures on 12/04/2020, suggesting only a small fraction of the total sequences have known structures. The PDB statistics, as on 12/04/2020, are shown in Table 7.1. Thus, the prediction of protein structures is essential to fill this significant gap.

Most wet-lab experimental methods provide structural information of proteins in static form, while practically proteins are highly dynamic (Dror et al. 2012). Molecular docking only provides a static pose, and it cannot illustrate the dynamics of the protein–ligand complex (Kalita et al. 2018b; Mamgain et al. 2015). In addition to the prediction of the dynamic behavior of biological systems, MD simulations can also help to explore the kinetic behavior and assemblies of molecules at the atomic level

Table 7.1 The PDB statistics as on dated 12/04/2020

Experimental method	Proteins	Nucleic acids	Protein/NA complex	Others	Total
X-Ray	135,436	2044	6562	460	144,502
NMR	11,344	1284	264	49	12,941
Electron microscopy	3638	35	1029	114	4816
Other	32	1	0	4	37
Hybrid methods	155	5	3	1	164
Total	150,605	3369	7858	628	162,460

(Alder and Wainwright 1959; Rajendran et al. 2018). The conformational dynamics in proteins cover large ranges in both magnitude and time-scale (Vogeli et al. 2012), and due to the conformational changes proteins can function in a variety of ways including acting as transporters, signaling molecules, sensors, and mechanical effectors and also interact with the substrate, drugs, and hormones through conformational changes (Sonkar et al. 2017; Pandey et al. 2017). Structural dynamics in protein conformations are fast, covering a magnitude from 10^{-11} to 10^{-6} m, within of time-scale between 10^{-12} s and 10^{5} s (Boehr et al. 2006; Wolf and Kirschner 2013; Krishnamoorthy 2012). The local motions: loop and side-chain motions take 10^{-15} s to 10^{-1} s to complete, while the helix, domain (hinge bending), and subunit motions take 10^{-9} s to 1 s. Processes, such as helix-coil transitions, association/dissociation, and folding/unfolding, come under large-scale motions and may take 10^{-7} s to 10^{4} s to complete. Practically, it is challenging to study such changes as they take place in a very short time, however, but by resembling the in vivo conditions computationally, these processes can be examined, visualized, and analyzed (Shukla et al. 2018a, d).

In MD simulation, a molecule can be understood as a series of charged points (atoms) linked by springs (bonds). Now, to describe the time evolution of bond lengths, bond angles, torsions, and also the non-bonding interactions between atoms, the force field is used. In MD simulation, an in-vivo like environment is created using protein and water molecules, and the atoms of protein and water move with a short time step (in the fs time duration), where forces of every atom are computed, and written in a file using a force field. This force field is a collection of equations and associated constants and includes the potential energy functions with bonded and non-bonded potential terms. It reproduces molecular geometry and selected properties of test structures. Cammon et al. performed the first MD simulation of biological macromolecules in 1977 at 9.2 ps for bovine pancreatic trypsin inhibitors. Before advances in MD simulations, experimental observations like hydrogen bond exchange were already studied (Berger and Linderstrom-Lang 1957). The role of thermal factor (B) in internal motions of proteins was investigated (Jeremy Smith et al. 1986; Brunger et al. 1985; Brooks and Karplus 1983) by that time. From the MD simulation data, we can also deduce and calculate principal components (PCs) and perform its analysis (Wolf and Kirschner 2013; David and Jacobs 2014). The calculated PCs are arranged according to their contribution to the total fluctuation along with the ensemble of conformations. The global and correlated motions can be predicted using the computed data. MD simulation also provides information on the conformational flexibility of macromolecules as well as in understanding the experimental results, such as the analysis of fluorescence depolarization (Frauenfelder et al. 1987), dynamics of NMR parameters (Brunger et al. 1987), and effect of solvent and temperature on protein stability (Nilsson et al. 1986; Colonna-Cesari et al. 1986). The simulated annealing is a widely used method for the refinement of X-ray structures (Harvey et al. 1984) and the determination of NMR structure (Case and Karplus 1979).

7.3 Principle of MD Simulation

The underlying principle of MD simulation is based on Newton's law for molecular mechanics (Adcock and McCammon 2006). For the computational study of biomolecular dynamics, MD simulation is the most important established technique (Adcock and McCammon 2006; Levitt and Warshel 1975; Karplus and Kuriyan 2005). In MD simulation, the interaction between the atoms and the molecules is examined for a time period by approximations of known physical attributes (Levitt and Warshel 1975; Karplus and Kuriyan 2005). Presently, significant progress has been made in the simulation of biomolecules. By now, we can examine the movement of atoms, the side-chain conformation of residues, and predict secondary structure and domains in a protein, as well as the binding pattern of nucleic acids and lipid membranes (Perilla et al. 2015). The binding free energy and conformational changes of systems can also be predicted by MD simulation as they are based on statistical mechanics. Due to this reason, the MD simulation has been used in the field of drug designing also (Paquet and Viktor 2015). Using the force field, the displacement of the particles, and energy values in each time step is calculated to define the new position of the atom (Adcock and McCammon 2006). The bonded (angles and atom bonds) and non-bonded contributions are included in the forcefield as an energy function in the classical MD simulation (Fig. 7.1).

The later contributions are made mainly by the van der Waals interaction, which is built by the Lennard-Jones 6-potential (Jones 1924). Coulomb's law is employed for the calculation of the electrostatic interaction (Cornell et al. 1995). Several algorithms involving Monte Carlo (MC) simulations, and Langevin dynamics, and

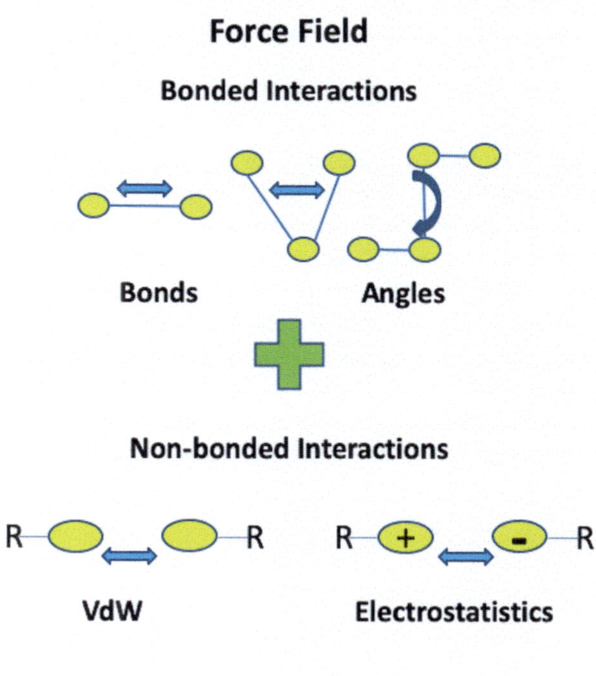

Fig. 7.1 The constituents of a force field, which represents bonded and non-bonded interactions

MD with their corresponding particularities and advantages have been reported (Adcock and McCammon 2006). The designed force field parameters and defined equations are well fitted and can reproduce the data from higher-level calculations or/and experiments. Most biomolecules, including proteins, nucleic acids, lipids, and sugars, are well parameterized in the force field for general use (Paquet and Viktor 2015). The parameters of force field for new ligands can be calculated using quantum chemistry treatment, along with many web servers, like PRODRG (van Aalten et al. 1996; Schuttelkopf and van Aalten 2004), ATP topology builder (Malde et al. 2011), and SwissParam (Zoete et al. 2010) that can generate the topology of the ligand. The bond lengths, bond dihedral angles, bond valence angles, and non-bonded interactions like van der Waals and electrostatic interactions contribute to the total energy of the systems (Hernandez-Rodriguez et al. 2016). Several force fields, like AMBER, GROMOS, and CHARMM, have also been developed (MacKerell et al. 1998). Every force field has a unique property, and a user defines the force field according to his choice based on the objective of the work (Ponder and Case 2003; Salsbury 2010). Once the force field and solvation of the proteins in an MD simulation are fixed, several parameters are also set by the users, which are defined below in brief.

7.3.1 Periodic Boundary Conditions

The periodic boundary condition (PBC) is an approach by which one can define a set of rules for the boundary of a simulation box so that atoms cannot move beyond the defined boundary during MD simulations. If the user does not define the PBC, the simulation box is repeated infinitely in every path and result in forming a lattice. For better computational efficiency, most MD simulations use this potential. Each particle interacts with adjacent images of the other particle in all these cut-off schemes (Holden et al. 2013). The long-range interactions are calculated in the case of molecular modeling and simulation by the isotropic periodic sum (IPS) method (Wu and Brooks 2009). Four significant advantages of using the IPS methods are as (1) it can eliminate the unnecessary symmetry artifacts, which originates in the PBC condition, (2) in the case of any functional form of the potential, it can be applied, (3) it can be used easily in the parallelized multi-processor computer, which indicates that it is computationally more efficient, and (4) it can predict the estimation of self-diffusion coefficient at the cut-off radius greater than 2.2 nm (Takahashi et al. 2010). Here, the long-range interactions are calculated based on the homogeneity of the simulation systems in the IPS method. Long-range interactions are represented by interactions with IPS images of a defined local region and can be reduced to short-range IPS potentials (Wu and Brooks 2009).

7.3.2 Ewald Summation Techniques

The calculation of long-range Coulombic interactions is time-dependent and labor-intensive in most of the MD simulation methods. Here, the Ewald summation method developed in 1921 for the prediction of long-range interactions is mostly used (Ewald 1921). Long-range interactions are estimated as sums that converge very slowly. Figure 7.2 shows that the conversion of the summation of two series of potential energy is the principal to obtain the Ewald sum in MD simulation.

7.3.3 Particle Mesh Ewald Method

In the particle mesh Ewald (PME) method, the potential energy is divided into two sums- Ewald's standard direct sum and the reciprocal sums. The classical Gaussian charge distributions are used in the PME method (Norberto de Souza and Ornstein 1999; Sagui and Darden 1999). The direct sum is computed directly utilizing cut-offs. In contrast, the reciprocal sum is determined by Fast Fourier Transform (FFT) with convolutions on a grid where charges interpolate in the grid points (Fig. 7.3) (Darden et al. 1993; Dessailly et al. 2007). Additionally, it does not interpolate while the forces are calculated by analytically differentiating energies. This significantly reduces the memory requirements for computation (Norberto de Souza and Ornstein 1999).

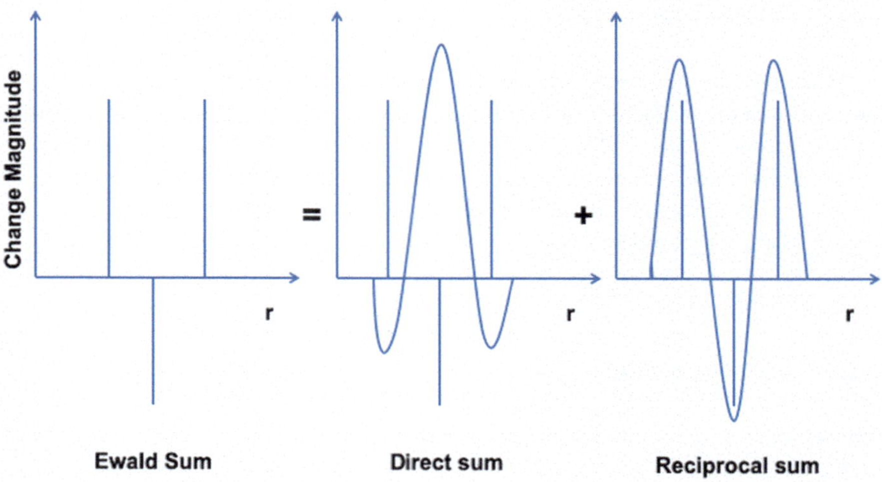

Fig. 7.2 Charge splitting into discrete and smeared distributions in the direct and reciprocal space

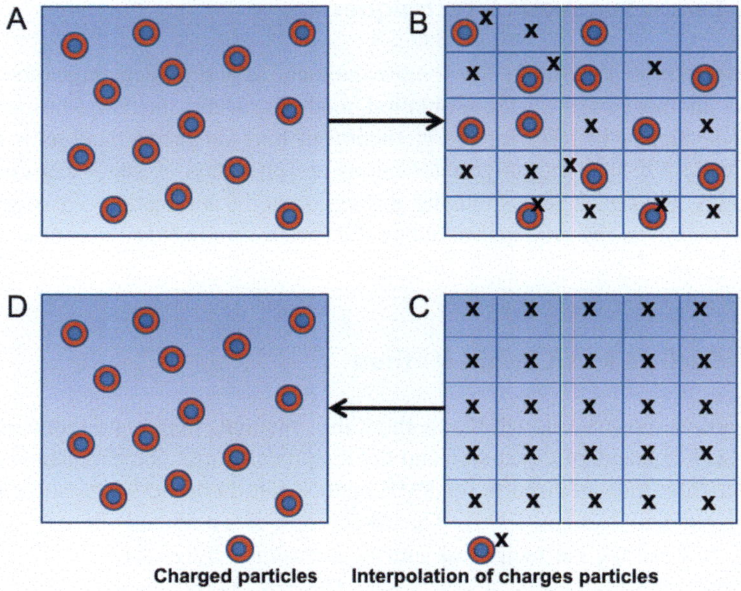

Fig. 7.3 The particle-mesh Ewald technique. A 2D representation, which is used by the majority of the Fourier-based methods. (**a**) The charged particle system, (**b**) Interpolation of charge in a 2D grid, (**c**) The forces and potential are computed at grid points by using FFT, (**d**) The coordinate updates and interpolate forces back to particles

7.3.4 Thermostats in MD

There are several thermostat methods for adding and removing the energy from the boundaries in the MD simulation system in a comparatively rational manner, for the approximation of classical ensemble (Fuzo and Degreve 2014). The number of particles (N), fixed volume (V), and the defined temperature (T) are conserved in the canonical ensemble. In a thermostat method, the energy exchange occurs between the endothermic and exothermic processes (Fuzo and Degreve 2014).

7.3.5 Solvent Models

Biomolecular MD simulation is performed in a realistic type water environment where the explicit solvent model is used (Nguyen et al. 2014). Several solvent models (mostly water models) are used in MD simulation as available: SPC/E, TIP3P, TIP4P, and TIP5P (Jorgensen and Tirado-Rives 2005). These water models are well optimized with one or many physical properties of water, such as density anomaly, diffusivity, and radial distribution function. To mimic the real cell-like environment, the MD simulation system contain explicit water molecules.

7.3.6 Energy-Minimization Methods in MD Simulations

There are different energy minimization approaches for MD structural data. By using a grid search method, the low energy regions are identified by the use of energy function in the zero-order method. In the case of gradient as an energy function, the conjugant gradient or steepest descent method is the first derivative technique. The Newton–Raphson algorithm uses the Hessian function for locating the energy minima as it is a second derived method (Kini and Evans 1991). Two main methods for energy minimization are (1) steepest descent method and (2) conjugant gradient algorithm method.

To remove the bad contacts and correction of bad geometries, the steepest descent method is widely used (Kini and Evans 1991). This method is particularly useful when the molecular system is farther from the minimum energy state, and it drifts down to the steepest slope on the potential energy surface by inducing minor structural modifications. In this method, the gradient is computed from its initial location and moves in the opposite direction to reach the minimum state. When the atoms are moving in a small increment pattern from one direction to another in coordinated systems, then for the initial geometry, the energy is computed. This process repeats for all the atoms, that eventually move to a new position downhill on the energy surface where every new step is at right angles to the one before it. This process occurs in the smaller steps to proceed down along a narrow valley and halts when the condition of a predetermined threshold value is achieved. The steepest descent method is used as the first rough and introductory run, followed by the subsequent minimization.

The conjugant gradient algorithm method is another method for energy minimization and is a primary order of minimization. This method performs minimization by using a mutually current gradient and preceding search command (Kini and Evans 1991). As compared to the steepest descent method, this method is congregated faster because it computes the search direction by using the history of minimization. It is the first derivative rate of change of the total energy in relation to the atomic positions with units of the gradient (kcal.mol^{-1} Å^{-1}). An array of directions is produced by which it succeeds over the oscillatory actions in the narrow valleys for the steepest descents method. In each minimization step, the gradient calculation is done for vector computations to predict the new direction for the minimization procedure as additional information (Feyfant et al. 2007). For the prediction of minimum energy, the direction is defined by each consecutive step. This method is preferred for larger systems (with a high number of atoms), and more storage space and calculation efforts are required. The expense of total computational and the long time period per iteration is compensated by efficient convergence to the minimum (Kini and Evans 1991). For illustrating convergence, there are various types of minimization procedures for molecular structures. For non-gradient minimizers, the augmentations in the energy and the coordinates can be measured to find the real geometry of the particular molecular system. All the gradient minimizers use atomic gradients.

7.4 Current Tools for MD Simulation

Several tools are available to investigate the atomic-level changes in the biomolecules using the MD simulation method (Khan et al. 2016). Some provide the graphical user interface like Desmond, while some run in command lines like GROMACS and AMBER. Some famous and widely used tools for MD simulation are GROMACS (Pronk et al. 2013; Oostenbrink et al. 2004), (AMBER) (Case et al. 2005; Salomon-Ferrer et al. 2013), Nanoscale MD (NAMD) (Phillips et al. 2005), and (CHARMM-GUI) (Brooks et al. 2009). For running such MD simulations, increased hardware power and software are essential components.

7.4.1 Recent Advances in Hardware to Run MD Simulation

Rapid development in computer hardware is a crucial part of MD simulation. Two reasons have an impact on trajectory analysis. The first one is the run of the long simulation result in GBs to TBs data storage, and the other is to develop the new rendering engines for the visualization effects using the latest video chipsets. Due to the advancement in the computer hardware, simulations can be performed from ns to μs with the help of GPUs (graphics processing units) that is configured with the molecular simulation suite (Hernandez-Rodriguez et al. 2016; Gotz et al. 2012; Salomon-Ferrer et al. 2013). The GPU cards are replacing the CPU (central processing unit) and becoming commodity software and play a crucial role in decreasing the time for MD simulation. The CUDA (Compute Unified Device Architecture) is a newly invented parallel computing platform, and its use in GPU increases the number of cores to run a long simulation within time (Zhou et al. 2012; Krieger and Vriend 2015; Ge et al. 2013). Due to the emergence of GPU-CUDA technology, vigorous and massively parallel clusters are developed, such as special-purpose supercomputer Anton and Blue waters (David et al. 2007). They are precise for running the MD simulation of biomolecules from μs to ms time-scale. But such resources are limited for limited researchers. To remove the time-scale gap, there is an urgent need to develop newer algorithms that allow enhanced sampling in the defined areas of conformational space and access long time-scale actions using necessary hardware. The purpose of this algorithm is to collect sufficient sampling that could result in the Boltzmann distribution of the diverse conformational states for the accurate calculation of the thermodynamic and kinetic properties of the system (Doshi and Hamelberg 2015). By the modification of the Hamiltonian method is to add a bias potential, several approaches have been developed like hyper dynamics (Voter 1997), local elevation (Huber et al. 1994; Voter 1997), and accelerated MD (Rodriguez-Bussey et al. 2016). In the case of hyper dynamics simulation, the identification of transition state required, but it is not necessary for classical MD simulation. Several tools are available to perform the MD simulation study with CPU or GPU. A few of the widely used tools are described in brief below.

7.4.2 GROMACS

GROMACS is the most widely used software for MD simulation. It is a freely available tool, and a brief tutorial of this tool can be accessed by this link (http://www.mdtutorials.com/gmx/) (Pronk et al. 2013). In the GROMACS simulation kit, MD simulation can be performed at various temperatures and pH values. In this simulation tool, several commands are available to perform a distinct function and calculate specific structural parameters. GROMACS, which is one of the MD simulation software, can read only the 20 natural amino acids, i.e., the non-standard amino acids are not read by GROMACS algorithms. Sometimes, there are force field limitations, for instance, Gromos and Amber cannot read the nicked DNA, but the same force field can read the same non-nicked DNA. The brief methodology for MD simulation using GROMACS is shown in Fig. 7.4.

To start, the user creates a box and fills the solvent (water). The solvent model depends on the force field. After placing the protein in the defined box in the solvent, the charge of the system is neutralized either by the addition of Na^+ or Cl^- ions; this is followed by the minimization of the system using the steepest descent method. Then, NVT (the constant Number of particles, Volume, and Temperature) simulation is run to maintain the volume and temperature of the defined system. The

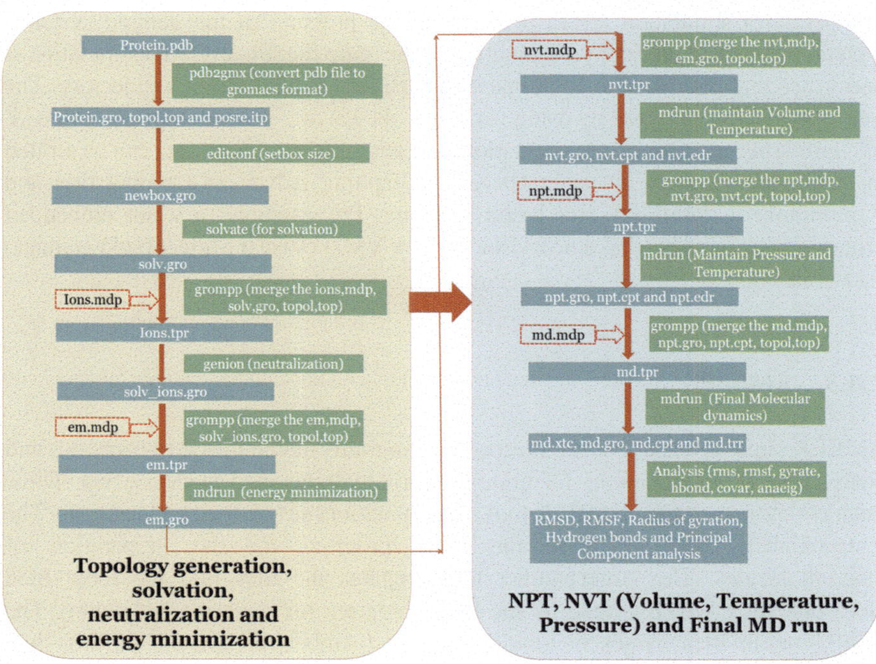

Fig. 7.4 Schematic representation showing the methodology of the MD simulation steps for GROMACS

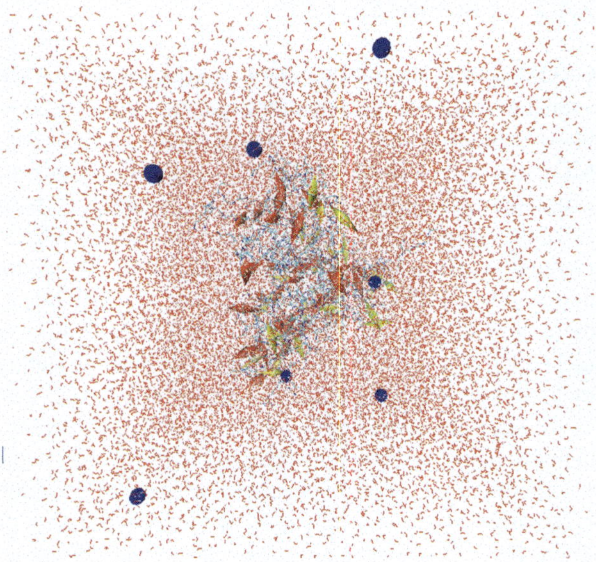

Fig. 7.5 A protein placed in a cubic box in the water system. The red color shows water molecules, while the blue color represents the ions

temperature of the system arises from 0 and attains the desired temperature that is set by the user. After that, NPT (the constant Number of particles, Pressure, and Temperature) simulation is run to maintain the pressure of the defined systems. Several parameters are set by the addition of the .mdp file. Finally, MD simulation is performed that provides the coordinates of each step in the form of a trajectory. The trajectory can be analyzed by using various tools that are embedded in GROMACS, like gmx–rms, gmx–rmsf, gmx–gyration, and gmx–hbond. These data can be plotted in an interactive form by using GRACE (Graphing Advanced Computation and Exploration of data), a Linux based software. For example, a water embedded protein molecule, placed in a box visualized by VMD (Visual Molecular Dynamics) shown in Fig. 7.5.

7.4.3 AMBER

AMBER simulation suite is a collection of programs that are used to carry out and analyze the MD simulations for proteins, carbohydrates, and nucleic acids. Three main components of the AMBER tool are preparation, simulation, and analysis. The Antechamber and LEaP are the main program for the preparation of macromolecules. The Antechamber tool prepares the files into the force filed descriptor files, which is read by the LEaP program for molecular modeling. The LEaP program then creates the topology files and Amber coordinates, which is then used in the MD simulation. The Sander program performs the MD simulation by fixing the temperature, pressure, and pH of the defined system. Lastly, the analysis part is performed by the ptraj module, which calculates the RMSD, RMSF, radius of gyration (Rg), H-bonds, and cross-correlation functions.

7.4.4 CHARMM-GUI

CHARMM-GUI is a simulation tool for the analysis of macromolecular dynamics and associated mechanical attributes. It performs standard MD simulations by using state-of-the-art algorithms for time stepping, long-range force calculation, and periodic images. Various analyses, such as energy minimization, crystal optimization, and normal mode analysis, can be performed using CHARMM.

7.4.5 NAMD

The simulations of much large biomolecular systems are performed using NAMD. The NAMD is available free of charge. The source code documentation including a set of compiled binary files configured with various parallel source software for calculations are freely available to the user. It supports massively parallel CUDA technology. NAMD can be used with graphical user interface software VMD. The simulation can be set and analyzed using the VMD as an interface. It is also compatible with AMBER and CHARMM (Khan et al. 2016).

7.4.6 Quantum-Mechanics/Molecular-Mechanics (QM/MM)

The QM/MM methods are a widely used approaches for biomolecular systems modeling (Groenhof 2013; Warshel and Levitt 1976). The various processes and charge transfer can be described using the QM method, but the QM methods are restricted to only a few hundred atom systems. Though simulations of a large system and a long-time period are performed by highly efficient force field-based MM methods. For modeling of the large biomolecules, the hybrid QM/MM method is very efficient (Senn and Thiel 2009). The QM/MM method is widely used with various applications, such as MD simulation, free energy calculation, geometry optimization, and computational spectroscopy (Groenhof 2013).

7.4.7 HyperChem

The HyperChem tool is also a tool for molecular modeling (Froimowitz 1993). It is an attractive commercial programming product manufactured by Hypercube Company and also given for free 30 days trial. It has a set of tools for molecular mechanics, quantum chemistry, and MD of the biological systems. The attractiveness of this software is attributed to the availability of complete documentation supported by examples, making this package ideal for studying the principle and practical approaches to molecular modeling (Gutowska et al. 2005). However, this program is comparitively slow as it cannot use multiprocessor support. An efficient strategy to use this tool is to employ it as an interactive molecular designer.

7.5 Other Advance Methods for MD Simulation

The MD simulation is a very progressive field. Several advancements have happened, and several other new methods are also introduced regularly to reduce the time complexity.

7.5.1 Metadynamics

It is an advantageous and powerful method of MD simulation to enhance sampling. The collective variables (CVs) define the free energy landscape reconstruction as a function of a few selected degrees of freedom. In this method, the sampling is accelerated by the history-dependent bias potential (Barducci et al. 2011). The space of collective variables is used for the adaptive construction of bias potential. In recent times, considerable improvements have been made to the actual algorithm, leading to a well-organized, flexible, and precise method that has found many successful applications in several domains. There are various examples of metadynamics study, and the most common umbrella sampling method is discussed below (Barducci et al. 2011).

The main challenge in computational biology is to predict the accurate binding free energy difference between two or more systems. For this problem, a new method umbrella sampling introduced, which is a biased MD simulation method, and it calculates free energy using the reaction coordinates. In this method, the system is driven from one thermodynamic state to other thermodynamics states. For example, reactant and product by using the bias potential reaction coordinate along with one or more directional (Kastner 2011). The intermediate steps are covered by a series of windows, at every stage of which an MD simulation is performed. Any functional form can represent bias potential. The harmonic potential is used in this method. Using the reaction coordinates, with the sampled distribution of a system, the free energy change in each window can be calculated. By using the umbrella integration or weighted histogram analysis method, the obtained windows are combined. The bias directly gives the free energy change between the two systems. The replica-exchange method can be used to improve the sampling of each window, either by replacing between successive windows or by running additional simulations at higher temperatures (Kastner 2011).

7.6 Analysis of MD Trajectories Through GUI-Based Software

The resulting output trajectories of any MD simulation can be visualized using GUI-based software. In the section below, we will discuss a few most popular software.

7.6.1 Visual Molecular Dynamics

VMD is developed at the University of Illinois (Hsin et al. 2008; Falsafi-Zadeh et al. 2012; Humphrey et al. 1996; Knapp et al. 2010). It is a potent tool for analysis and visualization of various biological systems such as proteins, nucleic acids, carbohydrates, and lipids. It is compatible with a large number of file formats, such as PDB and GROMOS. It can process a large amount of data for the visualization of trajectory movements (Falsafi-Zadeh et al. 2012). The molecules can be viewed in the animated form, and a movie can also be created from the input trajectory. It can be operated from a remote computer and also compatible with any operating systems with basic computer configuration. It is included with NAMD also. The additional functions of this tool are given below (Likhachev et al. 2016).

1. The visualization and analysis of macromolecules.
2. The atoms and amino acids can be selected.
3. Two structures can be aligned.
4. Support of user's action recording in the scripts.
5. The Raster3D format support (This format can give a high-quality image).
6. Ramachandran plots can be generated.
7. Support various types of molecular images.
8. Stereoscopic output.
9. Command-line support.
10. Working with arrays and vectors.
11. Support of JavaScript.

7.6.2 PyMOL

The PyMOL is among the most widely used software in the field of structural biology. It can accept various formats like PDB, Mol2, SDF, and several other file formats. The user can import the trajectory and analyze the simulation result. The user can generate a surface view model. Several additional plug-ins are also available to analyze the result of MD simulation. The user can create high-quality figures and animated movies from this tool.

7.6.3 Chimera

UCSF Chimera is an advanced software (Pettersen et al. 2004). It is widely used and freely available for academics. It was developed by Resource for biocomputing, visualization, and informatics (RBVI) and funded by the National Institutes of Health. The UCSF ChimeraX is also available, which is more advanced than Chimera. The Chimera 1.13.1 is released on 14-08-2018 and accepts the GROMACS and AMBER trajectory formats. After importing these trajectories, the user can make an animated movie with the time frame and generate high-quality pictures. The user can align two or more structures. The user can also generate the surface for cavity analysis during trajectory run. It has several features and supports the command line option also.

7.7 Structural Parameters for Analysis of MD Simulation

The MD simulation produces the result in the form of a trajectory. The trajectory contains all the parameters that were generated during each step movement of atoms. Various structural parameters that can be used to analyze the results are the following:

7.7.1 RMSD

The root mean square deviation (RMSD) is the most important and first parameter to analyze any MD trajectory (Kuzmanic and Zagrovic 2010). RMSD is used to measure the difference between the backbones of a protein from its initial structural conformation to its final position. The stability of the protein relative to its conformation can be determined by the deviations produced during the MD simulation process. RMSD is calculated with respect to the reference native conformation r_{ref} using the following formula:

$$\mathrm{RMSD}(t) = \left[\frac{1}{M}\sum_{i=1}^{N} m_i \left|r_i(t) - r_i^{\mathrm{ref}}\right|^2\right]^{1/2}$$

where $M = \sum_i m_i$ and $r_i(t)$ represents the atom, i position at the time t after least square fitting the structure to a reference structure.

The RMSD for all residues, backbone, side-chain, and Cα atoms can be calculated. RMSD is calculated with respect to the simulation time. Smaller deviations indicate a more stable protein structure and vice versa. In general, an RMSD value for a macromolecule should be less than 2 Å to extract any meaningful data. A comparative study of RMSD for native isocitrate lyase from *Mycobacterium tuberculosis* (MtbICL) and its mutant has been performed (Fig. 7.6a). A major difference in the RMSD of native (black) and mutant MtbICL (red) has been observed, and it indicates that a single mutation L148A in MtbICL protein ($\mathrm{MtbICL}_{\mathrm{L148A}}$) causes structural perturbations in the enzyme and reduces its stability (Shukla et al. 2018c).

7.7.2 RMSF

The root mean square fluctuation (RMSF) is the best way to study the residue-wise fluctuation of the protein from the MD trajectory. It describes the fluctuation of each residue or domain in a protein. The RMSF can be plotted as RMSF (nm) vs. residue

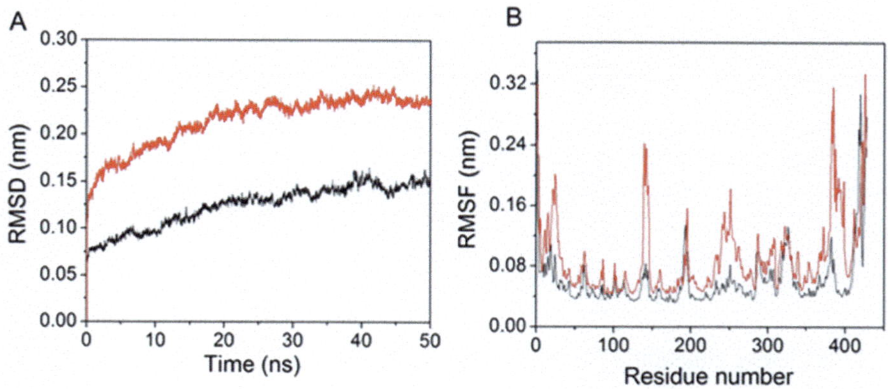

Fig. 7.6 (**a**) RMSD of native and mutant MtbICL. (**b**) RMSF of the Cα atoms calculated from the last 30 ns of the MD trajectory. Black and red lines represent MtbICL and $\mathrm{MtbICL_{L418A}}$, respectively

number (Kuzmanic and Zagrovic 2010). The RMSF of a protein is measured by the deviation between the position of particle *i* with its reference position, and *T* represents the time, and *riref* is the reference position of particle *i*.

$$\mathrm{RMSF}_i = \left[\frac{1}{T}\sum_{t_j=1}^{T}\left|r_i(t_j) - r_i^{\mathrm{ref}}\right|^2\right]^{1/2}$$

Well organized and rigid structures, like helix and sheets, show low RMSF, while loosely structure like bends and coils showed higher RMSF value because atom can have more fluctuation in the bends and coils as compared to helix and sheet. RMSF of the Cα atoms for native and mutant MtbICL was calculated from the MD trajectory to compare the residue-wise fluctuations (Fig. 7.6b). A high degree of residue-wise fluctuations observed in mutant MtbICL as compared to the native structure, and this happens due to a single mutation L148A, which causes more fluctuations and instability in the mutant. This mutation can also deform the shape of the binding site and which in turn can prevent the function of the enzyme (Shukla et al. 2018c).

7.7.3 Radius of Gyration

The radius of gyration (Rg) of a protein describes the compactness of the folded protein. For the same size proteins, ideal *Rg* value should less for the globular folded state, while in expanded form or protein form with more number of loops and turns, the *Rg* value should be relatively high. The *Rg* value of a structure is calculated from the following equation:

$$R_g = \left(\frac{\sum_i |r_i|^2 m_i}{\sum_i m_i} \right)^2$$

In this equation, the m_i represents the mass of atom i and r_i is the position of atom i with respect to the center of mass of the molecule. The system total mean energy is calculated by the following equation:

$$\langle E \rangle = \frac{1}{N} \sum_{i=1}^{N} E_i$$

In the above equation, E_i represents the energy of atom i.

7.7.4 Protein–Ligand Contacts

Non-covalent bonds play a significant role in determining the stability of the protein and ligand. These bonds play a pivotal role in protein structure folding and protein–ligand stability. During the folding, the hydrogen bonds provide a path for proper folding of a protein. Other interactions, such as hydrophobic bonds, make the inner patches and the catalytic triad of a biomacromolecule. The hydrogen bonds between residues can be calculated with respect to time during the simulation. In MD simulation analysis of apo-protein and protein–ligand, the hydrogen bonds, as well as other interactions, can be calculated during simulation time to find the potential residues necessary for ligand stabilization. The protein–ligand contact map is generated to find the residues that are in contact (interaction) with ligand during most of the simulation time.

7.7.5 SASA

The solvent accessible surface area (SASA) defines the area of the protein that interacts with the solvents in a simulation box (Mazola et al. 2015). For the same size proteins, the folded globular state shows lesser SASA value, while the expanded form of the protein shows higher SASA value. It is well-known that an increase in the temperature of the system will lead to protein unfolding, and the hydrophobic core of protein gets exposed toward the solvent. As a result, the SASA value increases upon unfolding.

7.7.6 Principal Component Analysis or Essential Dynamics

Essential dynamics (ED) reflect the overall increase of the motions in a protein during the time-scale of simulations (Maisuradze et al. 2009). The principal component analysis (PCA) predicts the collective motions of the protein and reveals the atomic fluctuations in the structure (David and Jacobs 2014). Every atom is related to each other in the MD simulation. The correlation motion analysis is very important for predicting the behavior of the molecules. The covariance matrix of atomic fluctuations is diagonalized for predicting the eigen values. The first eigenvectors play an essential role in the overall motions of the protein. The PCA analysis is used to compare the correlated motions of a protein under various conditions.

7.7.7 Secondary Structure Analysis

The secondary structure is a crucial parameter in the analysis of the MD results, and it describes the contents of secondary structure with respect to time. It clearly describes the structural contents to understand the stability of each domain. It is instrumental in mutational and protein unfolding studies, as it can explain the unfolding of a domain and its stability during a simulation.

7.8 Application of MD Simulation

MD simulation is a widely used technique for studying biological systems. It is applicable in several fields like mutational analysis of protein, protein–ligand complex stability prediction, protein unfolding studies, conformational protein stability prediction, etc. It is a potent tool that requires high computational power to reveal biological mysteries. In the following sub-sections, we will briefly discuss the applications of MD simulation in all these above-stated fields using case studies.

7.8.1 Mutational Analysis

MD simulation can be used to confirm the results of in vitro mutational studies, as it can predict the conformational movement of the atoms after and before mutation. It can also predict the alternate binding site. A case study of mutational analysis from our previous works is presented here. In these studies, the role of structurally distant amino acids (>10 Å) that are away from the active site signature motif (^{189}KKCGH^{193}) was studied (Fig. 7.7). We showed that point mutations brought the vast extent of conformation changes in the active site conformation, which results in loss of activity of the enzyme. Three mutations (F345A, L418A, and H46A) were introduced in the structure by the in silico approach, and then MD simulation was performed to see the effect of the mutation on the structure and function of the enzyme. Here, we take an example of several MD simulation analyses of an enzyme isocitrate lyase from *Mycobacterium tuberculosis* (MtbICL) and its three point mutants and extract structural and dynamic information from the MD simulation trajectories. Using MD simulation studies, we showed that distant point mutations at H46, F345, and L418, which are structurally distant (>10 Å) to the active site sequence (^{189}KKCGH^{193}), completely abrogates the enzyme activity.

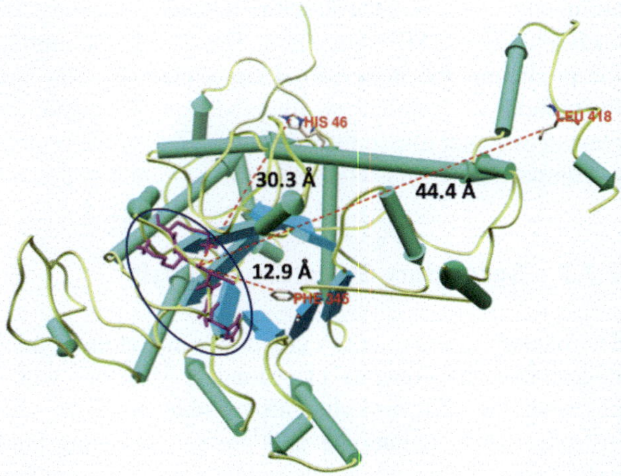

Fig. 7.7 Barrel structure of MtbICL. The distance of the mutated amino acids from the active site catalytic residues is shown in red lines

MtbICL is an important anti-tubercular drug target that catalyzes the first step of the glyoxylate shunt and is required for survival of the pathogen under dormancy condition. To understand the role of structurally distant amino acids in regulating the structural dynamics and conformational flexibility of MtbICL, several point mutants were created by site-directed mutagenesis, and their structure–activity relationship was studied. Three mutations, F345A (Shukla et al. 2018b), L418A (Shukla et al. 2018c), and H46A (Shukla et al. 2018f), were made. The experimental data revealed that these mutants were causing complete loss of enzyme activity, though these residues were not present within or near the active-site. Structurally, these residues were present more than 10 Å from the active site (Fig. 7.6).

To study the structural changes upon mutation and its effect on the dynamics of the enzyme, MD simulations were performed. A comparative study of RMSD for native and the three point mutants of MtbICL was done. The RMSD for the native, H46A, F345A, and L418A MtbICL mutant was found to be 0.12 nm, 0.13 nm, 0.14 nm, and 0.21 nm, respectively. The RMSD of F345A and H46A mutation shows an insignificant difference in the average RMSD values, thereby indicating that they do not significantly influence the overall stability of the protein. The RMSD value of L418A mutation suggests that it negatively affects protein stability, inducing destabilization in the mutant protein structure.

For understanding the effects of the mutations on the structural fluctuation of the entire protein and also on individual residues, the RMSF of Cα atoms was analyzed for all the proteins. RMSF of the native and the three mutants was calculated from the MD trajectories to compare the residues wise fluctuations. In H46A and F345A mutant, minor changes were observed in the RMSF (Shukla et al. 2018c; Shukla et al. 2018f), while the L418A mutant showed significant fluctuations (Shukla et al. 2018f). The data suggest that H46A and F345A mutation does not induce changes in the overall protein structure. In contrast, a high degree of residue wise fluctuations observed in the L148A mutant, which causes more fluctuations and instability in the structure (Shukla et al. 2018c).

H46A mutation did not show any global structural alternations; it caused changes in the catalytic site (Shukla et al. 2018f). The PCA and cross-correlation analysis showed that H46A mutation caused a change in conformational stability and collective motions of the protein, particularly in the active site region. The residue interaction network (RIN) analysis indicates that the active site geometry was disturbed in the H46A mutant. These results suggest that due to the mutation, the dynamic perturbation of the active site leads to enzyme transition from its active form to inactive form (Shukla et al. 2018f).

The mutation in F345 induced structural flexibility and conformational rearrangements near the active site. This mutation increased the collective motions and residual mobility of the enzyme, resulting in a decrease in the mutant enzyme stability. The result was confirmed by the lower free energy in the mutant enzyme indicating the destabilized structure (Shukla et al. 2018b).

The L418A mutation also nullifies the activity of the protein. The correlated motions, residual mobility, and flexibility in the enzyme increased upon mutation. Upon L418A mutation, the global conformational dynamics and the RIN of the protein changed. The RIN depicts that several hydrogen bonds, hydrophobic bonds were distorted due to the mutation. This alteration in RIN brings conformational changes in the active site leading to the loss of enzyme activity (Shukla et al. 2018f).

Ultimately, molecular docking data indicated that all three mutations affected the substrate interactions with the active site residues of MtbICL. These results reveal the internal dynamics of the enzyme structure and feature the importance of residue-level interactions in the enzyme.

7.8.2 Application in the Drug Designing

The protein–ligand docking is a significant phase in the field of drug designing. Nowadays, several software are available for structure-based virtual screening. The PDB entries are increasing due to newer structure determination methods. Most of the docking software considers the protein as a rigid body, while ligand is always considered as flexible. Some docking software recognizes the protein and ligand both as a flexible molecule, and they can produce a better pose with a binding affinity. The question arises now is: will this docking pose exist in the cell because every protein is dynamic? To solve this problem, MD simulation is an excellent approach used to predict the stability and dynamics of the protein–ligand complexes.

7.8.2.1 Inhibitor Designing Against MtbICL

Several MtbICL structures are available in the PDB in complex with inhibitor and substrates. An inhibitor-bound structure was retrieved from the PDB (PDB ID: 1F8I), and 167,674 compounds were screened in various runs. Following rounds of screening and docking refinement, 340 compounds were selected. For validation of the docking results and to study the dynamics of the system, MD simulation was performed as docking does not provide insight into the dynamics of the system. The MD simulation data of three compounds were compared and only one compound was found to have high potential to inhibit MtbICL. Thus, the MD simulation can be used to remove the false positive binders and provides information on the detailed mechanism of ligand-induced inhibition of enzyme function (Shukla et al. 2018e).

7.8.2.2 Inhibitor Designing Against *Fasciola gigantica* Thioredoxin Glutathione Reductase

Fasciola gigantica thioredoxin glutathione reductase (FgTGR) is a key drug target against fascioliasis caused by the helminth parasite *Fasciola gigantica*. We reported some novel inhibitors of FgTGR using the structure-based virtual screening approach, and the selected hits were validated by MD simulation. The compounds were screened against FgTGR in several runs. The selected compounds were evaluated through ADMET, and some compounds were chosen for further docking. Ultimately, three compounds were used for the MD simulation that resulted in one highly potent compound with a high affinity towards FgTGR. Thus, the MD simulation can play an important role in the screening of potential inhibitor against a target considering the physiological conditions of the interaction environment (Shukla et al. 2018b).

7.8.3 Unfolding Studies

MD simulation is a compelling technique to study the protein unfolding mechanism at an atomic level (Prakash et al. 2018; Sonkar et al. 2018). We can track the mechanism of unfolding as a result of chemical-, pH-, or temperature-induced denaturation using MD simulation, which provides a clear view of structural alternations taking place at a particular time-scale. Protein unfolding by pH, urea, GdnHCl, and temperature has already been performed using MD simulation.

7.8.3.1 Urea Induced Unfolding of FgGST1

Urea is a widely used denaturant to study the mechanism of protein unfolding. It has been proposed to disrupt the hydrophobic interactions, as a result of which the hydrophobic patches of proteins open and come into contact with water (Kalita et al. 2018a). We took an 8 M urea environment for analysis and performed 100 ns simulation at 300 K and 400 K temperature to understand the unfolding mechanism of a protein. The dynamics of protein were recorded at an interval of 40 ns; the data indicate that 8 M urea induces unfolding, and finally leads to the disruption of the complete protein 3D structure (Fig. 7.8). The RMSD, RMSF, Rg, and PCA analyses suggest that urea induces the unfolding of the protein. Different secondary structure alternations, such as loss of alpha-helices, loss of bends, and beta-sheets, were observed in the protein during MD simulation (300 K), which indicates the loss of native structure. In MD simulation at 400 K, a greater extent of unfolding in protein structure was observed.

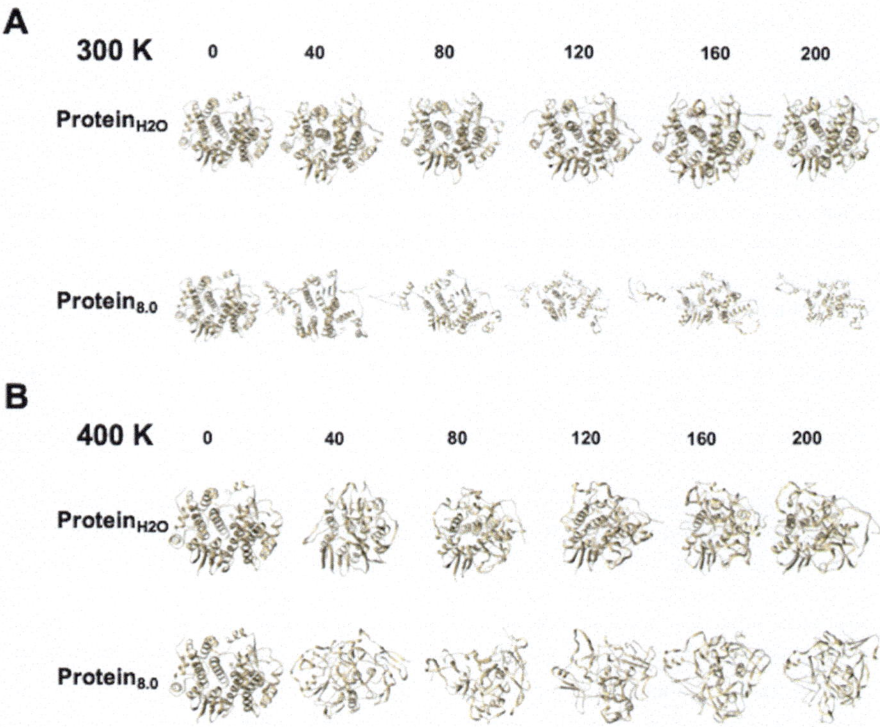

Fig. 7.8 Different time frame snapshots of unfolding at 40 ns time interval for a protein with water and urea at (**a**) 300 K (**b**) at 400 K temperature. Protein$_{H2O}$ and Protein$_{8.0}$ represent the protein in water and 8 M urea, respectively (Kalita et al. 2018a). The MD simulation was run for 200 ns

7.8.3.2 GdnHCl-Induced Unfolding Analysis

Syed et al. performed the denaturation study of a protein using GdnHCl at 300 K (Syed et al. 2018). 100 ns MD simulation was run to investigate the atomic-level changes. Various structural parameters such as RMSD, RMSF, Rg, SASA, the number of hydrogen bonds, and PCA were calculated and revealed the unfolding mechanism of the protein (Syed et al. 2018).

7.8.3.3 pH-Induced Effects on the Structure and Stability of the Protein

MD simulation can also be performed in different pH values to model the structural rearrangement pattern in different pH conditions. Syed et al. also used pH in MD simulation (2, 4, 6, 8, 10, and 12) to study protein unfolding. Their analyses on a particular protein suggest that it can maintain the secondary and tertiary structure in the alkaline pH. In contrast, in the acidic condition (pH 2.0–5.5), significant structural changes occur (Syed et al. 2015).

7.9 Conclusions

Proteins are dynamic entities, and the dynamic nature defines its function. The availability of an accurate 3D structure is essential to understand the protein dynamics and function. The 3D structure may be solved using X-ray crystallography, NMR, or computational methods. Although these methods provide detailed information on protein structure, they still fail to provide sufficient information on protein dynamics and motion. MD simulation has a history of more than 43 years. It is a widely used technique for predicting the dynamic picture of any biological system. Development of GPUs based high computational capability system is a milestone for the MD simulation to reduce the time for predicting the dynamics of biological molecules such as nucleic acids, proteins, carbohydrates, and many more and their molecular interactions with each other or with small molecules inhibitors. MD simulation is a potent tool to solve difficult biological problems, which happen in second to millisecond time-scales like molecular interaction of protein–protein, protein–ligand interaction, protein folding, and unfolding analysis. It considers the biological pH, the surrounding molecules for creating the cell-like environment like water and lipids, as well as co-enzymes, ions, and nucleic acid. MD simulations provide atomic-level details of atom interaction that are associated with the function of the molecules. The binding free energy, various energy constituents, and residue-wise binding contribution with a ligand can be predicted using the MM-PBSA tool. This information is beneficial for further improvement in the binding affinity. The QM and MM method implementation in the MD simulation has drastically changed towards the enhancement of accuracy of the binding free energy. By using these methods, we can easily find out the role of polarization and electronic effects in protein–protein and protein–ligand interactions. MD simulation can be used to reveal the chemistry and dynamics of a biological system by providing an appropriate model and physical conditions.

Competing Interest The authors declare that there are no competing interests.

References

Adcock SA, McCammon JA (2006) Molecular dynamics: survey of methods for simulating the activity of proteins. Chem Rev 106(5):1589–1615

Alder BJ, Wainwright TE (1957) Phase transition for a hard sphere system. J Chem Phys 27 (5):1208–1209

Alder BJ, Wainwright TE (1959) Studies in molecular dynamics. I. general method. J Chem Phys 31(2):459–466

Barducci A, Bonomi M, Parrinello M (2011) Metadynamics. Wiley Interdiscip Rev Comput Mol Sci 1(5):826–843

Berger A, Linderstrom-Lang K (1957) Deuterium exchange of poly-DL-alanine in aqueous solution. Arch Biochem Biophys 69:106–118

Boehr DD, Dyson HJ, Wright PE (2006) An NMR perspective on enzyme dynamics. Chem Rev 106(8):3055–3079

Brooks B, Karplus M (1983) Harmonic dynamics of proteins: normal modes and fluctuations in bovine pancreatic trypsin inhibitor. Proc Natl Acad Sci U S A 80(21):6571–6575

Brooks BR, Brooks CL 3rd, Mackerell AD Jr, Nilsson L, Petrella RJ, Roux B, Won Y, Archontis G, Bartels C, Boresch S, Caflisch A, Caves L, Cui Q, Dinner AR, Feig M, Fischer S, Gao J, Hodoscek M, Im W, Kuczera K, Lazaridis T, Ma J, Ovchinnikov V, Paci E, Pastor RW, Post CB, Pu JZ, Schaefer M, Tidor B, Venable RM, Woodcock HL, Wu X, Yang W, York DM, Karplus M (2009) CHARMM: the biomolecular simulation program. J Comput Chem 30 (10):1545–1614

Brunger AT, Brooks CL 3rd, Karplus M (1985) Active site dynamics of ribonuclease. Proc Natl Acad Sci U S A 82(24):8458–8462

Brunger AT, Kuriyan J, Karplus M (1987) Crystallographic R factor refinement by molecular dynamics. Science 235(4787):458–460

Case DA, Karplus M (1979) Dynamics of ligand binding to heme proteins. J Mol Biol 132 (3):343–368

Case DA, Cheatham TE 3rd, Darden T, Gohlke H, Luo R, Merz KM Jr, Onufriev A, Simmerling C, Wang B, Woods RJ (2005) The Amber biomolecular simulation programs. J Comput Chem 26 (16):1668–1688

Colonna-Cesari F, Perahia D, Karplus M, Eklund H, Braden CI, Tapia O (1986) Interdomain motion in liver alcohol dehydrogenase. Structural and energetic analysis of the hinge bending mode. J Biol Chem 261(32):15273–15280

Cornell WD, Cieplak P, Bayly CI, Gould IR, Merz KM, Ferguson DM, Spellmeyer DC, Fox T, Caldwell JW, Kollman PA (1995) A second generation force field for the simulation of proteins, nucleic acids, and organic molecules. J Am Chem Soc 117(19):5179–5197

Darden T, York D, Pedersen L (1993) Particle mesh Ewald: an N-log(N) method for Ewald sums in large systems. J Chem Phys 98(12):10089–10092

David CC, Jacobs DJ (2014) Principal component analysis: a method for determining the essential dynamics of proteins. Methods Mol Biol 1084:193–226

David ES, Martin MD, Ron OD, Jeffrey SK, Richard HL, John KS, Cliff Y, Brannon B, Kevin JB, Jack CC, Michael PE, Joseph G, Grossman JP, Ho CR, Douglas JI, John LK, Timothy L, Christine M, Mark AM, Rolf M, Edward CP, Yibing S, Jochen S, Michael T, Brian T, Stanley CW (2007) Anton, a special-purpose machine for molecular dynamics simulation. Paper presented at the proceedings of the 34th annual international symposium on computer architecture, San Diego, California, USA

Dessailly BH, Lensink MF, Wodak SJ (2007) Relating destabilizing regions to known functional sites in proteins. BMC Bioinf 8:141

Doshi U, Hamelberg D (2015) Towards fast, rigorous and efficient conformational sampling of biomolecules: advances in accelerated molecular dynamics. Biochim Biophys Acta 1850 (5):878–888

Dror RO, Dirks RM, Grossman JP, Xu H, Shaw DE (2012) Biomolecular simulation: a computational microscope for molecular biology. Annu Rev Biophys 41:429–452

Duan Y, Kollman PA (1998) Pathways to a protein folding intermediate observed in a 1-microsecond simulation in aqueous solution. Science 282(5389):740–744

Ewald PP (1921) Die Berechnung optischer und elektrostatischer Gitterpotentiale. Ann Phys 369 (3):253–287

Falsafi-Zadeh S, Karimi Z, Galehdari H (2012) VMD DisRg: new user-friendly implement for calculation distance and radius of gyration in VMD program. Bioinformation 8(7):341–343

Feyfant E, Sali A, Fiser A (2007) Modeling mutations in protein structures. Protein Sci 16 (9):2030–2041

Frauenfelder H, Hartmann H, Karplus M, Kuntz ID, Kuriyan J, Parak F, Petsko GA, Ringe D, Tilton RF (1987) Thermal expansion of a protein. Biochemistry 26(1):254–261

Froimowitz M (1993) HyperChem: a software package for computational chemistry and molecular modeling. BioTechniques 14(6):1010–1013

Fuzo CA, Degreve L (2014) Effect of the thermostat in the molecular dynamics simulation on the folding of the model protein chignolin. J Mol Model 18(6):2785–2794
Ge H, Wang Y, Li C, Chen N, Xie Y, Xu M, He Y, Gu X, Wu R, Gu Q, Zeng L, Xu J (2013) Molecular dynamics-based virtual screening: accelerating the drug discovery process by high-performance computing. J Chem Inf Model 53(10):2757–2764
Gotz AW, Williamson MJ, Xu D, Poole D, Le Grand S, Walker RC (2012) Routine microsecond molecular dynamics simulations with AMBER on GPUs. 1. Generalized born. J Chem Theory Comput 8(5):1542–1555
Groenhof G (2013) Introduction to QM/MM simulations. Methods Mol Biol 924:43–66
Gutowska I, Machoy Z, Machalinski B (2005) The role of bivalent metals in hydroxyapatite structures as revealed by molecular modeling with the HyperChem software. J Biomed Mater Res A 75(4):788–793
Harvey SC, Prabhakaran M, Mao B, McCammon JA (1984) Phenylalanine transfer RNA: molecular dynamics simulation. Science 223(4641):1189–1191
Heidari Z, Roe DR, Galindo-Murillo R, Ghasemi JB, Cheatham TE (2016) Using wavelet analysis to assist in identification of significant events in molecular dynamics simulations. J Chem Inf Model 56(7):1282–1291
Hernandez-Rodriguez M, Rosales-Hernandez MC, Mendieta-Wejebe JE, Martinez-Archundia M, Basurto JC (2016) Current tools and methods in molecular dynamics (MD) simulations for drug design. Curr Med Chem 23(34):3909–3924
Holden ZC, Richard RM, Herbert JM (2013) Periodic boundary conditions for QM/MM calculations: Ewald summation for extended Gaussian basis sets. J Chem Phys 139(24):244108
Hsin J, Arkhipov A, Yin Y, Stone JE, Schulten K (2008) Using VMD: an introductory tutorial. Curr Protoc Bioinformatics 24:5–7
Huber T, Torda AE, van Gunsteren WF (1994) Local elevation: a method for improving the searching properties of molecular dynamics simulation. J Comput Aided Mol Des 8(6):695–708
Humphrey W, Dalke A, Schulten K (1996) VMD: visual molecular dynamics. J Mol Graph 14 (1):33–38, 27-38
Jeremy Smith SC, Pezzeca U, Brooks B, Karplus M (1986) Inelastic neutron scattering analysis of low frequency motion in proteins: a normal mode study of the bovine pancreatic trypsin inhibitor. J Chem Phys 85(6):3636–3654
Jones JE (1924) On the determination of molecular fields. -II from the equation of state of a gas. Proc Roy Soc A 106(738):463
Jorgensen WL, Tirado-Rives J (2005) Potential energy functions for atomic-level simulations of water and organic and biomolecular systems. Proc Natl Acad Sci U S A 102(19):6665–6670
Kalita J, Shukla R, Tripathi T (2018a) Structural basis of urea-induced unfolding of Fasciola gigantica glutathione S-transferase. J Cell Physiol 234(4):4491–4503
Kalita P, Shukla H, Shukla R, Tripathi T (2018b) Biochemical and thermodynamic comparison of the selenocysteine containing and non-containing thioredoxin glutathione reductase of Fasciola gigantica. Biochim Biophys Acta Gen Subj 1862:1306–1316
Karplus M, Kuriyan J (2005) Molecular dynamics and protein function. Proc Natl Acad Sci U S A 102(19):6679–6685
Kastner J (2011) Umbrella sampling. Wiley Interdiscip Rev Comput Mol Sci 1(6):932–942
Khan FI, Wei DQ, Gu KR, Hassan MI, Tabrez S (2016) Current updates on computer aided protein modeling and designing. Int J Biol Macromol 85:48–62
Kini RM, Evans HJ (1991) Molecular modeling of proteins: a strategy for energy minimization by molecular mechanics in the AMBER force field. J Biomol Struct Dyn 9(3):475–488
Knapp B, Lederer N, Omasits U, Schreiner W (2010) vmdICE: a plug-in for rapid evaluation of molecular dynamics simulations using VMD. J Comput Chem 31(16):2868–2873
Krieger E, Vriend G (2015) New ways to boost molecular dynamics simulations. J Comput Chem 36(13):996–1007
Krishnamoorthy G (2012) Motional dynamics in proteins and nucleic acids control their function: revelation by time-domain fluorescence. Curr Sci 102(2):266–276

Kuzmanic A, Zagrovic B (2010) Determination of ensemble-average pairwise root mean-square deviation from experimental B-factors. Biophys J 98(5):861–871

Levitt M, Warshel A (1975) Computer simulation of protein folding. Nature 253(5494):694–698

Likhachev IV, Balabaev NK, Galzitskaya OV (2016) Available instruments for analyzing molecular dynamics trajectories. Open Biochem J 10:1–11

MacKerell AD, Bashford D, Bellott M, Dunbrack RL, Evanseck JD, Field MJ, Fischer S, Gao J, Guo H, Ha S, Joseph-McCarthy D, Kuchnir L, Kuczera K, Lau FT, Mattos C, Michnick S, Ngo T, Nguyen DT, Prodhom B, Reiher WE, Roux B, Schlenkrich M, Smith JC, Stote R, Straub J, Watanabe M, Wiorkiewicz-Kuczera J, Yin D, Karplus M (1998) All-atom empirical potential for molecular modeling and dynamics studies of proteins. J Phys Chem B 102 (18):3586–3616

Maisuradze GG, Liwo A, Scheraga HA (2009) Principal component analysis for protein folding dynamics. J Mol Biol 385(1):312–329

Malde AK, Zuo L, Breeze M, Stroet M, Poger D, Nair PC, Oostenbrink C, Mark AE (2011) An automated force field topology builder (ATB) and repository: version 1.0. J Chem Theory Comput 7(12):4026–4037

Mamgain S, Sharma P, Pathak RK, Baunthiyal M (2015) Computer aided screening of natural compounds targeting the E6 protein of HPV using molecular docking. Bioinformation 5:236–242

Mazola Y, Guirola O, Palomares S, Chinea G, Menendez C, Hernandez L, Musacchio A (2015) A comparative molecular dynamics study of thermophilic and mesophilic beta-fructosidase enzymes. J Mol Model 21(9):228

McCammon JA, Gelin BR, Karplus M (1977) Dynamics of folded proteins. Nature 267 (5612):585–590

McCammon JA, Karim OA, Lybrand TP, Wong CF (1986) Ionic association in water: from atoms to enzymes. Ann N Y Acad Sci 482:210–221

Nguyen TT, Viet MH, Li MS (2014) Effects of water models on binding affinity: evidence from all-atom simulation of binding of tamiflu to A/H5N1 neuraminidase. Sci World J 2014:536084

Nilsson L, Clore GM, Gronenborn AM, Brunger AT, Karplus M (1986) Structure refinement of oligonucleotides by molecular dynamics with nuclear overhauser effect interproton distance restraints: application to 5′ d(C-G-T-A-C-G)2. J Mol Biol 188(3):455–475

Norberto de Souza O, Ornstein RL (1999) Molecular dynamics simulations of a protein-protein dimer: particle-mesh Ewald electrostatic model yields far superior results to standard cutoff model. J Biomol Struct Dyn 16(6):1205–1218

Oostenbrink C, Villa A, Mark AE, van Gunsteren WF (2004) A biomolecular force field based on the free enthalpy of hydration and solvation: the GROMOS force-field parameter sets 53A5 and 53A6. J Comput Chem 25(13):1656–1676

Pandey T, Shukla R, Shukla H, Sonkar A, Tripathi T, Singh AK (2017) A combined biochemical and computational studies of the rho-class glutathione s-transferase sll1545 of Synechocystis PCC 6803. Int J Biol Macromol 94:378–385

Paquet E, Viktor HL (2015) Molecular dynamics, Monte Carlo simulations, and Langevin dynamics: a computational review. Biomed Res Int 2015:18

Perilla JR, Goh BC, Cassidy CK, Liu B, Bernardi RC, Rudack T, Yu H, Wu Z, Schulten K (2015) Molecular dynamics simulations of large macromolecular complexes. Curr Opin Struct Biol 31:64–74

Pettersen EF, Goddard TD, Huang CC, Couch GS, Greenblatt DM, Meng EC, Ferrin TE (2004) UCSF Chimera—a visualization system for exploratory research and analysis. J Comput Chem 25(13):1605–1612

Phillips JC, Braun R, Wang W, Gumbart J, Tajkhorshid E, Villa E, Chipot C, Skeel RD, Kale L, Schulten K (2005) Scalable molecular dynamics with NAMD. J Comput Chem 26 (16):1781–1802

Ponder JW, Case DA (2003) Force fields for protein simulations. Adv Protein Chem 66:27–85

Prakash A, Kumar V, Meena NK, Hassan MI, Lynn AM (2018) Comparative analysis of thermal unfolding simulations of RNA recognition motifs (RRMs) of TAR DNA-binding protein 43 (TDP-43). J Biomol Struct Dyn 37:178–194

Pronk S, Pall S, Schulz R, Larsson P, Bjelkmar P, Apostolov R, Shirts MR, Smith JC, Kasson PM, van der Spoel D, Hess B, Lindahl E (2013) GROMACS 4.5: a high-throughput and highly parallel open source molecular simulation toolkit. Bioinformatics 29(7):845–854

Rahman A (1964) Correlations in the motion of atoms in liquid argon. Phys Rev 136(2A):A405–A411

Rahman A, Stillinger FH (1971) Molecular dynamics study of liquid water. J Chem Phys 55 (7):3336–3359

Rajendran V, Shukla R, Shukla H, Tripathi T (2018) Structure-function studies of the asparaginyl-tRNA synthetase from Fasciola gigantica: understanding the role of catalytic and non-catalytic domains. Biochem J 475(21):3377–3391

Rodriguez-Bussey IG, Doshi U, Hamelberg D (2016) Enhanced molecular dynamics sampling of drug target conformations. Biopolymers 105(1):35–42

Sagui C, Darden TA (1999) Molecular dynamics simulations of biomolecules: long-range electrostatic effects. Annu Rev Biophys Biomol Struct 28:155–179

Salomon-Ferrer R, Gotz AW, Poole D, Le Grand S, Walker RC (2013) Routine microsecond molecular dynamics simulations with AMBER on GPUs. 2. Explicit solvent particle mesh Ewald. J Chem Theory Comput 9(9):3878–3888

Salsbury FR Jr (2010) Molecular dynamics simulations of protein dynamics and their relevance to drug discovery. Curr Opin Pharmacol 10(6):738–744

Schuttelkopf AW, van Aalten DM (2004) PRODRG: a tool for high-throughput crystallography of protein-ligand complexes. Acta Crystallogr D Biol Crystallogr 60(Pt 8):1355–1363

Senn HM, Thiel W (2009) QM/MM methods for biomolecular systems. Angew Chem Int Ed Engl 48(7):1198–1229

Shukla H, Khan SR, Shukla R, Krishnan MY, Akhtar MS, Tripathi T (2018a) Alternate pathway to ascorbate induced inhibition of Mycobacterium tuberculosis. Tuberculosis (Edinb) 111:161–169

Shukla H, Shukla R, Sonkar A, Pandey T, Tripathi T (2018b) Distant Phe345 mutation compromises the stability and activity of Mycobacterium tuberculosis isocitrate lyase by modulating its structural flexibility. Sci Rep 7(1):1058

Shukla H, Shukla R, Sonkar A, Tripathi T (2018c) Alterations in conformational topology and interaction dynamics caused by L418A mutation leads to activity loss of Mycobacterium tuberculosis isocitrate lyase. Biochem Biophys Res Commun 490(2):276–282

Shukla R, Shukla H, Kalita P, Sonkar A, Pandey T, Singh DB, Kumar A, Tripathi T (2018d) Identification of potential inhibitors of Fasciola gigantica thioredoxin1: computational screening, molecular dynamics simulation, and binding free energy studies. J Biomol Struct Dyn 36 (8):2147–2162

Shukla R, Shukla H, Sonkar A, Pandey T, Tripathi T (2018e) Structure-based screening and molecular dynamics simulations offer novel natural compounds as potential inhibitors of Mycobacterium tuberculosis isocitrate lyase. J Biomol Struct Dyn 36(8):2045–2057

Shukla R, Shukla H, Tripathi T (2018f) Activity loss by H46A mutation in Mycobacterium tuberculosis isocitrate lyase is due to decrease in structural plasticity and collective motions of the active site. Tuberculosis (Edinb) 108:143–150

Singh DB, Tripathi T (2020) Frontiers in protein structure, function, and dynamics. Springer, Singapore. https://doi.org/10.1007/978-981-15-5530-5. ISBN 978-981-15-5529-9

Sonkar A, Shukla H, Shukla R, Kalita J, Pandey T, Tripathi T (2017) UDP-N-Acetylglucosamine enolpyruvyl transferase (MurA) of Acinetobacter baumannii (AbMurA): structural and functional properties. Int J Biol Macromol 97:106–114

Sonkar A, Shukla H, Shukla R, Kalita J, Tripathi T (2018) Unfolding of Acinetobacter baumannii MurA proceeds through a metastable intermediate: a combined spectroscopic and computational investigation. Int J Biol Macromol 126:941–951

Syed SB, Shahbaaz M, Khan SH, Srivastava S, Islam A, Ahmad F, Hassan MI (2015) Estimation of pH effect on the structure and stability of kinase domain of human integrin-linked kinase. J Biomol Struct Dyn 37:156–165

Syed SB, Khan FI, Khan SH, Srivastava S, Hasan GM, Lobb KA, Islam A, Hassan MI, Ahmad F (2018) Unravelling the unfolding mechanism of human integrin linked kinase by GdmCl-induced denaturation. Int J Biol Macromol 117:1252–1263

Takahashi K, Narumi T, Yasuoka K (2010) Cutoff radius effect of the isotropic periodic sum method in homogeneous system. II. Water. J Chem Phys 133(1):014109

van Aalten DM, Bywater R, Findlay JB, Hendlich M, Hooft RW, Vriend G (1996) PRODRG, a program for generating molecular topologies and unique molecular descriptors from coordinates of small molecules. J Comput Aided Mol Des 10(3):255–262

Vogeli B, Kazemi S, Guntert P, Riek R (2012) Spatial elucidation of motion in proteins by ensemble-based structure calculation using exact NOEs. Nat Struct Mol Biol 19(10):1053–1057

Voter AF (1997) Hyperdynamics: accelerated molecular dynamics of infrequent events. Phys Rev Lett 78(20):3908–3911

Warshel A, Levitt M (1976) Theoretical studies of enzymic reactions: dielectric, electrostatic and steric stabilization of the carbonium ion in the reaction of lysozyme. J Mol Biol 103(2):227–249

Wolf A, Kirschner KN (2013) Principal component and clustering analysis on molecular dynamics data of the ribosomal L11.23S subdomain. J Mol Model 19(2):539–549

Wu X, Brooks BR (2009) Isotropic periodic sum of electrostatic interactions for polar systems. J Chem Phys 131(2):024107

Zhou Y, Liepe J, Sheng X, Stumpf MP, Barnes C (2012) GPU accelerated biochemical network simulation. Bioinformatics 27(6):874–876

Zoete V, Cuendet MA, Grosdidier A, Michielin O (2010) SwissParam: a fast force field generation tool for small organic molecules. J Comput Chem 32(11):2359–2368

Computational Approaches for Drug Target Identification

8

Pramod Katara

Abstract

It is assumed that due to the enormous investment in terms of time, money, human volunteers, and other resources, sometimes failure at the later stage mostly put pharmaceutical companies on the back foot. For the last two decades, pharmaceutical companies felt that the traditional drug designing process should be optimized to avoid huge financial loss and save time. Thus, despite its limitations, the use of computer-aided drug design (CADD) techniques in drug discovery and development process is successful. CADD approaches support almost all phases of the drug designing process, including drug target identification, lead identification, optimization of leads, and simulations. Drug target identification and characterization is a first and most essential step that begins with identifying the function of a possible molecular target (gene/protein) and its role in the disease. The availability of the huge amount of molecular data, i.e., big data, for human as well as pathogens with applications of knowledge-based data mining approaches can provide a list of probable drug targets which further can be validated through experiments can save time and cost of pharmaceutical companies and boost their research towards the development of new drugs. This chapter focuses on the computational approaches for drug target identification, which play a crucial role in the drug discovery and development process.

Keywords

Algorithm · Database · Drug target · Drug designing · Druggability · Biological network

P. Katara (✉)
Computational Omics Lab, Centre of Bioinformatics, University of Allahabad, Prayagraj, India

D. B. Singh (ed.), *Computer-Aided Drug Design*,
https://doi.org/10.1007/978-981-15-6815-2_8

8.1 Introduction

Drug designing deals with the discovery and development of therapeutic molecules for a drug target. The drug is a small molecule that has potential to modulate the function of drug targets, such as a protein and sometimes nucleic acid tool, i.e., regulatory RNAs (Dersch et al. 2017). Drug design involves the design of molecules that are complementary in shape to the chosen drug target and modulate in the desired manner (Zauhar et al. 2003). Nowadays various drug designing approaches are in practice, broadly they can be classified into two types: (1) traditional methods: traditional methods involve trial and error method of testing for chemicals on cultured or animals cell, and observe the outcome of treatments, and (2) rational drug design: this approach is based on the hypothesis that modulation of a specific biological target which will be considered as drug targets, may have therapeutic value. In this approach, a potential therapeutic target is identified and purified. The purified protein is used to develop a screening assay. In rational drug design, 3D structure of the drug target should be available. The small bioactive searched by screening libraries of a drug or bioactive compound. This can also be performed by the screening assay, which also known as chemical or wet screening assay.

Nowadays computational methods are also in practice to screen compounds virtually and are well known as virtual screening (McInnes 2007). After library screening, the molecules are subjected to biological screening to test toxicity and those who show positive screening enter into the clinical trials where they try on human volunteers/patients to check pharmacokinetics (ADMET) of the drug. In the case of the successful completion of the clinical trials, a molecule passes to the approval agency and then finally hits the market (Fig. 8.1). This whole drug designing process is very time consuming and expensive, and at any stage of the process, a lead molecule can fail. Failure of leads at a later stage is responsible for the loss of millions of dollars for pharmaceutical companies (Hughes et al. 2011).

To reduce the chance of later-failure and speed up the molecular screening process, computational approaches are in practice for the last one and a half decade. Nowadays, designing drug using computational approaches is well known as computer-aided drug designing (CADD). CADD involves various approaches such as QSAR, virtual screening, docking, etc. (Katsila et al. 2016). Computational approaches have speed up the process of drug discovery and have provided novel drug targets and lead structures (Katara 2013). The computational method can identify drug targets and leads against them, affinity and efficacy between them before clinical trials and saving enormous time and cost (Shekhar 2008; Katara 2017).

8.2 Drug Targets

The term drug target describes the native biomolecule in the human body whose function can be modulated by a drug molecule, which may have a therapeutic effect against the disease or some adverse effect. Mostly these drug targets are biological

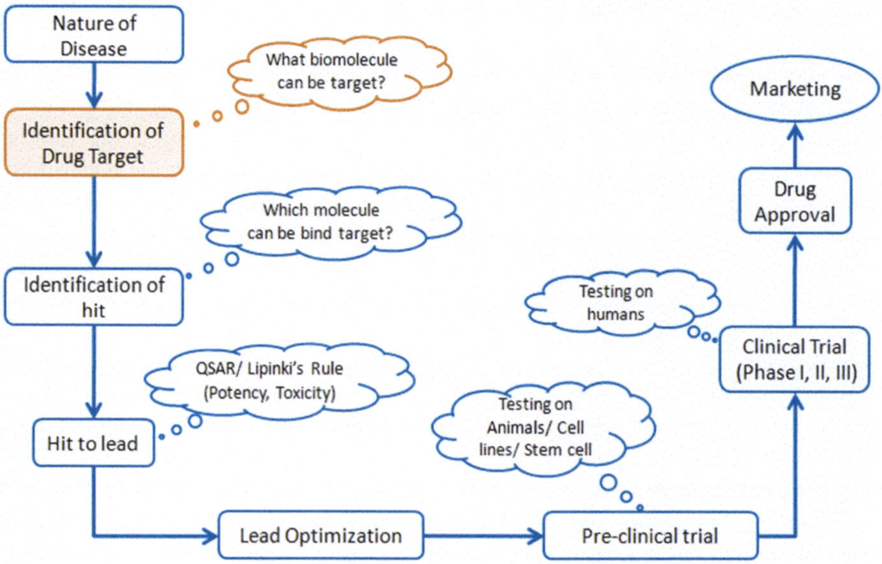

Fig. 8.1 The flow of drug designing process (Katara 2017)

Table 8.1 Details of frequently used drug target protein classes

S. No.	Target classes	Description
1	GPCRs	G protein-coupled receptors (GPCRs) play a central role in various signal transduction pathways responsible for cellular responses. Due to its indispensable role, GPCRs make up a large portion of the targets of approved drugs. Presently, more than hundreds of GPCRs are already in practice as targets of ~34% FDA approved drugs (Sriram and Insel 2018)
2	Ion channels	Ion channels play a very crucial role in controlling a very wide range of physiological processes in humans, and their dysfunction can lead to abnormalities, thus they are reported as one of the important drug targets (Kaczorowski et al. 2008)
3	Kinases	Kinase plays a pivotal role in the regulation of many cellular and biological processes. Abnormal kinase activity has been well reported to be linked with a variety of diseases and human cancers (Cohen 2002; Klaeger et al. 2017)
4	Proteases	Deficient or abnormal protease function is linked with many pathological conditions. An estimated 5–10% of all drugs under progress target the proteases (Docherty et al. 2003)

targets in nature. Various protein drug targets are currently utilized by available drugs, most of them belong to one of four major drug target protein classes (Table 8.1), in some cases, nucleic acids are also utilized by drugs as a target.

8.3 Drug Target Identification

After identifying the biological nature and origin of a disease, identification of potential drug targets is the first step in the discovery of a drug. Drug target identification follows the hypothesis that the most promising targets are tightly linked to the disease of interest, and have an established function in the underlying pathology, which can be observed with high frequency in the disease-associated population. By definition, it is not necessary for potential drug targets to be involved in the disease-causing process, or responsible for a disease, but they must be disease-modifying. Currently, various strategies are in practice for drug target identification, which is either based on experimental approaches or computational approaches.

Experimental approaches are mainly based on comparative genomics (expression profiling) and supplemented with the phenotype and genetic association analysis. Mostly, all experimental approaches provide reliable results, and theoretically, they should be the first choice methods for target identifications. Even though experimental approaches are more precise, they are suffering from some practical limitations, i.e., relatively high costs and intensive scientific labor required for experimental profiling of the full target space (>20,000 proteins, nucleic acid) of chemical compounds and they often end with few drug targets in hand. Due to all these limitations, mostly scientists and pharmaceutical companies utilize the computational methods for first-line research and then use the experimental approaches for further validation and other purposes.

8.4 Computational Approaches for Drug Target Identification

The development of bioinformatics has come up with various bioinformatics resources, including the database, algorithm, and software, which push the CADD in every aspect of the drug designing process (Table 8.2). One of the most important contributions is computational drug target identification, as discussed earlier that identification of the drug target is a very crucial and most decisive step of the drug designing process. In this regard, for the last one and half decades, various scientific studies carried out with the aim of drug target identification with the help of bioinformatics resources and proposed various approaches for drug target identifications. These approaches easily handle and deal with a huge amount of genomics, transcriptomics, and proteomics data, and also process it efficiently, and at the end provide potential drug targets in a short period at a low cost.

Currently, several computational approaches are available which utilized different molecular information, i.e., gene and genome sequence, molecular interaction information and protein 3D structure. Most of these approaches are interlinked. Still, based on their concept, they have broadly classified into two types: (1) homology-based approaches and (2) network-based approaches. The major features which are checked for drug target prediction are listed in Table 8.3 (Kim et al. 2017).

Table 8.2 Bioinformatics resources for drug target identification and CADD

S. No.	Database/ method	Description
1	DbMDR	It provides a collection of multidrug resistance genes and their orthologs, acting as potential drug targets (Gupta et al. 2011)
2	DEG	It contains all known essential genes from a different organism (Zhang et al. 2004)
3	DFVF	Collection of fungal virulence factors which collected>2000 pathogenic genes from a wide range of fungal sp. (Lu et al. 2012)
4	DrugBank	DrugBank is a richly annotated database, which provides detailed information about the drugs along with their target and drug action information (Wishart et al. 2008)
5	GEO	The database provides transcriptomics data (mainly array- and sequence-based) useful for functional genomics (Clough and Barrett 2016)
6	KEGG	KEGG offers information about the pathway, gene, and ligands in three different databases, i.e., Pathway, Gene, and Ligand (Kanehisa and Goto 2000)
7	MvirDB	Microbial protein toxins, virulence factors, and genes related to antibiotic resistance (Zhou et al. 2007)
8	PDTD	Database of potential proteins for in silico *drug* target identification (Gao et al. 2008)
9	TDR targets	Identification and prioritization of molecular targets for drug development (Magariños et al. 2012)
10	TTD	Publicly accessible cross-links database that provides inclusive information about known therapeutic targets with related information, i.e., pathway information and the corresponding drugs/ligands (Chen et al. 2002)
11	VFDB	Database contains virulence factors (VFs) of various medical significant bacterial pathogens (Chen et al. 2005)
12	Daspfind	Interactions between drugs and target proteins based on the similarities among them (Ba-Alawi et al. 2016)
13	iDTI-ESBoost	Evolutionary and structural feature-based model for identification of drug–target interactions (Rayhan et al. 2017)
14	NetCBP	Drug–target interaction prediction with the help of networks. It also predicts some new drugs without any known target interaction information (Chen and Zhang 2013)
15	SELF-BLM	It predicts drug–target interactions using a self-training support vector machine (SVM) based bipartite local model; SELF-BLM (Keum and Nam 2017)

8.5 Homology-Based Approaches

Homology-based approaches utilize sequence similarities among genes and proteins, further based on predicted homology, it takes the decision just like decision tree analysis. Mostly these methods consider the various level of homology test, which follows top-down direction. Each level of homology test scale down the data,

Table 8.3 Important features utilized in drug target identifications

S. No.	Features	Description
1	Essentiality of targets	To find out the indispensable nature of probable target for disease/pathogen
2	Gene ontology, biological process, involvement in pathways	To find out the biological process, pathways, and functional involvement of probable targets
3	Cellular localization	To find out the accessibility of probable target for a drug
4	Structural availability, druggability	To find out the binding pockets along with various physiochemical features involved in binding. It also helps to predict binding affinity and drug–target interaction mode
5	Gene expression patterns	Expression patterns play a significant role to check the availability of targets in given conditions. It also helps to predict the chance of adverse drug reaction, especially in the case of polypharmacological drugs

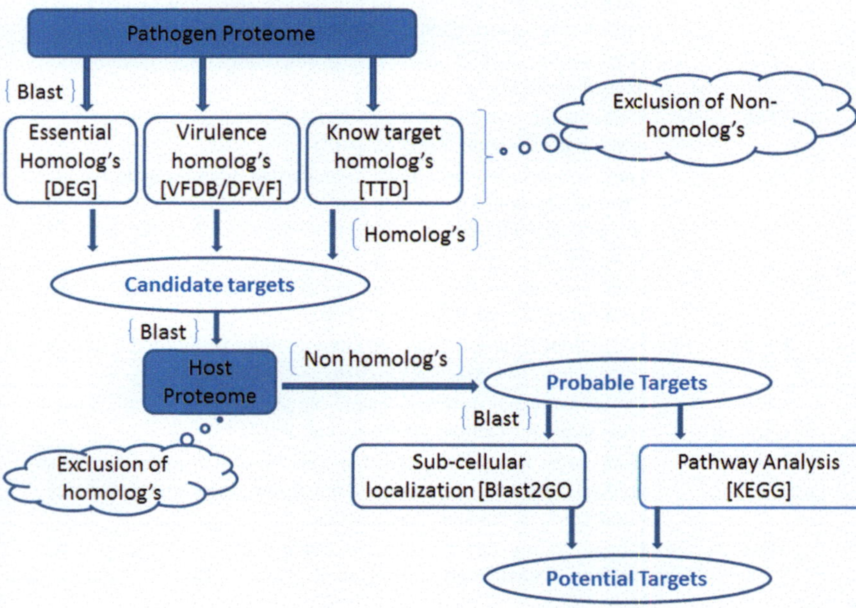

Fig. 8.2 Schematic diagram of the standard flowchart for drug target identification using homology-based approach

starting from complete genes or proteome, and step by step either eliminate those which fitted in "inappropriate" or select only those which fitted in "appropriate." Homology-based approaches always ended with countable potential drug targets (Fig. 8.2), and because of their scale down nature, these approaches are also known as subtractive (genomic or proteomic) approaches.

The term "inappropriate" and "appropriate" are conditional, and they are tested on various biological conditions that play a decisive role in target selection. The following are the major conditional tests that help to decide the further consideration of molecules for drug target identification.

8.5.1 Human Homologs

It is assumed that humans have various genes, and few of them are playing an indispensable biological role, considered as housekeeping genes. The use of human housekeeping genes or homologs of human housekeeping genes as a drug target can create lethal conditions and result in the death of human patients. To avoid such accidental use of the housekeeping gene as well as some important pathway-related gene as a drug target genes of the microbial pathogen are generally compared against the human, and those genes which show significant similarities with human housekeeping or crucial genes will be considered as "inappropriate" and mostly eliminate from rest of the process.

8.5.2 Human-Microbiome Homologs

The human body, especially, the gut has a lot of microbes that are already listed by the human microbiome project. Most of these microbes are involved in the biological process, which is beneficial for humans and thus considered beneficial microbes. Use of homologs from these beneficial microbes as a drug target can harm these bacteria, which can affect the related biological process in the human host, i.e., digestion, respiration process, etc., because of the above said reason, human-microbiome homologs are considered as "inappropriate" and eliminated from the further process.

8.5.3 Essentiality

Identification of drug targets against the microbial pathogen assumes that the essentiality of the target protein for pathogen-microbes is one of the advantageous and "appropriate" features. Without the function of essential proteins, microbial-pathogen will not able to survive. Various essential genes and proteins are identified by experimental approaches and enlisted in various databases. The database of essential genes (DEG) is one of the most active databases providing a collection of essential genes and protein sequences. Based on the above concept, those pathogenic genes/proteins which show homology with essential genes/proteins are considered as "appropriate" and include for the further process.

8.5.4 Virulence Factor Homologs

Those proteins whose role in virulence and pathogenicity is reported through the experiment are considered as virulence factors. Various such proteins are available, especially for microbes, and their molecular information is stored in various databases, i.e., virulence factor database (VFDB) and database of fungal virulence factors (DFVF). Genes/proteins of the pathogens that show homology with these virulence factors can be considered as "appropriate" and utilized as a potential drug target.

8.5.5 Drug Target Homologs

Information about known and explored drug/therapeutic targets is available, i.e., therapeutic target database (TTD). Homology mining with TTD is in practice, and those candidate molecules which show significant homology with these known targets are considered as "appropriate" and included for further exploration.

8.5.6 Cellular Location

The cellular location of the target protein is one of the very important features and plays a crucial role in target selection. In a homology-based approach, sequence-based gene ontology (GO) and annotation are in practice to look at the sub-cellular location along with the cellular component, biological process, and molecular function. Generally, those targets whose access is easy are preferable over others.

8.5.7 Role in the Biological Pathway

Biological pathways are responsible for the synthesis or metabolism of various bio-products. Few of these pathways are very important and unique, and they are solely responsible for their processes and products. The blockage of these pathways creates a scarcity of their products and finally reduces the chance of survival of the pathogen. Various pathway databases are available to conduct such checks. Current literature shows that the KEGG pathway is one of the richest and preferable pathway databases utilized for this purpose. Those pathways which are unique for pathogen are considered as appropriate pathways, and gene/proteins involved in them were considered for the further process. In contrarily those pathways which are also shared by human/host and their gene/proteins are "inappropriate" and excluded from further consideration.

It has been observed that homology-based approaches are very fast and almost cover the entire target space, and it only needs sequence information as input. Available reviews suggest that uses of homology-based approaches are very common for microbial disease and generally restricted with them only. Their use for other types of infection or disease is not in common practice.

8.5.8 Case Study: Subtractive Approach for Drug Target Identification

The subtractive approach is one of the very famous approaches that have been utilized for target identification against various pathogens. In 2011 Katara et al. presented a subtractive approach exploiting the knowledge of global gene expression along with sequence comparisons to predict the potential drug targets in *Vibrio cholerae*, cholera causing bacterial pathogen, efficiently. Their analysis was based on the available knowledge of 155 experimentally proved virulence genes (seed information) (Fig. 8.3). For target identification, they utilized co-expression based gene mining and multilevel subtractive approach. At the end, they reported 36 gene products as a drug target, to check the reliability of the predicted targets they also performed gene ontology through Blast2GO. They observed these targets for their involvement in a crucial biological process and their cellular location. They found all these 36 gene products as reliable targets and conclude them as potential drug targets.

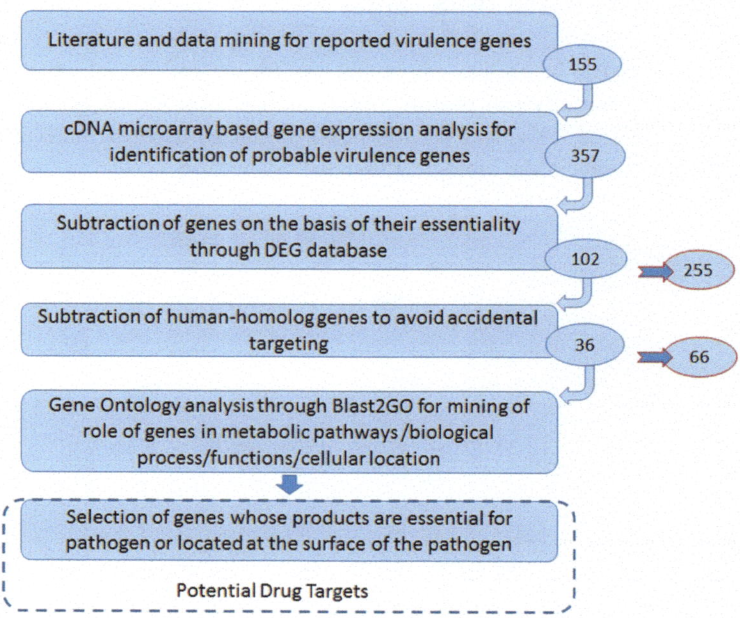

Fig. 8.3 Subtractive approach for drug target identification

8.6 Network-Based Approaches

It examines the effects of drugs in the context of molecular networks (i.e., protein–protein interactions, gene networks, transcriptional regulatory networks, metabolic networks, and biochemical reaction networks). In molecular network models, molecules refer as nodes, and each edge corresponds to an interaction between two molecules, based on the direction and importance of interaction between nodes, sometimes edges also mention the direction and weight (Fig. 8.4). Drug target identification through the network is based on the fact that networks have many important nodes that are vulnerable and can be targeted in many ways. Most of the time, these nodes are very crucial, and sometimes essential for the whole network structure, inhibition of such nodes can reduce their efficiency and damage of these nodes can shut down the complete network. Network inhibition process follows one of the following two models: (1) partial inhibitions: Partial knockout of the interactions of the target nodes, and (2) complete inhibition: all interactions around a given target node are eliminated.

In the drug designing process, these target nodes can be considered as potential drug targets. Various molecular networks (Table 8.4), including protein-interaction networks, regulatory, metabolic, and signaling networks individually or in integrated form can be subjected to a similar analysis (Imoto et al. 2007; Sridhar et al. 2008; Kotlyar et al. 2012; Shin et al. 2017).

8.6.1 Centrality Based Drug Target

Network centrality can be used as a potential tool for network-based target identification. Network centrality can prioritize proteins based on the network centrality measures (i.e., degree, closeness betweenness). It can be used to characterize the importance of proteins in the biological system.

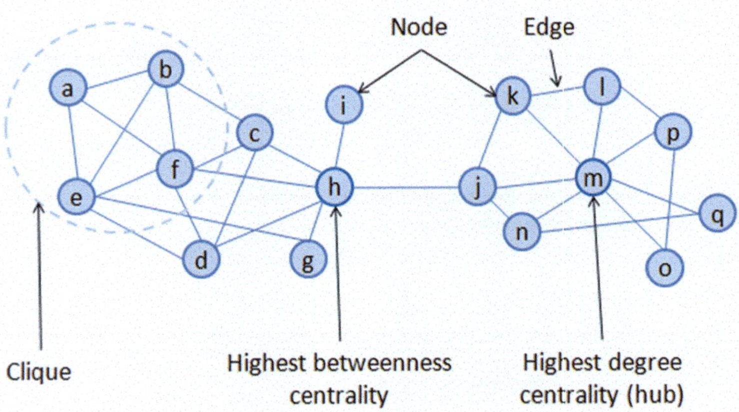

Fig. 8.4 Various components of a standard network

Table 8.4 Types of the biological network for drug target identification

S. No.	Network	Description
1	Protein–protein interactions (PPIs)	Here, proteins are nodes, and their interactions are edges. Proteins with high degrees of connectedness are likely to be more crucial than proteins with lesser degrees (Zheng et al. 2013; Shin et al. 2017; Verma et al. 2020)
2	Gene regulatory networks (GRN)	Transcription factors bind to multiple binding sites in a genome. As a result, all cells have complex networks between transcription factors (with respect to their target gene) that form a GRN (Imoto et al. 2007)
3	Gene co-expression network (GCN)	GCN is an undirected graph network that shows connectivity between co-expressed genes that supposed to be regulated by the same transcriptional regulatory system (Cheng et al. 2012; Yang et al. 2014)
4	Metabolic networks	The network of biochemical reactions is called metabolic network. Flux-balance analysis of these networks provides information about potential targets (Sridhar et al. 2008)
5	Cell signaling networks	Signaling networks represent connectivity between cellular signals typically, by combining PPIN, GRN, and metabolic networks (Behar et al. 2013)
6	Composite network	Composite cellular (transcriptional, signaling, PPI) networks identify the susceptible nodes which can act as a potential target (Pinto et al. 2014)

8.6.1.1 Hubs as Target

Real-world networks almost show a scale-free degree distribution, which means that in these networks, some nodes have a tremendous number of connections to other nodes (high degree), whereas most nodes have just a few. Here, nodes with a great number of connections than average called hubs. It assumes that the functionality of such scale-free networks heavily depends on these hubs, and if these hubs are selectively targeted, the information transfer through networks gets hindered and results in the collapse of the network (Pinto et al. 2014).

8.6.1.2 Betweenness Centrality Based Target

Hubs are the centers of local network topology, thus only provide the local picture of the network. Betweenness centrality is another approach that can be used to explain network centre, unlike, hub it provides central elements of the network in the global topology, thus, provide a global picture of network connections. Conceptually, betweenness is the number of times a node is in the shortest paths between two other nodes (Fig. 8.4), thus higher the betweenness means more importance of the node in quick network communication. Such higher betweenness centrality nodes can be utilized as a potential target against drugs (Melak and Gakkhar 2015).

8.6.1.3 Mesoscopic Centrality Based Target

Considering the advantage of both local and global centers of network topology for drug target identifications, the third class of centrality called mesoscopic centrality has also been reported. Mesoscopic centrality is neither fully based on local information (such as hubs) nor global information (such as betweenness centrality) on network structure. It mainly considers long-range connections between high degree nodes, which make a profound effect on small-world networks.

8.6.1.4 Weight-Based Drug Target

Recently, the weighted-directed network is also reported for drug target identification studies (Wang et al. 2013). The weighted-directed network is closer to the real, cellular scenario, where PPIs are characterized by their affinity and dominance (link weight) as well as direction (e.g., in form of signaling), as mentioned in Fig. 8.5. It has been assumed that the deletion of the links with the highest weighted centralities is often more disturbing to network behavior than the removal of the most central links in the similar un-weighted network topology.

Utilization of the complex structural information of real-world networks to measure the centrality is not an easy task, and it requires more sophisticated methods to overcome these challenges. Bioinformatics provides various tools to support network construction, visualization, and network-based analysis, i.e., weight, centrality, interaction directions (Table 8.5).

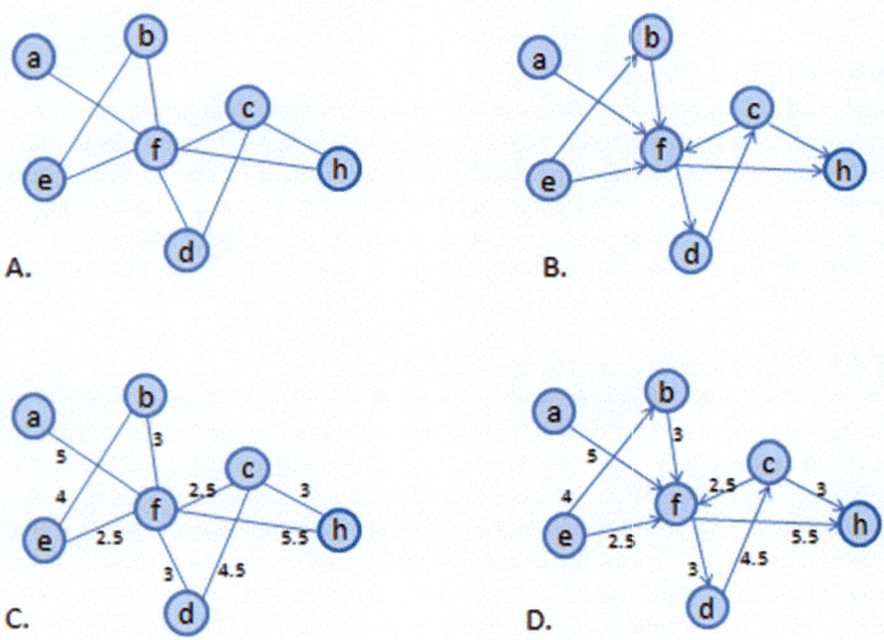

Fig. 8.5 Molecular network with a different type of connectivity between nodes (**a**) undirected (**b**) directed (**c**) weighted, and (**d**) weighted directed

Table 8.5 Tools supporting molecular network analysis for drug target identification

S. No.	Resource	Description
1	BioGRID	It is a repository of biological network information that can be visualized by Cytoscape (Oughtred et al. 2019)
2	BioMart	It contains data, software, and provides data services to facilitate scientific interactions and drug target discovery (Haider et al. 2009)
3	Connectivity map	It is a collection of genome-wide expression data from bioactive treated cultured human cells. It provides transcriptome based functional connections between drugs, genes, and diseases (Lamb et al. 2006)
4	MetaboAnalyst	It is an analysis tool for high-throughput metabolomics data, including data processing, biomarker discovery, and pathway analysis (Xia et al. 2015)
5	Netpredictor	Netpredictor is an R package that dealing with a unipartite or bipartite network. It can utilize to explore interactome and enrichment analysis for disease pathway and ontology (Seal and Wild 2018)

8.6.2 Limitations

Drug target identification through the biological network is an empirical approach, which relies on available information on molecular networks. However, numbers of molecular interaction databases are available, and most of them suffer from uncertainties, false-positive entries, and the average probability of particular interaction along with nomenclature as well as interpretation problems. However, to overcome these issues, recently, PPI databases are linked with protein structure data, which provides more reliable and validated interactions. At the same time, scientists also propose some alternative, i.e., use of the curated database and low-resolution network to surmount the above-mentioned problems (De-Alarcón et al. 2002).

8.7 Properties of an Ideal Drug Target

Identification of potential drug targets is not the last step. Nowadays, through various computational approaches, a huge number of probable targets are reported against different diseases and are available in databases and literature (Katara et al. 2011). It is not a good idea to recommend them directly for testing, its recommendation that first, we check them for an ideal property (Table 8.6), and then for druggability. Only those targets which fulfill most of them are considered as an ideal drug target and recommended them for further validation and testing (Gashaw et al. 2011).

Table 8.6 Important properties to assess the ideal drug targets

S. No.	Property	Detail
1	Disease-modifying	Target should be disease-modifying with proven function in disease pathophysiology
2	Disease specific modulation	Modulation of the target must be explicit to the targeted disease, should not affect standard physiology in normal or other disease conditions
3	Druggability assessment	Target druggability should be observable
4	Assay ability	Target should have favorable assay ability, specifically through high-throughput screening
5	Tissue-specific expression	Target expression should be tissue-specific, it should not affect unrelated tissue or organs

8.8 Druggability of Drug Target

In drug designing process, the potential of any target is defined by its druggability (affinity of the target to bind with drug-like molecules), thus the target must be druggable (Fauman et al. 2011). Biomolecules (i.e., protein, nucleic acid) with an activity that can be modulated by a drug are considered as a druggable target. These targets must have binding sites with typical structural and physicochemical properties that favor binding interaction with high affinity and specificity.

8.8.1 Importance of Druggability

Despite technological advancement in the drug designing process, most drug discovery projects fail because of the druggability problem. To avoid the failure of a drug discovery project, which is mostly very expensive, it is very important to understand the difficulties associated with a potential target. Druggability has become part of the target identification and validation process, more significantly in the case where targets do not belong to traditional classes (Finan et al. 2017).

8.9 Computational Methods for Druggability Assessment

To date, various targets are reported and documented through various methods, and few of them are already in practice (drugs are available against them), such targets are druggable. If no drug available for a target, then predict druggability is required. Various computational methods are available to evaluate the druggability of target protein, mainly rely on either sequence-based or 3D-structure based properties of proteins (Fauman et al. 2011).

8.9.1 Sequence-Based Methods

A protein is druggable if its other family members are known to be targeted by drugs. For such analysis, sequence alignment can be used to predict sequence similarity (homology) between probable target (query) proteins and database of known druggable targets (Finan et al. 2017). The sequence-based concept provides a significant approximation of druggability, but it suffers from the following limitations: (1) its predictions are limited to known drug target families, it does not attempt for those potential targets, which belong to the novel "un-drugged" protein family; and (2). It assumes that all members of the protein family are equally druggable, which is not true.

8.9.2 Structure-Based Methods

Structure-based methods rely on the availability of 3D structure information, thus only can apply to those proteins whose structures are available. Along with experimentally determined 3D structures, it also considers high-quality structure models through homology modeling. Several structure-based methods are available for the assessment of target druggability, irrespective of their different algorithms; all of them consist of the following three common components.

8.9.2.1 Identifying Cavities and Binding Pockets

Many computational methods and tools have been developed for binding pocket identification, which scans 3D surface and interior of the target protein for potential cavities (possess suitable properties for binding a ligand) that can act as binding pockets. These tools mainly tend to look for cavities with suitable size, shape, and composition to accommodate drug-like molecules.

Working of binding pockets detection methods depends on either energy-based or geometry-based detection algorithms (Nisius et al. 2012; Zheng et al. 2013). Energy-based detection predicts pockets by computing the interaction energy between atoms of protein and a probe molecule (Ghersi and Sanchez 2011). Geometry-based detection predicts the solvent accessible area that is embedded in the protein surface. Comparative studies suggest that both types of detection algorithms have good performance and advantages (Schmidtke et al. 2010). It has been observed that geometry-based detections are more suitable for large-scale pocket detection. Their inherent advantages, i.e., high speed and robustness against structural variations or missing atoms and residues in the input structures, provide the edge over an energy-based detection algorithm (Schmidtke et al. 2010). With the increasing availability of binding cavity information, recently, one new class of methods called information-based detection methods are developed. These methods utilize available cavity information from its neighbor and similar proteins whose binding cavities are known.

8.9.2.2 Druggability of Binding Pocket

This second step aims to calculate the physicochemical and geometric properties of the pocket to check whether these properties are complementary with the properties of drug-like molecules. Lipinski's rule of five (RO5) connects the physicochemical properties of a drug with its pharmacokinetic properties (Lipinski 2000). It is a well-known fact that the physicochemical properties of the druggable pocket should be the mirror image of the physicochemical properties of the drug-like molecule itself. This analogy gave the concept of a druggable pocket. Therefore, the complementary properties of the pockets reflect the Lipinski's rule of five of "drug-likeness" (H-bond donors >5, H-bond acceptors = 10, molecular weight > 500, and the Log P (CLog P) is >5).

The major features which define and affect the druggability of pockets are pocket descriptors. Characteristic features of a binding site play a very crucial role in druggability calculation, and the selection of those descriptors, which are crucial for binding drug-like molecules, needs to be described as accurate as possible. Observations suggest that none of the individual pocket descriptors is sufficient for druggability explanation, and a group of descriptors is required to describe and calculate pocket druggability. Both physiochemical and geometrical features play a crucial role as descriptors. Physiochemical descriptors and frequently used physiochemical pocket descriptors include size, shape, electrostatics, hydrogen bonding, hydrophobicity, polarity, amino acid composition, rigidity, and secondary structure (Halgren 2009; Krasowski et al. 2011). Geometrical descriptors: Along with physicochemical properties, geometrical properties, i.e., the shape and size of the binding pocket, play a crucial role in suitable interactions with a small molecule (Zheng et al. 2013). The following are the major geometrical features involved in pocket druggability measurement.

Position of the Atoms

It has been observed that the position of the atoms in pockets affects the contribution of an atom in interaction. Atoms located at the contact surface considerably give a major contribution in contact energy (hydrophobic interaction) than those who lie outside of the surface, i.e., within the bulk of the protein cavity.

Cavity Size

Large spherical cavities are more exposed to the solvent, thus not suitable for binding, especially with small drug molecules. Narrow (micro) cavity pockets are less exposed to the solvent and offer more van der Waals contact, thus they are more druggable. These micro-cavities are also defined as hot spots, which are characteristic of highly druggable targets.

8.9.2.3 Target Specificity Assessment

Drug target must be specific; structure similarity of drug target molecules with other unwanted molecules will create problems in the drug development process. Structural similarity of the binding sites could make the design of selective inhibitors difficult. During target selection, it is very important to assess the structural landscape of the primary binding sites of the target to confirm the druggability. Sequence and structural alignment based computational methods are available to perform specificity assessment.

Sequence Alignment Based Assessment
It is based on the sequence information of binding sites of the target protein. It assumes that when the degree of conversation between the two sequences is sufficiently high, then identical amino acids in the sequence will likely correspond to identical binding site structure.

Structure Alignment Based Assessment
These methods are based on either structural superposition or pharmacophore features. Structural superposition generally utilizes a 3D grid force field around the binding sites, which can be calculated using various types of energy terms, i.e., electrostatic, hydrophobic, and hydrogen bonding. In the grid approach, the field potentials can be calculated for each suspicious protein and are used for comparing their binding sites. The structural similarity between a pair of proteins can be studied by correlation functions of the various molecular interaction fields (MIFs) of the two grids or by utilizing the Fourier transformation of correlation functions or related approaches. Another approach consists of identifying pharmacophore features that generally summarized with the help of surface chemical features (SCF), including hydrophobic centers, H-bond donors and acceptors, positive and negative charges, and aromatic centers, etc. This SCF based on the consideration can be determined on the whole protein surface or a chosen cavity. Binding sites with the highest SCF matches show the highest similarity with the query binding site. Various computational tools are already available, which provide the facilities to evaluate binding site similarities and assess the specificity (Table 8.7). Almost all tools rely on the available entries at the protein structural database.

8.9.3 Quantification of Druggability

Quantification of druggability could provide the best criteria for target selection, but till now, none of the standard explanation is available for this purpose. Each method has its measures for druggability, thus a druggability score of a specific target might vary. However, irrespective of an individual's weaknesses and strengths, all major druggability measures can classify targets into druggable, non-druggable, medium druggable, and difficult-druggable.

8.9.4 Major Concern

8.9.4.1 Size of Training Sets

Most of the druggability assessment methods are based on the machine learning algorithm, thus highly dependent on available training sets (ChEMBL, BindingDB,

Table 8.7 Bioinformatics resources for druggability detection and evaluation

S. No.	Tool/ algorithm	Description
1	CavityPlus	Protein cavity detection and functional analyses (Xu et al. 2018)
2	Dr. PIAS	A druggability assessment system. Along with druggability, it also provides functional annotation of interacting proteins (Sugaya and Furuya 2011)
3	DrugEBIlity	It evaluates the druggability of targets. The server can search with a sequence, PDB id, or uploaded structure (https://www.ebi.ac.uk/chembl/drugebility)
4	DrugPred	Structure-based druggability predictor that relies on the affinity between known drugs and their target proteins (Krasowski et al. 2011)
5	IsoCleft	Detection of local geometric and chemical similarities between potential binding cavities for small molecules (Kurbatova et al. 2013)
6	IsoMIF finder	Detection and comparison of binding site molecular interaction field (MIF) (Chartier et al. 2016)
7	MultiBind	Recognize the common spatial chemical binding patterns along with shared physicochemical binding site properties (Shulman-Peleg et al. 2008)
8	PockDrug-server	Pocket druggability with and without ligand proximity information. In both cases, it provides consistent druggability results using different pocket estimation methods (Hussein et al. 2015)
9	SiteAlign	Align, compare druggable ligand-binding sites, and to measure distances between druggable protein cavities (Schalon et al. 2008)
10	SiteMap's	Provide prediction of the target's binding sites with druggability. It also provides quantitative and graphical information about the target (Halgren 2009)

PubChem, etc.) used to train them. The size and quality of the available datasets in databases directly affect the reliability and scope of the assessment methods.

8.9.4.2 Binding Site Flexibility

The identification of the binding cavity in a rigid target is based on the assumption that the cavity already exists. There are some proteins whose binding pockets do not exist in their native structure, and their active pockets behave like inducible allosteric sites, which only revealed after protein conformational changes. In such a case, it is very difficult to assess the binding pockets, and this situation is considered as a binding site "flexibility problem." The presence of multiple X-ray conformers for a specific target can help us to handle binding site flexibility. Multiple conformers allow us to assess the relative variability of certain residues within the binding site pockets. Based on such relative variability information, it is possible to assess the plasticity of the binding site.

8.10 Target-Based Drug Discovery

As discussed, drug targets are the most crucial element of the drug designing process, and selection of the targets decides the fate of the drug designing process that it will succeed or get fail at a later stage. For several decades, pharmaceutical companies are successfully using well established one drug-one target approach for drug designing purposes. By realizing the scenario, the central dogma of the drug designing process has now shifted from one drug-one target to one drug-multi-target concept and considers multiple targets for a single drug.

8.10.1 Multi-Target Drug Designing

Computational approaches specifically those, which are based on system biology concepts are very crucial in the identification of multi-targets, thus play a major role in the success of the multi-target-based drug designing (Vasaikar et al. 2016). Multi-target-based drug designing approach is, to some extent, similar to single target-based drug designing, but it initiated with the set of targets multi-targets (Fig. 8.6). The following are the main steps of multi-target drug designing.

8.10.1.1 Identification of a Set of Targets "Multi-Targets"

This is the most crucial step which decides the fate of the whole following process. System biology-based molecular networks are in practice to identify multi-targets.

8.10.1.2 Generation of Multi-Target Pharmacophore

Computational methods are available to design multi-target (structure) based pharmacophore, which utilizes combinatorial algorithms (Kumar et al. 2018; Ramsay et al. 2018). The most common steps in multi-target pharmacophore generation include (1) interaction profiling (MIFs) of all targets, (2) identification of common MIFs/features, and (3) multi-target specific and selective ensembles development.

8.10.1.3 Virtual Screening

Pharmacophore generation is followed by virtual screening of chemical libraries to find suitable compounds against multi-target pharmacophore.

8.10.1.4 Generation or Selection of Multi-Target Compound

Multi-target compounds are generated through the integration of pharmacophore of above-selected molecules (already known drugs or drug candidates).

8.10.1.5 Evaluation and Optimization of Multi-Target Specific Compound

Evaluation and optimization process mainly includes multi-target specific interaction assay (to avoid off-targeting), QSAR, and degree of modulation. Though multi-target drugs seem promising and designing of these compounds is not a

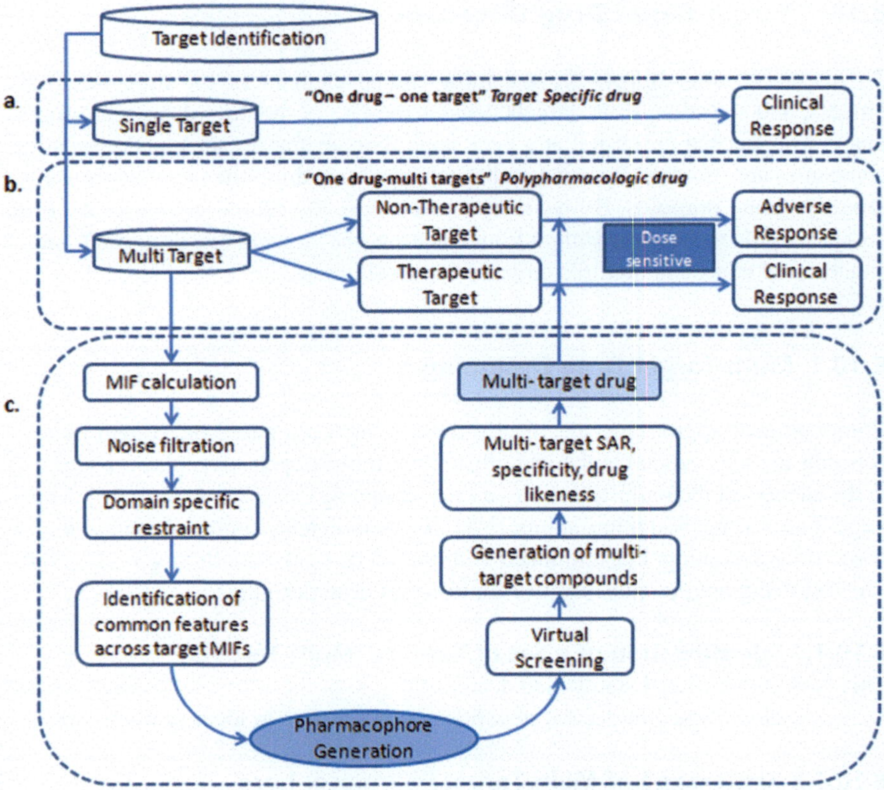

Fig. 8.6 Target-based drug designing (**a**) single target-based drug designing, (**b**) multi-target-based drug designing and (**c**) major steps involve in multi-target-based drug designing

straightforward task. It needs to deal with various crucial issues, i.e., right target-sets selection, balanced activity towards them, and excluding activity at off-target(s), while at the same time retaining drug-like properties (Hopkins 2008; Bottegoni et al. 2012). Available experimental methods are not enough to handle these issues, thus the feasibility of multi-target drugs profoundly depends on computational approaches and resources. Various databases are also there, i.e., DrugBank, STITC H, BindingDB ZINC, PubChem, KEGG DRUG, which provide required information about molecular pathways, 3D structure, chemical reactions, side effects, and known drug targets, thus help in the success of poly-pharmacologic drugs.

8.11 Summary

Now day's computational biology becomes an indispensable tool for almost every aspect of biology and related fields, and drug designing is not an exception. CADD is now a mature field, and its success influenced by its first and pivotal step that is the identification of drug targets. This chapter provides an overview of various computational approaches available for drug target identification. It also discusses various bioinformatics resources, i.e., database, methods, and software, which can be handy for drug target identification purposes.

Competing Interest The author declares that there are no competing interests.

References

Ba-Alawi W, Soufan O, Essack M, Kalnis P, Bajic VB (2016) DASPfind: new efficient method to predict drug-target interactions. J Cheminf 8:15

Behar M, Barken D, Werner SL, Hoffmann A (2013) The dynamics of signaling as a pharmacological target. Cell 155(2):448–461

Bottegoni G, Favia AD, Recanatini M, Cavalli A (2012) The role of fragment-based and computational methods in polypharmacology. Drug Discov Today 17(1–2):23–34

Chartier M, Adriansen E, Najmanovich R (2016) IsoMIF Finder: online detection of binding site molecular interaction field similarities. Bioinformatics 32(4):621–623

Chen H, Zhang Z (2013) A semi-supervised method for drug-target interaction prediction with consistency in networks. PLoS One 8(5):e62975

Chen X, Ji ZL, Chen YZ (2002) TTD: therapeutic target database. Nucleic Acids Res 30 (1):412–415

Chen L, Yang J, Yu J, Yao Z, Sun L, Shen Y, Jin Q (2005) VFDB: a reference database for bacterial virulence factors. Nucleic Acids Res 33:D325–D328

Cheng F, Liu C, Jiang J et al (2012) Prediction of drug-target interactions and drug repositioning via network-based inference. PLoS Comput Biol 8(5):e1002503

Clough E, Barrett T (2016) The gene expression omnibus database. Methods Mol Biol 1418:93–110

Cohen P (2002) Protein kinases—the major drug targets of the twenty-first century? Nat Rev Drug Discov 1(4):309–315

De-Alarcón PA, Pascual-Montano A, Gupta A, Carazo JM (2002) Modeling shape and topology of low-resolution density maps of biological macromolecules. Biophys J 83(2):619–632

Dersch P, Khan MA, Mühlen S, Görke B (2017) Roles of regulatory RNAs for antibiotic resistance in bacteria and their potential value as novel drug targets. Front Microbiol 8:803

Docherty AJ, Crabbe T, O'Connell JP, Groom CR (2003) Proteases as drug targets. Biochem Soc Symp 70:147–161

Fauman EB, Rai BK, Huang ES (2011) Structure-based druggability assessment--identifying suitable targets for small molecule therapeutics. Curr Opin Chem Biol 15(4):463–468

Finan C, Gaulton A, Kruger FA, Lumbers RT, Shah T, Engmann J, Galver L, Kelley R, Karlsson A, Santos R, Overington JP, Hingorani AD, Casas JP (2017) The druggable genome and support for target identification and validation in drug development. Sci Transl Med 9(383):eaag1166

Gao Z, Li H, Zhang H, Liu X, Kang L, Luo X, Zhu W, Chen K, Wang X, Jiang H (2008) PDTD: a web-accessible protein database for drug target identification. BMC Bioinf 9:104

Gashaw I, Ellinghaus P, Sommer A, Asadullah K (2011) What makes a good drug target? Drug Discov Today 16(23–24):1037–1043

Ghersi D, Sanchez R (2011) Beyond structural genomics: computational approaches for the identification of ligand binding sites in protein structures. J Struct Funct Genom 12(2):109–117

Gupta S, Mishra M, Sen N, Parihar R, Dwivedi GR, Khan F, Sharma A (2011) DbMDR: a relational database for multidrug resistance genes as potential drug targets. Chem Biol Drug Des 78 (4):734–738

Haider S, Ballester B, Smedley D, Zhang J, Rice P, Kasprzyk A (2009) BioMart Central Portal—unified access to biological data. Nucleic Acids Res 37:W23–W27

Halgren TA (2009) Identifying and characterizing binding sites and assessing druggability. J Chem Inf Model 49(2):377–389

Hopkins AL (2008) Network pharmacology: the next paradigm in drug discovery. Nat Chem Biol 4 (11):682–690

Hughes JP, Rees S, Kalindjian SB, Philpott KL (2011) Principles of early drug discovery. Br J Pharmacol 162(6):1239–1249

Hussein HA, Borrel A, Geneix C, Petitjean M, Regad L, Camproux AC (2015) PockDrug-Server: a new web server for predicting pocket druggability on holo and apo proteins. Nucleic Acids Res 3(W1):W436–W442

Imoto S, Tamada Y, Savoie CJ, Miyano S (2007) Analysis of gene networks for drug target discovery and validation. Methods Mol Biol 360:33–56

Kaczorowski GJ, McManus OB, Priest BT, Garcia ML (2008) Ion channels as drug targets: the next GPCRs. J Gen Physiol 131(5):399–405

Kanehisa M, Goto S (2000) KEGG: Kyoto Encyclopedia of genes and genomes. Nucleic Acids Res 28(1):27–30

Katara P (2013) Role of bioinformatics and pharmacogenomics in drug discovery and development process. Netw Model Anal Health Inform Bioinf 2(4):225–230

Katara P (2017) Stem cell: a key to solving the drug screening enigma. In: Verma V, Singh MP, Kumar M (eds) Stem cells from culture dish to clinic. Nova Science, New York, pp 257–268

Katara P, Grover A, Kuntal H, Sharma V (2011) In silico prediction of drug targets in Vibrio cholerae. Protoplasma 248(4):799–804

Katsila T, Spyroulias GA, Patrinos GP, Matsoukas MT (2016) Computational approaches in target identification and drug discovery. Comput Struct Biotechnol J 14:177–184

Keum J, Nam H (2017) SELF-BLM: prediction of drug-target interactions via self-training SVM. PLoS One 12(2):e0171839

Kim B, Jo J, Han J, Park C, Lee H (2017) In silico re-identification of properties of drug target proteins. BMC Bioinf 18(Suppl 7):248

Klaeger S, Heinzlmeir S, Wilhelm M et al (2017) The target landscape of clinical kinase drugs. Science 358(6367):eaan4368

Kotlyar M, Fortney K, Jurisica I (2012) Network-based characterization of drug-regulated genes, drug targets, and toxicity. Methods 57(4):499–507

Krasowski A, Muthas D, Sarkar A, Schmitt S, Brenk R (2011) DrugPred: a structure-based approach to predict protein druggability developed using an extensive nonredundant data set. J Chem Inf Model 51(11):2829–2842

Kumar P, Kaalia R, Srinivasan A, Ghosh I (2018) Multiple target-based pharmacophore design from active site structures. SAR QSAR Environ Res 29(1):1–19

Kurbatova N, Chartier M, Zylber MI, Najmanovich R (2013) IsoCleft Finder—a web-based tool for the detection and analysis of protein binding-site geometric and chemical similarities. F1000Res 2:117

Lamb J, Crawford ED, Peck D et al (2006) The Connectivity Map: using gene-expression signatures to connect small molecules, genes, and disease. Science 313(5795):1929–1935

Lipinski CA (2000) Drug-like properties and the causes of poor solubility and poor permeability. J Pharmacol Toxicol Methods 44:235–249

Lu T, Yao B, Zhang C (2012) DFVF: database of fungal virulence factors. Database 2012:bas032

Magariños MP, Carmona SJ, Crowther GJ et al (2012) TDR targets: a chemogenomics resource for neglected diseases. Nucleic Acids Res 40:D1118–D1127

McInnes C (2007) Virtual screening strategies in drug discovery. Curr Opin Chem Biol 11 (5):494–502

Melak T, Gakkhar S (2015) Comparative genome and network centrality analysis to identify drug targets of Mycobacterium tuberculosis H37Rv. Biomed Res Int 2015:1. https://doi.org/10.1155/2015/212061

Nisius B, Sha F, Gohlke H (2012) Structure-based computational analysis of protein binding sites for function and druggability prediction. J Biotechnol 159(3):123–134

Oughtred R, Stark C, Breitkreutz BJ et al (2019) The BioGRID interaction database: 2019 update. Nucleic Acids Res 47(D1):D529–D541
Pinto JP, Machado RS, Xavier JM, Futschik ME (2014) Targeting molecular networks for drug research. Front Genet 5:160
Ramsay RR, Popovic-Nikolic MR, Nikolic K, Uliassi E, Bolognesi ML (2018) A perspective on multi-target drug discovery and design for complex diseases. Clin Transl Med 7(1):3
Rayhan F, Ahmed S, Shatabda S et al (2017) iDTI-ESBoost: identification of drug target interaction using evolutionary and structural features with boosting. Sci Rep 7(1):17731
Schalon C, Surgand JS, Kellenberger E, Rognan D (2008) A simple and fuzzy method to align and compare druggable ligand-binding sites. Proteins 71(4):1755–1778
Schmidtke P, Le Guilloux V, Maupetit J, Tufféry P (2010) Fpocket: online tools for protein ensemble pocket detection and tracking. Nucleic Acids Res 38:W582–W589
Seal A, Wild DJ (2018) Netpredictor: R and shiny package to perform drug-target network analysis and prediction of missing links. BMC Bioinf 19(1):265
Shekhar C (2008) In silico pharmacology: computer-aided methods could transform drug development. Chem Biol 15(5):413–414
Shin WH, Christoffer CW, Kihara D (2017) In silico structure-based approaches to discover protein-protein interaction-targeting drugs. Methods 131:22–32
Shulman-Peleg A, Shatsky M, Nussinov R, Wolfson HJ (2008) MultiBind and MAPPIS: webservers for multiple alignment of protein 3D-binding sites and their interactions. Nucleic Acids Res 36:W260–W264
Sridhar P, Song B, Kahveci T, Ranka S (2008) Mining metabolic networks for optimal drug targets. Pac Symp Biocomput 13:291–302
Sriram K, Insel PA (2018) G protein-coupled receptors as targets for approved drugs: how many targets and how many drugs? Mol Pharmacol 93(4):251–258
Sugaya N, Furuya T (2011) Dr. PIAS: an integrative system for assessing the druggability of protein-protein interactions. BMC Bioinf 12:50
Vasaikar S, Bhatia P, Bhatia PG, Chu Yaiw K (2016) Complementary approaches to existing target based drug discovery for identifying novel drug targets. Biomedicines 4(4):E27
Verma Y, Yadav A, Katara P (2020) Mining of cancer core-genes and their protein interactome using expression profiling based PPI network approach. Gene Rep 18:100583
Wang W, Yang S, Li J (2013) Drug target predictions based on heterogeneous graph inference. Biocomputing 2013:53–64
Wishart DS, Knox C, Guo AC et al (2008) DrugBank: a knowledgebase for drugs, drug actions and drug targets. Nucleic Acids Res 36:D901–D906
Xia J, Sinelnikov IV, Han B, Wishart DS (2015) MetaboAnalyst 3.0—making metabolomics more meaningful. Nucleic Acids Res 43(W1):W251–W257
Xu Y, Wang S, Hu Q et al (2018) CavityPlus: a web server for protein cavity detection with pharmacophore modelling, allosteric site identification and covalent ligand binding ability prediction. Nucleic Acids Res 46(W1):W374–W379
Yang Y, Han L, Yuan Y, Li J, Hei N, Liang H (2014) Gene co-expression network analysis reveals common system-level properties of prognostic genes across cancer types. Nat Commun 5:3231
Zauhar RJ, Moyna G, Tian L, Li Z, Welsh WJ (2003) Shape signatures: a new approach to computer-aided ligand- and receptor-based drug design. J Med Chem 46(26):5674–5690
Zhang R, Ou HY, Zhang CT (2004) DEG: a database of essential genes. Nucleic Acids Res 32: D271–D272
Zheng X, Gan L, Wang E, Wang J (2013) Pocket-based drug design: exploring pocket space. AAPS J 15(1):228–241
Zhou CE, Smith J, Lam M, Zemla A, Dyer MD, Slezak T (2007) MvirDB—a microbial database of protein toxins, virulence factors and antibiotic resistance genes for bio-defence applications. Nucleic Acids Res 35:D391–D394

Computational Screening Techniques for Lead Design and Development

9

Pramodkumar P. Gupta, Virupaksha A. Bastikar, Alpana Bastikar, Santosh S. Chhajed, and Parag A. Pathade

Abstract

Virtual screening is a computational screening technique used to screen drug-like compounds from vast libraries of chemicals based on the binding energy of compounds with a target. To discover new plausible drug candidates, computational chemistry tools are used for studying the absorption, distribution, metabolism, excretion, and toxicity (ADMET) of potential drugs, as well as also decipher the mechanisms of drug action and its interaction mode with the target. Many drug designing tools are available, which can assist in the design and discovery of new drugs for the treatment of diseases with fewer or no side effects. The objective is to evaluate several millions of compounds saving time and cost of discovery. The quantitative structure-activity relationship (QSAR) analysis has made it possible to theoretically correlate the biological activity of a compound with its physicochemical properties, and the predictive equation has been derived for the assessment of the biological response of a compound using molecular

P. P. Gupta (✉)
School of Biotechnology and Bioinformatics, D Y Patil Deemed to be University, Navi Mumbai, Maharashtra, India
e-mail: pramod.gupta@dypatil.edu

V. A. Bastikar
Amity Institute of Biotechnology, Amity University, Mumbai, Maharashtra, India

A. Bastikar
Department of Computer Aided Drug Design, Navin Saxena Research and Technology Pvt. Ltd, Gandhidham, Gujarat, India

S. S. Chhajed
Department of Pharmaceutical Chemistry, MET Institute of Pharmacy, Nashik, Maharashtra, India

P. A. Pathade
Department of Pharmaceutical Chemistry, KBHSS Trust's Institute of Pharmacy, Malegaon, Maharashtra, India

D. B. Singh (ed.), *Computer-Aided Drug Design*,
https://doi.org/10.1007/978-981-15-6815-2_9

descriptors. Bioinformatics and Cheminformatics database in houses several million of compounds with similar architecture and biological properties. Screening and identification of potential candidates for the appropriate term and target is a difficult mining task in terms of cost and time. Virtual screening tools developed by various bioinformatics and cheminformatics groups are contributing to the field of open source drug discovery project. Online computational resources in drug discovery research are a helping hand to the community, with remote and free access to the resources.

Keywords

ADME · Virtual screening · QSAR · Molecular modeling · Toxicity

9.1 Introduction

Biological "trial and error" experiments trying huge gatherings of small molecules for a precise pharmacological consequence is the conventional way to discover new lead complexes, which consequently assist as models for additional optimization in medicinal chemistry platforms. In the course of last few decades, the introduction of recombinant DNA technologies joint along through developed assay methods and high-performance laboratory mechanization intensely altered the pharmacological screening progression. Nowadays, high-throughput screening (HTS), QSPR, QSAR, ADME, and toxicity profiling have been increased by many considerations. The drug design process makes use of computational methods to develop drugs at a faster rate with low cost. The most fundamental goal in drug design is to predict whether a given molecule will bind to a target and if so how strongly. Molecular mechanics or molecular dynamics is most often used to estimate the strength of the intermolecular interaction between the small molecule and its biological target. These methods are also used to predict the conformation of the small molecule and to model conformational changes in the target that may occur when the small molecule binds to it. Semi-empirical, ab initio quantum chemistry methods, or density functional theory are often used to provide optimized parameters for the molecular mechanics calculations and also provide an estimate of the electronic properties (electrostatic potential, polarizability, etc.) of the drug candidate that will influence binding affinity.

9.2 High-Throughput Screening

The quick rise in the field of molecular biology brings about an expansion of quick, effective, drug analyzing schemes. Approaches developed through schemes remain jointly recognized as high-throughput screening (Thomas 2007). High-throughput screening techniques gives a precise conclusion even after vastly minute quantities of scrutinize substance are existing. Though if it is to be used in an economic approach as well as proficiently, this technology needs the fast manufacture of a big amount of materials for examining which cannot be seen through the old-fashioned method toward organic synthesis, frequently adapt to manufacture of single synthesis blend at a time.

Fundamentally it is a procedure of selection and examining the massive quantity of living modulators as well as effectors contrary to selected and accurate marks. The codes as well techniques of HTS discover along with its claim aimed at the selection of combinatorial harmony, genetics, protein, and peptide archives. The principal effort of this performance remains toward accelerating drug finding procedure by the selection of great complex archives through quickness, which might exceed a somewhat thousand compounds in the single day time or for every week. For some assay or screening by HTS to be productive certain stages similar object recognition, compound management, component composition, test expansion also high-throughput assortment selection ought to remain approved available by highest attention and accuracy.

HTS is a very huge part of the study and progress having unlimited openings comprising enzyme analysis, entire organ analysis, and even complete animal examination over cassette dosing. Cassette dosing is a technique for HTS permitting to speedily evaluate the pharmacokinetics of great figure drug applicants. Unlike, additional methods to measure pharmacokinetics, in these technique solitary animals are given instantaneously, and blood trials composed to evaluate the same. The foremost lead is pharmacokinetics of an unlimited extent of complexes can be measured speedily and precisely. Nevertheless, the crucial drawback is that concurrent administration can lead to drug–drug interaction. No doubt, HTS is a new technique aimed at drug discovery. However, it is not the lone technique, and it also benefits the development of current drug moieties to enhance their activity. This too relates to screening the continuously growing compound libraries coming up to be selected owing to upsurge equivalently also combinatorial compound blend. An investigation is correspondingly approved available to expurgate the drug development expenses. Therefore, industries hold up-to-date through increasing competition. It is expected that along with the overview of human genomes as latent applicants the compound library will be as big as 100 million applicants, which would necessitate around 1012 assays to create their structure-activity relationship (SAR). Primarily the assays stood approved in 96-well plates nevertheless through development currently nearby 1586-well plates accessible. Distinctive HTS sequencers have great capacities for screening up to 10,000 complexes per day, although certain research laboratories with ultra-high-throughput screening (UHTS) can attain 100,000 assays for every day (Carnero 2006).

9.2.1 Assay Design

HTS as an explanation computed toward quick partially computerized prompt chief selection through giant figures of blends used for vigorous complexes. Procedure for the usage of bio micro-assays has grounded, which are speedy to perform and also involves usage of a small number of chemicals and trial compounds. Such tests remain approved available through 96-along with larger-well plates using specific treatment equipment. They are constructed on trial composite interrelating through an object, like protein molecules, chemical messenger, and ligand-activated transcription factors, which remain connected with the ailment phase, which is underneath the inspection (Thomas 2007). Subsequently, it might be essential to recognize, purify, and separate target earlier a reference library stay curtained. Additionally, tests are recurrently as well predominantly proposed aimed at study, therefore they need to remain confirmed. Such early explorations would be expensively collected in terms of money and period. Contribution by living objects in tests said so tests remain frequently accepted in water media. Accordingly, test determination simply is operative, if a considerable amount of product below examination liquefies in aqueous solution. Therefore, maximum tests remain conducted in water. Dimethyl-sulfoxide (DMSO) is recurrently combined to test combinations for upsurge solubility of the test compound in water (Thomas 2007).

The price of specialized chemicals used for screening big libraries of lonely mixtures can be costly. Therefore, frequent companies and researchers reduce funds by screening enormous archives using combinations of compounds. Though, this could prime for misleading consequences. For case, improper assertive consequences might ascend afterward combination under test covers a huge quantity of distinct compounds through a frail activity (Tate and Ward 2004). As a consequence, the mixture delivers a decent overall reply to trial. Therefore outcome would be erroneously unspecified through specialist by way of a mixture having a sturdily vigorous composite. This avoids active complexes obligatory to object accordingly, allowing a decent test reply. Due to the strength of the trial composite, this incorporated in the test compound leads to incorrect acceptances and rejections. As spending furthermore great application of trial compound similarly contributes incorrect confidence since attentiveness determined non-discerning obligatory to object. Equally, unreasonably minor deliberation can stretch an untruthful deleterious after an insufficient amount of dynamic trial complex been extant toward fixation of the object. Supplementary fundamentals of precisions like-wise ascend in particular kinds of micro-assay.

The micro-assays castoff in HTS might remain categorized as aimed at suitability as any biochemical founded assays. Biochemical examines are depend on the collaboration of trial complex over certain biochemical things inaccessible after cells like protein molecules, chemical messenger and ligand-activated transcription factors, whereas complete cell tests remain founded through the usage of integral compartments. Though, it's emphasized that HTS is castoff as a chief monitor for vigorous mixtures. Some vigorous mixtures (hits) that appear routine of extra inspection requirement remain exposed to a wider diversity of action trials earlier which might be measured for experimental expansion (Broach and Thorner 1996).

9.2.2 Biochemical Assays

These are also called as mechanism founded tests generally, it is established on obligatory of ligand to a receptor or the reserve of an enzyme-catalyzed response by an object which has been recognized using existence relevant to stated ailment phase (Singh et al. 2006). Target has been typically separated as of cell and is not at all extended portion of a cell. The experimental complex has been bind to object, which is counted through the usage of radioactive elements plus conventional investigative approaches like spectroscopic means of a variety of protocols similar by determining fluorescence in scintillation proximity assays (SPA). SPA usage resin globules whose exterior has remained contrived, therefore it accomplished binding to the wide variability of constituents. Bead too contains a sparkle that only fluorescence at the time of low-slung energy radioactive source ascends inside about 20 mm of the exterior of the beads (Glickman et al. 2008). The radioactive elements cast off in SPA tests release little energy discharges that need exact small trails in water solution.

Numerous SPA enzyme-based assays are conducted using radio-labeled ligands. Assume, for instance, an enzyme reserve assay established on the practice of a substrate A-B aimed at enzyme wherever, B as a serving of substrate which comprises the radioactive element and A comprises a termed capture cluster, which covers constructions bind to SPA globules. The action of substrate A-B through an enzyme along with certain significant co-enzymes in the absenteeism of inhibitor, consequences in the breakdown of entirely substrate A-B. After SPA beads remain, further not any fluorescence is detected as individual non-radioactive A binds to the beads (Acker and Auld 2014). Superior the reticence, fewer the fragment of substrate A-B. Subsequently, during SPA beads are added, the non-reacted radioactive substrate A-B and the non-radioactive A fix to beads. Subsequently, the strength of fluorescence is in a linear relationship with bead-bound A-B. Simply it means, maximum fluorescence, maximum the mark of enzyme inhibition of substrate A-B by the trial complex (Acker and Auld 2014).

9.2.3 Whole-Cell Assays

When the condition of disease is not been clear, in such cases, whole-cell assays have favored, which also gives many other benefits upon biochemical investigations. This test might recognize a composition that can work on sites except for the target site. Such test carried out below situations those are further identical come across if the test sample remained used in a patient. Those test compounds having a more hydrophobic property, which leads to toughly binding properties with serum albumin and in such case compound not crosses the cell membrane until it's been active. So it has been comparatively easy to find these compounds and eradicate them from the examination. Besides, poisonous test compounds are frequently recognized for their consequence on the cells used in the test.

Cell-based assays aimed at HTS can be classified below subsequent classes (Du et al. 2016). Signal transfer through activated cell-surface receptors is regulated by the second messenger assay. The speed of measurement of signals infraction of seconds is done by the second messenger assay. Second messenger assays typically measure fast passing fluorescent signals that arise in a fraction of seconds or milliseconds. Several luminous molecules are recognized to reply variations in intracellular concentration of calcium ion, and several other parameters, henceforth they are used for receptor stimulus and ion-channel initiation in an extension of second messenger assays. The improvement of hydrophobic voltage-sensitive probes has been serving the progression of the screening method for ion-channel in the discovery of a new drug. A response from the transformation level for a cell has been regulating by reporter gene assays. It specifies the occurrence or deficiency of a genetic factor creation that imitates variations in an indication transduction corridor. Such measurement of the reporter gene is generally conceded through biological approaches just like by determining the activity of the enzyme.

Reagents perform always vital part during the chemical synthesis or testing of compound, HTS is also no exemption in this regard. Characterization and optimization of reagents are essential before usage. The study revealed that nucleic acid, aptamers which bind to further molecules with greater affinity. Therefore this nucleic acid may be used as useful chemicals in competition binding HTS assays to recognize and improve minor ligands to protein targets. Speed of aptamer identification as compared to other more aptamer-protein interaction surfaces, the greater affinity of other protein targets towards aptamer. These are a few important benefits of aptamers usage in HTS assays. Aptamers could be predominantly beneficial in HTS assays through protein targets, which have no recognized obligatory followers like orphan receptors.

A hit arises after the action of a test composite showing significance better than the random smallest worth fixed through the agents by utilizing that assay. Like an enzyme reserve test, a hit can record during the action of the enzyme is reserved through the appeared worth of 60 percent. It's significant to establish standards intended for a hit formerly conduct the assay subsequently hit tolls remain repeatedly cast off as per the extent of the rationality of test method. Hit duties remain distinct as the number of vigorous mockups exposed by test stated using the percentage of whole models cast off now that monitor. Tests through standards of around 0.2% hits are generally stared as existence usable. Nevertheless, great hit tolls might be of use after emerging an assay (Du et al. 2016).

9.2.4 Automatic Methods of Library Generation and Robotics in HTS

Conventional carbon-based mixture and tests remain work rigorous. Beginning of HTS and combinatorial chemistry proceeds engaged to amplified usage of mechanization in drug-related chemistry through numeral companies generating "of the self" involuntary complex synthesizers then HTS analyzers, also norm constructed machines. Synthesis is regularly conceded about in sequence of computerizing monitored phases using the suitable arrangement of reaction vessels at stand-alone work locations (Michael et al. 2008). Operations in the synthesis like filling the reaction containers with chemicals and diluents, pipetting, washing, warming, separation, etc. usually maintained by a mechanical limb fixated through a software governing synthesizer. Mechanical arms and pathway organizations are cast-off toward rearrangement of microplates among work locations. Auto analysis remains supported in an identical overall way, which was, at work stations using mechanical supports, etc. to hand over models and chemicals.

Robotic HTS systems recurrently need humidified CO2 incubators and are surrounded by tissue culture work. Like to gathering track manufacturing, microplates are accepted route in a consecutive way toward successive dispensation components. Every one component has its individual modest preference and keeps the robotic arm (permit plates to the subsequent unit) and microplate treating scheme. Consequently, by every component, a unique phase of the assay is accomplished. Such preparation, attached through Windows NT™ (Microsoft, Redmond) along with ethernet TCP/IP linkage amongst components, delivers a considerable modest and extra firm stage than robot-centric HTS systems (Li Pira et al. 2010).

9.2.5 Profiling

The first objective of HTS is the identification of little authenticated HITS per big compound libraries. The conclusion as to whether a specific hit is a rate pursuing as a chemical lead in a drug discovery development rest on numerous influences, significant ones present chemical features and its pharmacodynamics and pharmacokinetic properties. The technology involved in reduction, mechanization, and assay data desired for HTS is ongoing to improve quickly, and as it does so, the laboratory provisions associated with HTS facilities are progressively developing their capabilities external their key determination of recognizing hits (Zhong et al. 2015). As this occurs, it turns out to be conceivable for HTS methods to be applied to additional miscellaneous compound profiling assays linking not only to the target discrimination of the compound libraries but also to their pharmacokinetic characteristics. Progressively, hence, initial compound summarizing jobs on hit compounds are actuality conceded out in the HTS laboratory wherever the essential technical proficiency is focused. Minor devoted robotic workstations are desired, somewhat than the fast but, stubborn factory-style robotic assembles used for large-scale HTS. It is clear that pharmacological outlining will be a growing activity of HTS units in upcoming, and will help to add more cost in the drug discovery sequence (Hann and Oprea 2004).

9.2.6 Screening Expense and Outsourcing Screening

Uncommonly, a certain corporation desires to screen 100,000 compounds each day domestic. The motive behind this comprises boundaries of several drug discovery procedures, apparatus/robotic necessities, infrastructure investment, and insufficient pre-condition to participate in altering technologies. Certain definite charges connected to screening are assay chemical charges (chemicals, cell culture overheads, etc.), microplates expenses, pipette tip box charges, screening worker expenses, record management/investigation period, databank budgets, robot earning charges, and workroom interplanetary expenses. Owing to combine difficulties of overhead, a rising amount of arrangement screening corporations are developing (such as Tropix, Panlabs, and Evotec). The facilities provided by these corporations regularly consist of assay expansion with screening, statistics examination, besides additional archive establishment necessities for HTS (Wildey et al. 2017). Contract screening corporations are also being used for their capability to deliver assay data over-accurate fast development period. They attain this by running 24-hr changes and using HTS robotic technologies. This retains the advanced value, additional registered minor screening internally, and permits the upkeep of a high ratio of hit production resulting since the authorized primary screening. The cost of total screening such a large compound library for a single assay might amount to over $ 300,000 (Liu et al. 2004; Mayr and Bojanic 2009).

9.3 QSAR Theories

QSAR studies have a very important application in modern chemistry and biochemistry. QSAR helps in finding the compounds with desired properties using chemical information and its association with biological activity. The physicochemical properties such as partition coefficient, and presence or absence of certain chemical features are taken into consideration (Roy et al. 2015). QSAR attempts to correlate structural, chemical, statistical, and physical properties with biological potency using various methods. QSAR models are used to predict and classify the biological activities of new chemical compounds. QSAR guides the process of lead optimization and also used as a screening and enrichment tool to remove the compounds that do not possess drug-likeness properties or predicted toxic (Tropsha 2010).

9.4 Molecular Descriptors Used in QSAR

Molecular descriptors are a numerical representation of chemical information present within a molecule (Caruthers et al. 2003). There are many parameters such as hydrophobic, electronic, and steric parameters, as well as associated descriptors used for QSAR (Roy and Das 2014). Descriptors associated with hydrophobic parameters are Partition coefficient (log P), Hansch's substituent constant (π), hydrophobic fragmental constant (f), distribution coefficient (log D), apparent log P, capacity factor in HPLC (log k, log kW), and solubility parameter (log S). Hammett constant (σ, σ+, σ -), Taft's inductive (polar) constant (σ*), ionization constant (pKa, ΔpKa), and chemical shifts are the descriptors used to define electronic parameters. Similarly, steric parameters are defined by Taft's steric parameter (Es), molar volume (MV), Van der Waals radius and volume, molar refractivity (MR), and Parachor. Atomic net charge (Qσ, Qπ), super delocalizability, energy of highest occupied molecular orbital (EHOMO), energy of lowest unoccupied molecular orbital (ELUMO) are known as quantum chemical descriptors. Spatial descriptors such as Jurs descriptors, shadow indices, radius of gyration, and principle moment of inertia are also used in developing a QSAR model. The information about molecular descriptors depends on the representation of a molecule and algorithm used for calculations of descriptors.

9.5 Methods of QSAR

Several methods are available for QSAR analysis which depends on the following criteria or factors related to study (Fig. 9.1):

1. Structural features or parameters (2D structure of a chemical compound to 3D conformations) that are derived from a series of molecules.

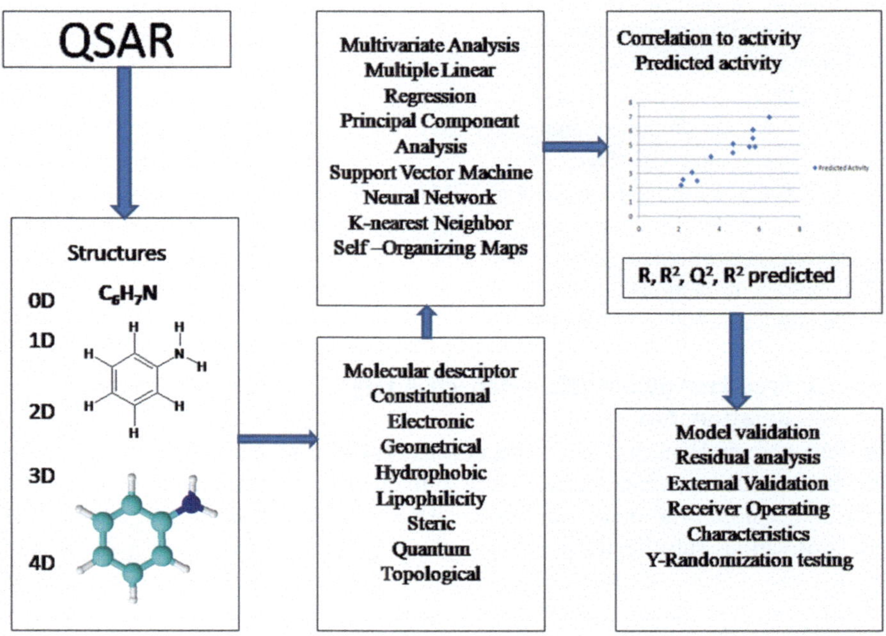

Fig. 9.1 QSAR methodology, mathematical models, and validation procedures

2. Mathematical approach used for establishing the relationship between the structural parameters and biological activity. Figure 9.1 explains the methodology of QSAR employed for any general QSAR type. Structures are fragmented to establish their relevant descriptor properties. With the help of various mathematical analysis tools, the data is trained to establish a mathematical QSAR model, which will correlate with the biological activity. The model developed is validated by various validation methods and tested for external prediction. Finally, a robust QSAR model is established that takes into account the relevant parameters for the biological activity for the given set of compounds.

9.5.1 2D QSAR Methods

There are many approaches used for 2D QSAR models. Different types of models can be used for the generation of any 2D QSAR. This approach can predict the biological activity and other molecular properties of compounds without having any experimental data related to the activity. QSAR analysis is based on computational and mathematical approaches, and it requires no use of in vivo and in vitro experiments, which are costly and time-taking. There are many related QSAR models as given below:

1. Free energy models—Hansch analysis: linear free energy relationship (LFER).
2. Mathematical models.
   (a) Free Wilson analysis.
   (b) Fujita-Ban modification.
3. Other statistical methods.
   (a) Discriminant analysis (DA).
   (b) Principal component analysis (PCA).
   (c) Cluster analysis (CA).
   (d) Combine multivariate analysis (CMA).
   (e) Factor analysis (FA).
4. Pattern recognition.
5. Topological methods.
6. Quantum mechanical methods.

9.5.1.1 Free Energy Models—Hansch Analysis Linear Free Energy Relationship

In 1969, Corwin Hansch extended the concept of LFER to describe the effectiveness of a biologically active molecule. It is used to quantify the therapeutic response of a drug molecule on the biological system (Hansch 1969). It takes into account the effect of various substituents in electronic, steric, hydrophobic, and dispersion data in the non-covalent interaction of a compound and target. In this approach, a set of parameters (descriptors) derived from a series of compounds and then used for activity prediction. The distribution of a drug depends on the partition coefficient (lipophilicity, log P) of the drug molecules. Hansch proposed that the biological activity of a drug molecule depends on two parameters, i.e. bulk of substituent groups (steric factor) and electron density on an interacting group (electronic factor) (Roy and Das 2014).

9.5.1.2 Mathematical Model

Free Wilson Analysis

It is a structure-activity analysis method that considers the contribution of various structural fragments to the biological activity of a molecule. It is used to find the presence or absence of a particular structural feature in a molecule that can be used as a parameter for the determination of biological activity. This mathematical model considers the symmetry equation to minimize linear dependency between variables. In this method, the structural features of a molecule are used to predict biological activity. But in the case of Hansch analysis, physicochemical parameters of the molecules are used to predict the biological activity (Kubinyi 1988).

Statistical Methods

Statistical methods such as multivariate analysis, classification, and regression analysis are used for interpretation and theoretical prediction of biological activity for new compounds. Statistical methods are very useful in finding a correlation between variables, building a model between associated variables, and also in assessing its accuracy. Regression generates a model in the form of an equation which represents a relationship between dependent variables or output variable (usually activity) in terms of independent variables or input variable (descriptors). This equation can be used to predict the biological activity of unknown set which can be further helpful in the screening of potential compounds with a good predicted activity (Everitt and Dunn 1992).

Discriminant Analysis

Discriminant analysis is used to separate molecules based on their constituent classes. It finds a linear combination of factors that best discriminates between different constituent classes. In this method, molecules are categorized as active and inactive based on the value of their biological activity parameters (Fisher 1936).

Cluster Analysis

Clustering is the process of dividing a set of objects into groups so that each cluster contains highly similar objects, and object in one cluster are dissimilar objects of other clusters. When cluster analysis is applied on a compound data set, the number of clusters provides information about the number of structural types present in a compound set. A diverse subset of compounds can be prepared by taking one or more compounds from each cluster (Kriegel et al. 2011). It is applied to sample diverse subset of compounds from a larger compound dataset. Hierarchical clustering, k-means clustering, and non-hierarchical clustering are the methods used for compound clustering.

9.5.1.3 Principal Component Analysis (PCA)

The number of variables used to describe an object is known as dimensionality. PCA is used to reduce the dimensionality of data set when a significant correlation exists between some or all of the variables (descriptors). PCA gives information about the significant principal components and represents most information on independent variables (Shaw 2003).

9.5.1.4 Quantum Mechanical Methods

- Quantum mechanical methods are used to analyze the electrostatic potential and ionization potential. In this method, electronic descriptors are derived from molecular wave function and used in the QSAR analysis (Roy and Das 2014). Hence there is a need for more advanced QSAR methodologies. There are certain demerits of predicting biological activity of compounds from QSAR analysis as it does not provide detailed and accurate knowledge about the mechanism of biological response and may also lead to wrong predictions associated with the biological activity of a compound.

9.5.2 3D-QSAR

3D-QSAR generates the quantitative relationship between the biological activity of a set of compounds and their 3D structural properties (Fig. 9.2). 3D-QSAR uses a

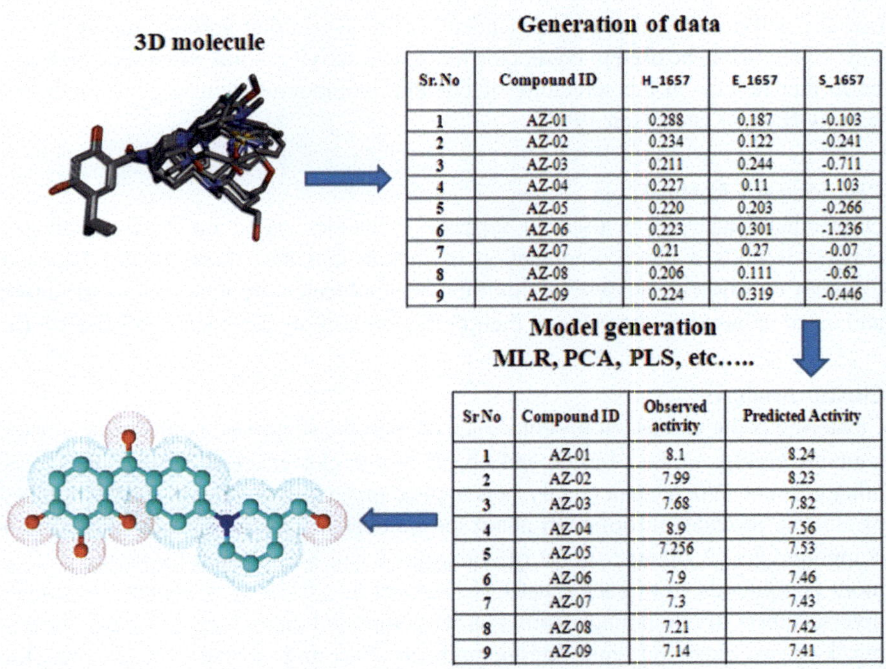

Sr. No	Compound ID	H_1657	E_1657	S_1657
1	AZ-01	0.288	0.187	-0.103
2	AZ-02	0.234	0.122	-0.241
3	AZ-03	0.211	0.244	-0.711
4	AZ-04	0.227	0.11	1.103
5	AZ-05	0.220	0.203	-0.266
6	AZ-06	0.223	0.301	-1.236
7	AZ-07	0.21	0.27	-0.07
8	AZ-08	0.206	0.111	-0.62
9	AZ-09	0.224	0.319	-0.446

Sr No	Compound ID	Observed activity	Predicted Activity
1	AZ-01	8.1	8.24
2	AZ-02	7.99	8.23
3	AZ-03	7.68	7.82
4	AZ-04	8.9	7.56
5	AZ-05	7.256	7.53
6	AZ-06	7.9	7.46
7	AZ-07	7.3	7.43
8	AZ-08	7.21	7.42
9	AZ-09	7.14	7.41

Fig. 9.2 3D-QSAR methodology: molecular alignment, generation of descriptors and model building

probe to determine values of 3D properties such as steric and electrostatic of molecules and then correlate and build a relationship model between 3D descriptors of molecules and its biological activity (Verma et al. 2010).

9.5.2.1 Molecular Shape Analysis (MSA)

MSA is an approach that includes conformational flexibility and molecular shape data in 3D QSAR analysis. In MSA, the 3D structure of many compounds is superimposed to find the commonly overlapping steric volume, and common potential energy fields between superimposed molecules are also identified to establish a correlation between the structure and activity of a set of compounds. This analysis also provides structural insight into the shape and size of the receptor-binding site.

9.5.2.2 Self-Organizing Molecular Field Analysis (SOMFA)

SOMFA divides the entire molecule set into actives (+) and inactive (−), and a grid probe maps the steric and electrostatic potentials onto the grid points. The biological activity of molecules is correlated with steric and electrostatic potentials using linear regression.

9.5.2.3 Comparative Molecular Field Analysis (CoMFA)

CoMFA is a grid-based 3DQSAR technique (Cramer et al. 1988). It assumes that in most cases, the drug–receptor interactions are governed by non-covalent interaction. COMFA considers that a correlation exists between steric and electrostatic fields of molecules and their biological activity. Here, the steric and electrostatic fields of the ligands at the various grid points in a 3D lattice are calculated. Partial least square (PLS) analysis is used to correlate steric and electrostatic fields with biological activities of molecules.

9.5.2.4 Comparative Molecular Similarity Indices Analysis (CoMSIA)

In COMSIA, molecular similarity indices serve as a set of field descriptors. This technique of 3D QSAR is used to determine the common features that are important for binding with the target molecule. Here, not only steric and electrostatic features, but also hydrophobic fields, hydrogen bond donors, and hydrogen bond acceptors are also taken into account for predicting the biological activity of a compound.

9.5.2.5 3D Pharmacophore Modeling

In pharmacophore modeling, the features governing the biological activity are determined from a set of known drugs that binds to a specific target. The entire structure of a molecule is not responsible for carrying out the biological activity. It is the only pharmacophore, which decides the biological response. Pharmacophore modeling is used for searching new potential drugs that share the same pharmacophore as available in other biologically active drugs of the same target. Pharmacophore models are hypothesis on the 3D arrangement of structural features such as hydrophobic groups, aromatic rings, hydrogen bond donor, and acceptor. Structurally diverse molecules bind with the receptor in a similar pattern, and their pharmacophore interacts with the same atom or functional groups of the receptor

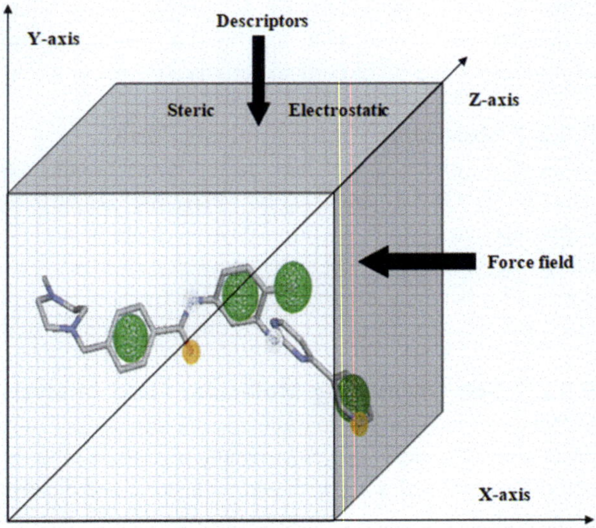

Fig. 9.3 3D-QSAR: expressing each molecule in its steric and electrostatic force field

molecule (Virupaksha and Alpana 2012) (Fig. 9.3). In the 3D QSAR model, molecules are aligned and superimposed with the core structure, and the molecular descriptors are calculated based on their conformation in the 3D space. The descriptors are correlated with biological activity, and a mathematical model is established. The descriptors in the 3D QSAR are the steric properties of the molecules, electrostatic forces, and force field descriptors.

Advantages of 3D QSAR over 2D QSAR

1. No dependent on experimental values.
2. Can be applied to molecules of unusual substituents.
3. Not restricted to molecules of same structural class as in case of pharmacophoric mapping.
4. Predictive ability.

9.5.3 4D-QSAR

The 4D-QSAR considers the conformational and alignment flexibility for training sets of structure-activity data by performing ensemble averaging. The fourth dimension in 4D-QSAR considers that each molecule can be represented by an ensemble of conformations, orientations, and protonation states (Damale et al. 2014). The 4D QSAR takes into account the conformational analysis of the molecule (Fig. 9.4). The different conformations of the molecules in the 3D space are designed and based on those conformations, the 3D descriptors are calculated.

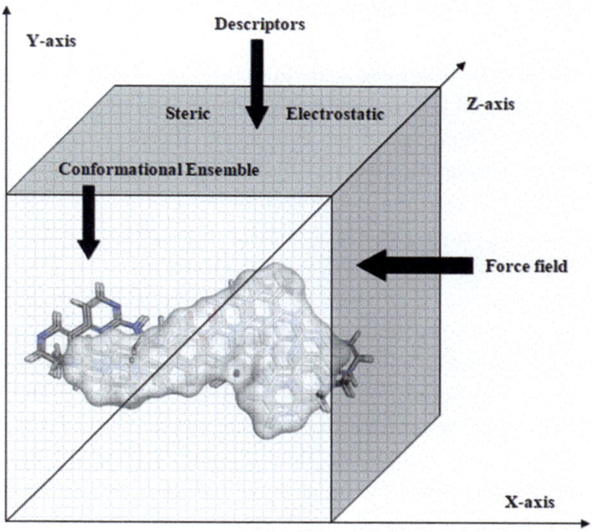

Fig. 9.4 4D QSAR: generating different conformations of molecule in its steric and electrostatic force fields

9.5.4 5D-QSAR

The fifth dimension in 5D-QSAR is the possibility to represent an ensemble of up to six different induced-fit models (Ducki et al. 2005). The 5D QSAR adds a fifth dimension to the QSAR models by taking into account the docked complexes of the molecules with their receptors along with their different conformations. This leads to the development of active site specific QSAR models that help to identify active site molecular fragments that are responsible for the biological activity of the molecule.

9.5.5 4D vs 5D-QSAR

The 4D-QSAR and 5D-QSAR are the multidimensional QSAR used in drug discovery. The performance of 5D-QSAR is evaluated using the analyzing residual test. In 4D and 5D-QSAR, multiple representations are used as ensembles and native conformations are chosen from a set of conformations using the concept of genetic algorithm (Damale et al. 2014). It is one of the best theoretical approaches to assess the relationship between substituent's physicochemical property and biological activity. Much advancement has occurred in the field of QSAR study, and a lot of molecular descriptors have been identified. COMFA and COMSA are the 3D-QSAR approaches, which are used for studying a large number of 3D descriptors for molecules to find a QSAR model. 3D-QSAR approaches have a certain limitation, which can be overcome by the use of 4D, 5D, and 6D-QSAR. 5D-QSAR analysis can be performed using Biographics Laboratory 3R, Quasar, and Raptor. 6D-QSAR has been developed to include one more parameter/dimension in the analysis, i.e. salvation function for different salvation states (Damale et al. 2014).

9.6 ADME Screening

ADME is well defined and studied terminology in pharmacology and describes the disposition of pharmaceutical compounds in an organism. In pharmacology, the term Pharmacokinetic gives an idea about ADME/T of a drug molecule. More than 50% of drug candidate falls out during the clinical trial experiments due to their poor ADME properties. Modern techniques and advancements to the entire drug discovery process have generated enormous potential therapeutic molecules and are in the pre-clinical ADMET assessment (Balakin et al. 2005; Morya et al. 2012). The complex route of any new chemical entity (NCE) to influence its target and achieving an optimum therapeutic index, commonly engages with the passage through numerous barriers and survival into complicated biological systems too. This collective process determines the bioavailability of NCE and several factors influence its pharmacokinetic properties (Bocci et al. 2017; Prentis et al. 1988). Pre-clinical ADME studies help in the identification of poor performer from the pool of chemical entities and reduce the breakdown induced by PK, yet the drug toxicity continues a big issue to the filtered one (Schuster et al. 2005). Non-optimal ADME and toxicity (ADMET), in a pair may end up with late-stage collapsation, accounts for a massive desecrate of time, money, and resources, unlucky cases like rofecoxib (Vioxx) and troglitazone (Rezulin) persuades the paradigm to fail early, fail cheap (Bocci et al. 2017; McNaughton et al. 2014).

To reduce the late-stage failure of a drug candidate, and early reach in the market, extensive studies of ADME processes is conceded out at an initial stage of drug discovery phases. Computational approaches are still pursued by biopharmaceutical investigators to foresee the consequences of drugs in the organism, and to determine the premature liability of toxicity (Singh and Dwivedi 2019). Prediction and simulation of various ADME properties are significantly less in cost than in vitro screening (Czarnik and Mei 2007; Van de Waterbeemd and Testa 2007). Currently, numerous pharmaceutical industry, third-party consultancy services, research groups, and academic institutions have their web-based chem-informatics services, private databases of compounds, ADME prediction tools and online servers deployed via the mode of standalone applications, Internet of Things and cloud-based applications for the users and giving medicinal chemists easy access to web-based and standalone based easy screening.

9.6.1 Absorption

How much of the drug is absorbed and how quickly? (Bioavailability). Absorption defines the amount of drug is absorbed, and its time taken to be absorbed, the amount of drug reaching the systemic circulation in an unchanged form is called bioavailability. Drugs cross biological membranes by facilitated diffusion down the concentration gradient, which involves two types of membrane transport proteins: carrier proteins and channel proteins (Bocci et al. 2017; Jenkinson 1991). Factors affecting the absorption of the drug are as discussed below;

9.6.1.1 Biologic Factors

Several drugs need to cross one or more cell membranes to reach their effective site of action. A most common feature of all cell membranes is the phospholipids layer about 10 mm in thickness, a lipophilic region facing inside and a hydrophilic head outside, giving a sandwich effect. The cell membranes are semi-permeable with lipoid sieve region and aqueous channels and a variety of special carrier molecules (Jenkinson 1991).

9.6.1.2 Passive Diffusion

Molecules less than 150–200 molecular weights were supported to pass through channels using passive aqueous diffusion in tissues. In an exceptional case, where endothelial capillary linings with large pore size help in a molecule with 20–3000 molecular weight to pass. Due to the absence of these large size pores in the brain capillaries, it makes difficult to big size molecule pass through. Passive lipid diffusion is the uttermost significant and widely used adsorptive mechanism. Lipid soluble drugs get dissolve into these membranes and are induced by concentration gradient across the membrane (Cocucci et al. 2017).

9.6.1.3 Carrier-Mediated Facilitated Transport

Drugs those are structural analogs of endogenous compounds for which specific carrier membrane is well established and known uses carrier-mediated assisted transport system. For example, anticancer drug methotrexate is structurally similar to folic acid and is successfully transported by the folate membrane transport system (Pratt et al. 1990).

9.6.1.4 Local Blood Flow

As the local blood flow continuously maintains the concentration gradient, important for passive diffusion hence an important factor for the rate of absorption. In the case of orally administered drugs, the blood supply demanding the gut passes via the liver before entering the systemic circulation and achieving its maximum bioavailability. Since the liver is the most vital site for the drug metabolism, and this first-pass reaction may diminish the quantity of drug reaching the target tissue and in achieving its therapeutic challenges. In certain cases, the first-pass effect may result in the activation of prodrugs, too (Jenkinson 1991; Pratt et al. 1990).

9.6.1.5 Gastric Emptying Time

It varies from patient to patient and adds significantly to intersubject uncertainty in drug absorption. Numerous factors such as meal composition and consistency, phase of the menstrual cycle, body position, smoking, gender, and time of day the study is performed are dependent variables and used to calculate the gastric emptying values (Vasavid et al. 2014).

9.6.1.6 pH-Partition Theory

For an ideal drug candidate, it should be able to cross the membrane barrier and must be soluble in both the phases, i.e. lipid layer and aqueous phase. For better absorption and distribution, a drug should contain polar, non-polar characteristics or either weak acids or bases. The solubility of the drugs is controlled by the following factors such as drugs with either weak acids or weak bases, pKa of a drug, pH of the bloodstream, pH of GI tract fluid, and the membrane lining of the GI tract (Shore et al. 1957). In 1957, the pH-partition theory was explained, and the extent of drug transfer or drug absorption under the influence of GI, pH, and drug pKa. The ration D is calculated by Eq. (9.1).

$$D = \frac{\text{Total Concentration in Blood}}{\text{Total Concentration in GI tract}} \tag{9.1}$$

9.6.1.7 Ion Trapping

It is the condition, which favors the non-ionized form of the drug in enhancing its drug permeation (absorption). In the case where weak acidic drug (aspirin; pKa 3.5) crosses the gastric mucosa at pH 2.0, further reverts to an ionized form within the cell at pH 7.0, and therefore deliberately it passes to the extracellular fluid (Sharma and Sharma 2011).

9.6.1.8 Chemical Modifications Affect the Absorption

Modification of drug structure without altering its pharmacological activity is one of the profitable ways to improve the absorption of a drug. The modification has been commonly used to modify the physicochemical property of a drug such as molecular weight, molecular size, pKa, solubility, lipophilicity (hydrophobicity), hydrophilic nature, and related activity. For example, chemical modification of salmon calcitonin to elcatonin (bond replacement from (C–N) to (S–S)) shows better bioavailability of elcatonin over salmon calcitonin (Liu et al. 2019).

9.6.1.9 Optimizing Absorption

Absorption of a molecule is highly dependent on above discussed factors, optimizing each factor including chemical modification is time-consuming and tedious task. Many computational methods were developed to understand and support in optimizing the drug absorption rate. Log Po/w estimation a classical descriptor was designed with a various recital on diverse chemical sets. In SwissADME, five predictive models help in optimizing the compound lipophilic activity i.e. XLOGP3 using a knowledge based library. WLOGP is an atom based method for lipophilicity prediction that uses the fragmental system of Wildman and Crippen. MLOGP prediction is based on the topological method, whereas the SILI COS-IT is based on a hybrid method that uses fragments and topological descriptors both. Another lipophilicity prediction approach, iLOGP depends on free energies of solvation in n-octanol and water. Optimizing all the properties is a major and tedious task and may fail to achieve the desire pharmacological target (Daina et al. 2017). Considering the optimizing lead molecules physicochemical properties like solubility, lipophilicity (hydrophobicity), pKa, and hydrophilic is a trial and error based approach, Computational optimization helps in time saving and reduce in false landing to the desire target.

9.6.2 Distribution

A drug is distributed through the blood system to reach a target site, they may also distribute to different muscles and organs. The distribution of drugs mainly depends on the binding and free bound form from enzymes and proteins present in the bloodstream. The effective distribution of a drug is affected by its binding to the plasma proteins. If a less amount of drug binds to plasma proteins, then a higher amount of drug disseminates across cell membranes and achieve higher bioavailability. Human serum albumin, lipoprotein, glycoprotein, and α, β, and γ globulins are the most common/typical proteins into the blood that drugs bind (Bocci et al. 2017). The rate of distribution is highly affected by the molecular size, smaller ones pass through easily and have a high rate of distribution, whereas one with larger molecular size finds it difficult to cross biological membranes and hence have a lower rate of distribution. Polar drugs (e.g. penicillin class) are not capable to cross biological barriers, except they are taken by pinocytosis (e.g.: insulin) or either with the help of carrier proteins (Bocci et al. 2017; Thomas 2008). There are numerous factors that affect drug distribution.

1. Tissue permeability of the drug.
   (a) Physiochemical property of a drug.
   (b) Physiological barriers to diffusion: simple capillary endothelial barrier, blood–brain barrier, simple cell membrane barrier, placental barrier, cerebrospinal fluid barrier, and blood–testis barrier.
2. Organ/tissue size and perfusion rate.
3. Binding of drugs to tissue components: binding to blood components, and extravascular tissue proteins.
4. Miscellaneous factors: age, obesity, diet, pregnancy, and drug interaction (Pavan 2013).

9.6.2.1 Optimizing Distribution

Appropriate absorption and distribution are prerequisites for a chemical entity to act as a drug. Poor bioavailability and pharmacokinetics are major concerns and hurdles in drug design processes. Gastrointestinal absorption and blood–brain barrier permeability are two vital pharmacokinetic aspects which are essential to estimate at different stages of the drug discovery process (Daina et al. 2017; Daina and Zoete 2016). Computationally Boiled-Egg method helps in the prediction of gastrointestinal absorption and blood–brain barrier (BBB) permeation (Daina and Zoete 2016) and Lipinski filter helps in stabilizing the drug-like properties (Lipinski et al. 2001). Factor like solubility is responsible for solute dissolution in an aqueous/solvent medium to achieve a homogenous system. It is a vital parameter to study and optimize the drug concentration in systemic circulation for an anticipated pharmacological response or to achieve its desired bioavailability. Technically solubility of any compound can be improvised by the following ways and they are categorized into physical modification, chemical modifications, and other techniques. In physical modification: particle size reduction akin to nanosuspension and micronization. Crystal modification such as cocrystallization, polymorphs, and amorphous is form of a compound. The use of a buffer, chemical derivatization, complexation, salt formation, and change of pH comes under chemical modification. Miscellaneous methods include the use of novel adjuvant like surfactants, solubilizers, co-solvency, and supercritical fluid process (Savjani et al. 2012). 2D and 3D Quantitative structure-property relationship (QSPR) and quantitative structure-activity relationship (QSAR) based models help us to identify the suitable functional group for desire activity and their possible interaction in optimizing drug-related affinities. Significant advancement has been carried out in computational based methods to predict the solubility of a compound by considering numerous models from high-throughput assay based model, highly accurate small scale thermodynamics based measurements using pure water or in the non-complex buffer, molecular dynamics based simulations to study and understand the compound behavior in the complex system (Bergström and Larsson 2018).

9.6.3 Metabolism

Once the drug is entered into the body of an organism, the process of catabolism and anabolism commonly known as metabolism is started with the help of numerous enzymes supported by various chemical moieties, and solvent systems. The main objective of the drug metabolism is to convert these drug compounds into a more polar, water-soluble intermediates, or final products that can be easily excreted from the organism's body (Thomas 2008). The metabolism phase is divided into two parts.

9.6.3.1 Phase I

Usually, the liver is the prime site for drug metabolism, including other organs such as lungs and kidneys that also carry out drug metabolism. Phase I metabolism reaction is carried out in the liver where oxidation, reduction, hydrolysis, cyclization, and decyclization process are done. In phase I reaction, the C–H bond converts into a C–OH, transforming an inactive compound to an active form, exhibiting pharmacological actions including conversion of non-toxic compounds into a toxic one. A variety of enzymes binds the drug or NCE as their substrate and introduces reactive and polar groups into it. The most common class of enzyme cytochrome P-450 includes all the following processes listed below (Akagah et al. 2008; Guengerich 2001; Schlichting et al. 2000).

Oxidation

Reactions resulting in the removal of hydrogen and, or addition of oxygen is known as an oxidation reaction. Cytochrome P450 monooxygenase, flavin-containing monooxygenase, alcohol dehydrogenase, aldehyde dehydrogenase, monoamine oxidase, and peroxidases enzyme system carry out the oxidation process in an organism.

Reduction

Reactions resulting in the removal of oxygen and, or addition of hydrogen are known as a reduction process, the following enzyme carries out a reduction process in an organism.

NADPH-cytochrome P450 reductase.
Reduced (ferrous) cytochrome P450.

Hydrolysis

Hydrolysis is a reaction in the presence of water where one compound breaks into two separate compounds. Simultaneously the water molecule splits in two and transfers the hydrogen atom to one compound, and the hydroxide group to another is known as hydrolysis. In a cytochrome P-450, a reliant mixed-function oxidase system catalyzes a hydroxylation process within its substrates as a common modification. The enzyme complexes act to integrate an oxygen atom into non-activated hydrocarbons, i.e. the addition of hydroxyl groups or N–, O–, and S-dealkylation of substrates (Schlichting et al. 2000).

$$O_2 + NADPH + H^+ + RH \rightarrow NADP^+ + H_2O + ROH$$

9.6.3.2 Phase II—Conjugation

Phase II is commonly known as the conjugation process, and the reactions include stimulated xenobiotic metabolites are associated with charged species like as glutathione (GSH), sulfate, glycine, or glucuronic acid. The phase II reactions produce less reactive products compared with their substrates. Phase I products associate the inclusion of highly polar molecules to a functional group and form more soluble compounds, which can be easily eliminated (Akagah et al. 2008; Guengerich 2001; Schlichting et al. 2000).

9.6.3.3 Factors Affecting the Metabolism of a Drug

Numerous factors may affect the influence of drug metabolism such as:

1. Chemical factors: (a) Enzyme induction, (b) Enzyme inhibition, (c) Environmental chemicals.

2. Biological factors: (a) Age, (b) Diet, (c) Sex difference, (d) Species difference, (e) Strain difference, (f) Altered physiological factors.
3. Physiochemical properties of a drug (Jenkinson 1991; Pratt et al. 1990).

9.6.3.4 Optimizing Metabolism

Drug metabolism is an important link between chemistry and biological reactions, yet a complex mechanism. Drug molecule goes under a series of chemical transformations like in Phase I oxidation, reduction, hydroxylation, dealkylation to conjugation process of reaction followed by in phase II reactions predominantly conjugation reactions like glucuronide, sulfate, and acetate conjugation.

The optimization process of metabolism reaction can be counteracted by using the prodrug concept and alteration of the physicochemical properties of a lead compound (Singh 2018). Chemical alteration, such as the introduction of the ester group can be performed to increase the metabolic rate. *N*-dealkylation can be prevented by replacing *N*-methyl group by *N*-t-butyl group and oxidation of aromatic rings be reduced by the introduction of –Cl, –NR3, –COOH and sulfate to decrease the metabolic rate. Prodrugs are compounds that are biologically inactive (or maybe less active) but are metabolized to a bioactive molecule known as a metabolite. A number of prodrugs have also been designed to be site specific and in order to improve the biopharmaceutical, pharmacokinetic, and poor drug-like properties (Sanches and Ferreira 2019). Salicylic acid is one of the oldest analgesics known. However, its use can cause gastric irritation because of free carboxylic acid functionality present. Masking of the carboxylic group will be carried by acetylation reaction to form an acetylated prodrug known as aspirin. Aspirin is known for producing less degree of gastric irritation (Mahfouz et al. 1999). This reduces the amount of salicylic acid in contact with the gut wall lining. RH1 anticancer drug, RH1 therapy is based on the prodrug conversion by cancer cells into the more active by enzyme NQO1 (Gupta et al. 2017; Parkinson et al. 2013).

9.6.4 Excretion

The process of removal or elimination of unwanted products or metabolic waste from an organism's body is called excretion. In vertebrates, the process is principally carried out by the lungs, kidneys, and skin (Beckett 1987). The metabolized or un-metabolized components should be excreted from the organism's system. The overall complex process of elimination is carried out via the kidneys as urine, feces, and sometimes through sweats.

9.6.4.1 Factors Affecting ADME Properties and Modeling Process

To be an effective and pocket-friendly drug, a sufficient amount of a drug must reach and modulate the drug target with minimal toxic effect, including the low cost in terms of time, money, resources to the innovator. Traditional methods including in vitro testing and late phase drug failure cause numerous loss in terms of subject health hazards and loss of overall resources including time. In silico ADME-Tox prediction plays an essential role in assisting the selection of experimental ligands or drugs by pharmaceutical industries before initiating an expensive clinical trial (Alqahtani 2017; Sliwoski et al. 2014). As numerous groups are working on the modeling of the ADME process, considering the major factor of drug profit and loss business focuses point. Collateral study and comparison of various ADME factors, including gene, protein, and reactions in silico modeling, and prediction are useful in an early phase drug retraction. Molecular and physiochemical properties persuade both pharmacokinetic and pharmacodynamic processes, including drug safety. The theories and concept of drug likeness characterize the borderline of fundamental properties of a drug to aid the medicinal chemists in superiorizing the drug candidates. Core competence is considered for the Molecular weight and lipophilicity of an NCE as it changes widely according to its size, and the nature of the functional group attached to the scaffold (Vallianatou et al. 2015).

9.6.4.2 Drug Likeness

Drug Likeness qualitatively assesses the chemical entity to develop into an oral drug with due recognition to its bioavailability (Daina et al. 2017). The assessment was customized from a structural or physicochemical examination of the selected compounds. This concept is regularly engaged to perform filtering of chemical libraries to suspend out the molecules with properties most apparently unsuitable with an acceptable pharmacokinetic profile (Daina et al. 2017). Online tools or web servers such as SWISS ADME, ADMETSAR, and related are used as a standard tool for filtering the compounds (Yang et al. 2018). SWISS-ADME provides access to five different rule-based filters. The most prominent approaches used as Lipinski filter are (Pfizer) (Lipinski rule of five) (Lipinski et al. 2001) and followed by Ghose (Amgen) (Ghose et al. 1999), Veber (GSK) (Veber et al. 2002), Egan (Pharmacia) (Egan et al. 2000), and Muegge (Bayer) (Muegge et al. 2001).

9.6.4.3 Lipophilicity

Lipophilicity is defined as the competence of a compound to dissolve or diffuse in fats, oils, lipids, non-polar solvents like hexane and toluene. As the cell membranes are composed of lipid bilayers, including phospholipids and glycolipids, hence it is necessary to study and analyze the lipophilic activity of the drug, and the environment of the system (Schroeder 2018). The affinity of a drug for a lipid environment is known as the Lipophilicity log P equation and is a critical parameter to study the pharmacokinetic and pharmacodynamic parameters of a drug and receptor interactions. Log P, one of the primitive and model-based descriptors for lipophilicity are considered as a partition coefficient among *n*-octanol and water (log Po/w) (Arnott and Planey 2012; Mannhold et al. 2009). As octanol represented the most optimum behavior to cell membranes, tissue lipids, and other lipophilic components of cells, hence it has not been yet replaced by cyclohexane and artificial lipids.

The Log P measurable range generally lies between −2 and 6, but these range values are not constant and can be varied depending on the detection techniques were used (Stoner et al. 2008). Numerous in silico tools for log P calculation have been developed from the large chemical dataset. Cheminformatics tools such as ACD Chemsketch, Marvin sketch, Molecular design suite (V Life sciences), QikProp (Schrodinger) are among the few that give a good prediction over the set of molecules Eq. (9.2).

$$\text{Log } P = \frac{\text{Concentration of Compound in Octanol}}{\text{Concentration of Compound in Water}} \tag{9.2}$$

9.6.4.4 Solubility

The absorption and distribution activity of a drug is significantly supported by the aqueous solubility of a compound. A low soluble compound exhibits poor absorption. Soluble compounds facilitate the drug development process, including the ease of handling, formulation, and support discovery projects aiming an oral and parenteral administration (Ottaviani et al. 2010; Ritchie et al. 2013). Online tools such as Osiris, SWISS-ADME, databases such as PubChem, drug bank, and related provides the information for the typically entered compound.

9.6.4.5 Pharmacokinetic Process

In the early phase of the drug discovery process, one should know the compounds that may be a substrate, non-substrate, or an inhibitor to numerous enzymes that play an essential role in drug transportation and metabolism. CYP and P-gp operate small molecules synergistically to improve the protection of tissues. Around, 50–90% of drugs act as a substrate for major CYP isoforms (CYP1A2, CYP2C19, CYP2C9, CYP2D6, CYP3A4), whereas P-gp plays an important role in protecting central nervous system (CNS) from xenobiotics. Inhibition of CYP superfamily isoenzymes may widely affect drug elimination through metabolic biotransformation, and drug–drug interactions (Montanari and Ecker 2015).

Lower clearance or accumulation of the drug or its metabolites results in a toxic or adverse effect (Sharom 2008; Szakács et al. 2008). While evaluating a large data size of compounds, one has to be cautious about the selection of drug and CYP family isoenzyme interaction selections. Some compounds might act as an inhibitor to one or more CYP isoenzyme, and might be a selective substrate for the other (Testa and Krämer 2007; van Waterschoot and Schinkel 2011; Wolf et al. 2000). Hence, it is important to select the most optimum CYP interactions that would be beneficial for the process of discovery (Huang et al. 2008; Kirchmair et al. 2015; Veith et al. 2009). Tools used for ADME screening of compounds are listed in Table 9.1.

Table 9.1 Online and offline tools for ADME screening

S No.	Tools	Description	URL
1	ADMETlab	Systematic ADMET using ADMET database	http://admet.scbdd.com/
2	ADMET predictor	ADMET property estimation	https://www.simulations-plus.com/software/admetpredictor/
3	ADVERPred	Prediction of adverse effects of drugs	http://www.way2drug.com/adverpred/
4	eMolTox	Prediction of molecular toxicity	http://xundrug.cn/moltox
5	LIVERTOX	Hepatotoxicity	https://livertox.nih.gov/
6	Molinspiration	Molecular properties	http://www.molinspiration.com/
7	MouseTox	Cytotoxicity assessment for small molecules	http://enalos.insilicotox.com/MouseTox/
8	PreADMET	ADME properties	https://preadmet.bmdrc.kr/
9	Pred-hERG	Predict hERGcardiotoxicity	http://labmol.com.br/predherg/
10	Pred-skin	Chemically-induced skin sensitization	http://labmol.com.br/predskin/
11	QikProp	Schrodinger tool for ADMET	https://www.schrodinger.com/qikprop
12	SOM prediction	Knowledge based method for prediction	http://www.scfbio-iitd.res.in/software/drugdesign/som.jsp
13	SwissADME	ADME parameters	http://www.swissadme.ch/
14	vNN	ADMET predictions	https://vnnadmet.bhsai.org/vnnadmet/login.xhtml

9.7 Toxicological Screening

Toxicity is an evaluation of the measure of any unwanted or unfavorable effect of any chemical or substance on the human body or environment. It can be quantifiable (like LD 50-lethal dose) or qualifiable (toxic or non-toxic). The toxicity is calculated in terms of various endpoints such as Genotoxicity, carcinogenicity, skin sensitization, irritation, ecotoxicity, etc. The studies identify the effects of chemicals on humans, animals, plants, or the environment through single dosing or multiple dosing (Stephens 2010). Several factors that determine the toxicity of chemicals are taken into consideration while calculating the toxicity such as route of exposure like oral, dermal, inhalation, amount of dose, frequency of exposures like single or multiple exposure, duration and time of exposure, ADME properties of the drug, biological characteristics of the patients, and chemical properties (Raies and Bajic 2016).

In silico toxicity testing uses computer-based methods, software, algorithms to analyze and predict the various toxicity endpoints for a given chemical. This information can be used to modify or discard a given chemical entity. It helps in designing and developing molecules that may have no or less toxic effects, reduce the cost and time of in vitro and in vivo toxicity studies. Also, it has an added advantage of predicting the toxicity of a given chemical moiety even before it is synthesized, just by knowing its structure, thus also reducing the cost of the synthesis of such molecules whose toxicity is high.

Various tools are involved in the toxicity prediction. Generally, the prediction consists of the following points:

1. Databases that store the toxicity information of various endpoints can be manually curated and literature referenced.
2. Software that generate molecular properties and descriptors.
3. Simulation and systems biology tools for molecular dynamics and computational drug design.
4. Modeling algorithms for toxicity prediction.
5. Statistical packages for prediction model generation.
6. Expert systems for reasoning and analysis.
7. Visualization tools.

Unique toxicity prediction software comprises of the above tools and properties to give a comprehensive overview of a particular prediction of a unique molecule (Gramatica 2013). The molecule prediction methods consist of 6 main steps to develop a prediction model (Fig. 9.5).

1. Collecting biological information that can associate the chemical structure to the molecule with the toxicological endpoints.
2. Comparing the chemical with similar chemicals in the database.

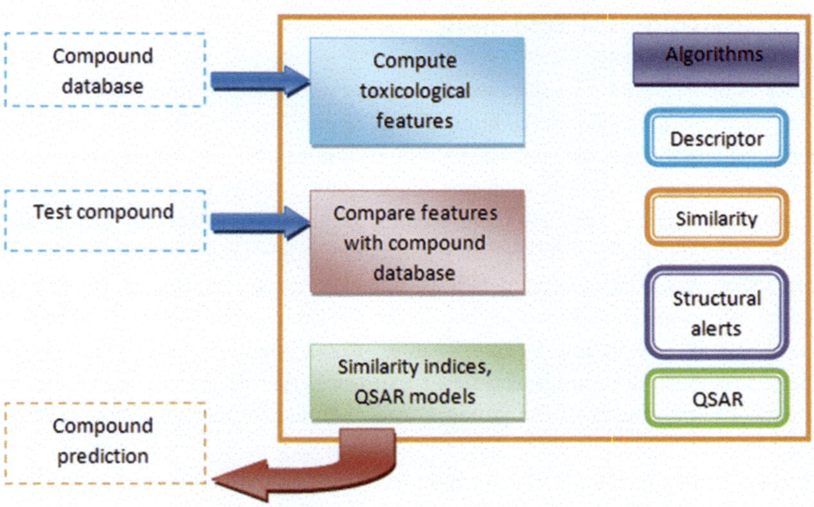

Fig. 9.5 Steps in toxicity prediction models

3. Calculate molecular descriptors for the chemical.
4. Generate a prediction for the chemical.
5. Evaluating the accuracy of the prediction.
6. Interpreting the model.

A compound database is created containing all the possible toxicological features, from the literature or through user input. The data is curated and stored. A query compound is searched against all the features in the database using various algorithms and similarity searches. Rule-based structural alerts are matched with the fragments in the query structure, and statistical-based QSAR predictions done based on the descriptors and similarity indices. Both methods give their respective predictions, and the compound is categorized accordingly.

9.7.1 Acute Systemic Toxicity

Acute systemic toxicity testing is the evaluation of the dangerous potential of the chemical by short-term exposure by determining its systemic toxicity. The results are given as median lethal dose LD50 for acute oral exposures and as median lethal concentration LC50 for inhalation exposures (Botham et al. 2002).

9.7.2 Toxicological Endpoints

The following are some of the toxicity endpoints predicted by various prediction software (Richard et al. 2008).

1. Carcinogenicity—cancer-induced due to harmful chemicals may be due to genotoxic compounds or non-genotoxic compounds.
2. Dermal penetration- the rate at which a chemical enters the skin.
3. Ecotoxicity—harmful effects on different life species in the environment due to chemicals. Commonly studies are conducted on fish, systemic, dietary, and reproductive systems as well as bioaccumulation.
4. Eye irritation/corrosion—reversible or irreversible eye damage caused due to chemicals.
5. Genotoxicity—induced mutations or changes in the structure, number or content of the DNA, or segregation of the genetic material caused due to harmful chemicals that may or may not lead to carcinogenicity.
6. Neurotoxicity—harmful effects on the brain, brain tissue, spinal cord, or any part of the nervous system caused due to chemicals.
7. Phototoxicity—toxic changes induced in the substance due to exposure to light or skin irradiation.
8. Organ toxicity—these are caused due to repeated daily exposure to harmful chemicals.

9. Reproductive and developmental toxicity—harmful effects on fertility, the sexual function of an individual caused due to chemicals.
10. Skin irritation/corrosion—skin damage that can be reversible or irreversible due to harmful chemicals.
11. Skin sensitization—allergic reactions and responses due to contact with any harmful chemicals.

9.7.3 Structural Alerts and Rule-Based Method

Structural alerts are also known as toxicophores or toxic fragments. These are small molecules fragments present as a part of the root structure that is toxic. These toxic fragments attribute to the toxic nature of the overall root compound (Table 9.2). Different structural alerts are associated with different endpoints. Thus if a particular structural alert giving rise to genotoxicity is present in a compound, then it is likely that the entire compound is genotoxic. The following are some of the examples of common toxicophores associated with genotoxicity (Plošnik et al. 2016).

Various software uses structural alerts or rule-based methods to predict toxicity endpoints. These software incorporate structural alerts manually curated by human experts. The structural alert list for skin sensitization was published in 1982 by Dupius and Benezra (Payne and Walsh 1994). Another structural alert list to predict carcinogenicity and mutagenicity was developed by Ashby and Tenant in 1988 (Ashby and Tennant 1988). One of the most developed and widely used structural alert lists is of carcinogenicity and mutagenicity proposed by Benigni et al. (2008). Other lists of alerts and rule-based methods are developed for endpoints like hepatotoxicity, cytotoxicity, irritation, corrosion of skin and eye, and skin sensitization. There are many software that employs rule-based methods such as oncologic cancer expert system (OCES), toxtree, Derek Nexus, Hazard Expert, and Meteor (Raies and Bajic 2016).

9.7.4 Read Across Methods Using Chemical Category

A chemical category is a cluster of molecules whose toxicity effects follow a similar pattern. The chemicals or molecules are grouped into categories depending on a particular property such as ADME or mechanism of action, physicochemical properties, interactions, or structural similarity (Berggren et al. 2015). The structural similarity is regarded as a starting point to group the chemicals into a similar category. The toxicity of an unknown molecule is predicted using these similar chemicals whose toxicity is known. Read across models are developed using two approaches-

- Analog approach—one molecule is compared to one or a few molecules.
- Category approach—one molecule is compared to many molecules (Raies and Bajic 2016).

Table 9.2 Genotoxicity structural alerts and compounds

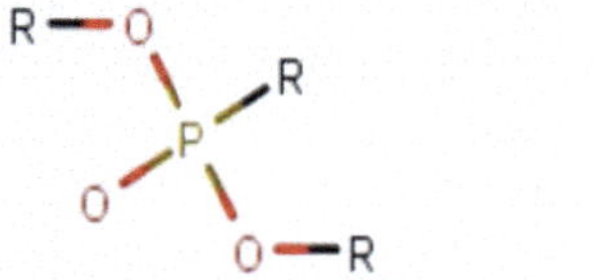 Alkyl esters of phosphonic acids	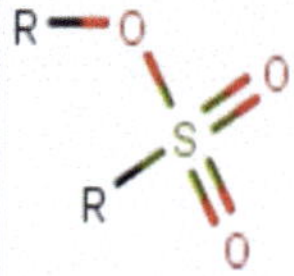 Alkyl esters of sulfonic acids	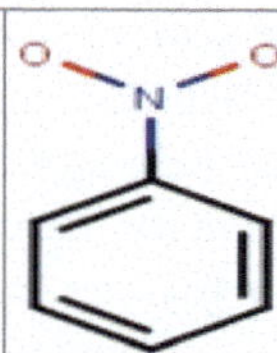 Aromatic nitro groups
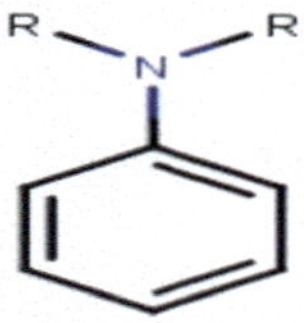 Aromatic mono- and dialkylamino groups	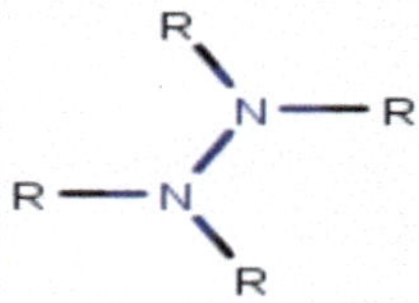 Alkyl hydrazines	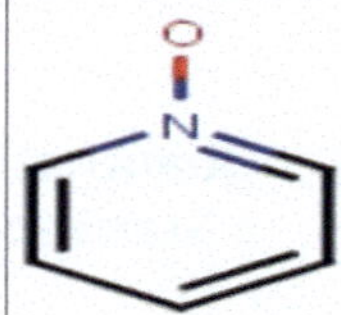 Aromatic *N*-oxides
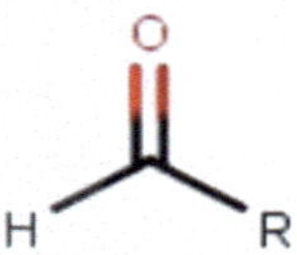 Simple aldehydes	 *N*-methylol derivatives	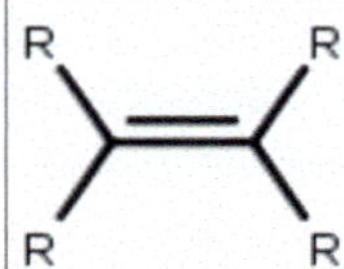 Monohaloalkenes
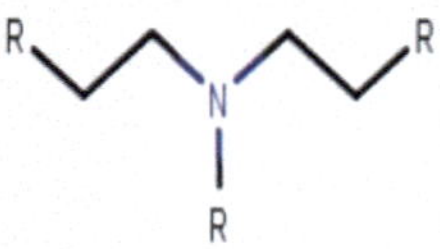 *N*-mustards	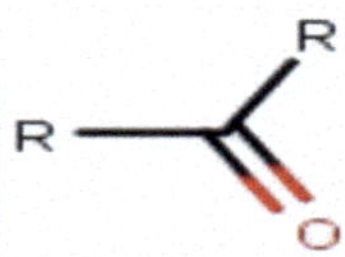 Acyl halides	 *S*-mustards
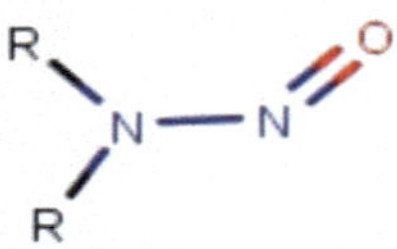 Alkyl and aryl *N*-nitroso group	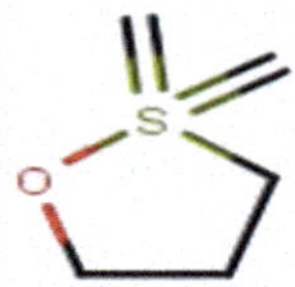 Propiolsultones	 Epoxides
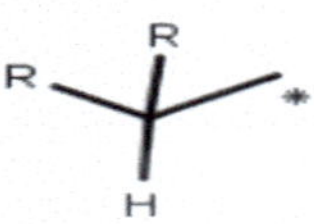 * = Halogens Aliphatic halogens	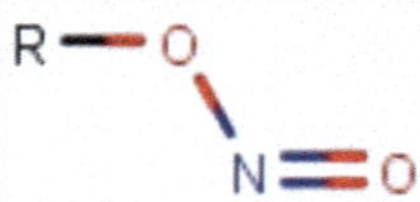 Alkyl nitrite	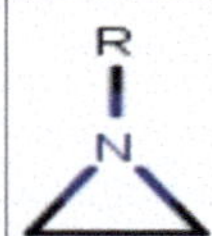 Aziridines
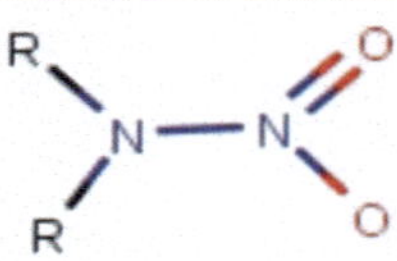 Aliphatic *N*-nitro group	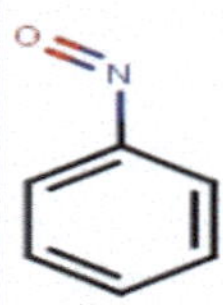 Aromatic nitroso group	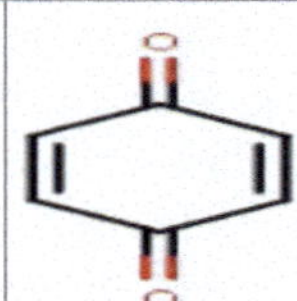 Quinones

A read across approach can be both qualitative as well as quantitative. The major advantages of read across methods are that it is transparent, easy to deduce and execute. A broad range of descriptors or parameters can be calculated in the read across the technique. Read across methods are applied for various endpoints such as carcinogenicity, hepatotoxicity, reproductive toxicity, skin sensitization, and environmental toxicity (Berggren et al. 2015). Various tools and software that implement read across techniques are OECD QSAR toolbox, ToxMatch, Toxtree, AMBIT, AMBITdiscovery, AIM, and DSSTox.

9.7.5 Quantitative Structure Activity Relationship Model Using a Statistical Method

This model is also known as the Quantitative structure-toxicity relationship model (QSTR) as it deals with the prediction of toxicity for a given compound. Just as in the case of traditional QSAR, a mathematical equation is derived to correlate the toxicity of the compound with its structure. The statistical method gives a probability score as to the compound being toxic or non-toxic in case of qualitative endpoints, and the model predicts a value in case of a quantitative toxicological endpoint. The statistical QSTR models are used in predicting the toxicity of various aromatic nitro compounds, nitrobenzene compounds, cytotoxicity of TIBO derivatives, and carcinogenicity of sulfa drugs (Morales et al. 2006). The advantages of QSAR models are that they are easy to interpret and more meaningful. They can model endpoints with the help of molecular descriptors. Various software, which uses the statistical QSAR method are OECD QSAR Toolbox, Topkat, Sarah nexus, Hazardexpert, VEGA, and METEOR.

9.7.6 Organization for Economic Cooperation and Development (OECD) Guidelines

Many new chemicals that are used as drugs, pesticides, food additives, and biotechnological products are launched into the market every year, and they require safety testing all over the world. The OECD has developed certain guidelines for the chemical testing, and safety of the chemicals (Fritsche et al. 2017; Milstein and Schreyoegg 2016; Sakuratani et al. 2018). They cover safety testing concerning physical and chemical properties, the effect on biological systems, environmental fate (degradation, accumulation), and health effects on individual living beings. Each organization, before launching its chemical product into the market has to abide by these rules and have to submit a complete data report of the chemical testing of these molecules. To reduce the time and cost of toxicity studies, the regulatory bodies accept the in silico toxicity reports of the chemicals generated through validation and evaluation. These reports should be generated based on the OECD principles laid down by the OECD committee. For the submission of a genotoxicity report for a particular molecule, the studies should comply with the ICH M7 guidelines and should satisfy the OECD principles. To consider a QSAR model for the regulatory purpose, it should satisfy the following points:

Principle 1: any QSAR model should possess a well-defined endpoint, which can be the physicochemical, biological, or environmental effect.

Principle 2: the algorithm used for the development of the prediction model should be unambiguous, which ensures transparency.

Principle 3: the prediction model should have a definite domain of applicability, ensuring that the predicted molecule does not go beyond the domain of prediction.

Principle 4: the QSAR model should be robust and have an internal performance training set, and external predictivity test set.

Principle 5: the model should have a mechanistic interpretation wherever possible and applicable.

The ICH guidelines recommend a QSAR prediction of toxicity by both rule-based and statistical approaches. A joint report of these predictions, along with an expert review is accepted by the regulatory authority. The software used for such predictions should be duly validated and accepted by regulatory bodies. Thus the field of in silico toxicity is under development, and novel methods are introduced and applied. If used appropriately, these tools are effective to predict the toxicity of a chemical. However, these models need continuous evaluation, validation, and improvement. Hence, there is a need to understand their current strengths, weakness, and their specific applications.

9.7.6.1 Optimizing Toxicity

In silico approaches can rapidly screen and optimize pharmacokinetics and toxicity profile of a drug like compounds. Several toxic substructures have been reported from experimental studies, and information about these toxic alerts should be kept in mind while designing a derivative or analog of compounds. We should avoid the addition of toxic group or substructure on a pharmacophore. The toxicity of a compound can be predicted using computational toxicity models generated from the known dataset. If a compound possesses very high toxicity, then it needs to be modified by either changing the scaffold or by changing the substituents of the molecules (Singh 2018).

9.8 Limitations and Future Scope

In silico techniques have supplied numerous and powerful toolbox for screening a large set of compounds, target identification and validation, optimization of compounds and lead molecules, toxicity modeling, and evaluation. Still, it limits further validation and can be only assessed by in vitro and in vivo experiments. Due to poor or unreliable data sets as an input to ADME or drug toxicity may limit the accuracy of outcomes. With the advancement in techniques and reliable datasets as a training set to the prediction, systems will surely increase the reliability of models.

Futuristic drug discovery techniques will highly depend on the accuracy of data models where input will comprise from disease-specific genomic and proteomic expressions, information regarding potential drug targets, active natural compounds, pharmacophore, and physiochemical properties, QSAR, QSPR (Quantitative Structure-Property Relationship), and ADMET models will help us to avoid the causes of drug failure.

9.9 Conclusions

Traditionally computers were extensively used in solving various biological complex problems. Numerous computational tools used in drug discovery approach suggest that the chance of improvisation is always open. New data descriptors based modeling and precise toxicity prediction will help to identify better tolerable drug candidates to market. Novel technologies and computational algorithms are required to move the computer-aided drug designing forward, as new developments are likely to lead to tools for disease identification and the screening of potential lead compounds.

Competing Interest The authors declare that there are no competing interests.

References

Acker MG, Auld DS (2014) Considerations for the design and reporting of enzyme assays in high-throughput screening applications. Perspect Sci 1:56–73

Akagah B, Lormier AT, Fournet A, Figadère B (2008) Oxidation of antiparasitic 2-substituted quinolines using metalloporphyrin catalysts: scale-up of a biomimetic reaction for metabolite production of drug candidates. Org Biomol Chem 6:4494–4497

Alqahtani S (2017) In silico ADME-Tox modeling: progress and prospects. Expert Opin Drug Metab Toxicol 13:1147–1158

Arnott JA, Planey SL (2012) The influence of lipophilicity in drug discovery and design. Expert Opin Drug Discovery 7:863–875

Ashby J, Tennant RW (1988) Chemical structure, Salmonella mutagenicity and extent of carcinogenicity as indicators of genotoxic carcinogenesis among 222 chemicals tested in rodents by the U.S. NCI/NTP. Mutat Res 204:17–115

Balakin KV, Ivanenkov YA, Savchuk NP, Ivashchenko AA, Ekins S (2005) Comprehensive computational assessment of ADME properties using mapping techniques. Curr Drug Discov Technol 2:99–113

Beckett BS (1987) Biology: a modern introduction. Oxford University Press, Oxford

Benigni R, Bossaa C, Jeliazkovab N, Netzevac T, Worthc A (2008) The Benigni/Bossa rulebase for mutagenicity and carcinogenicity – a module of Toxtree. https://ec.europa.eu/jrc/en/publication/eur-scientific-and-technical-research-reports/benigni-bossa-rulebase-mutagenicity-and-carcinogenicity-module-toxtree. Accessed May 2020

Berggren E, Amcoff P, Benigni R, Blackburn K, Carney E, Cronin M, Deluyker H et al (2015) Chemical safety assessment using read-across: assessing the use of novel testing methods to strengthen the evidence base for decision making. Environ Health Perspect 123:1232–1240

Bergström CAS, Larsson P (2018) Computational prediction of drug solubility in water-based systems: qualitative and quantitative approaches used in the current drug discovery and development setting. Int J Pharm 540:185–193

Bocci G, Carosati E, Vayer P, Arrault A, Lozano S, Cruciani G (2017) ADME-space: a new tool for medicinal chemists to explore ADME properties. Sci Rep 7:6359

Botham PA, Hayes AW, Moir D (2002) The international symposium on regulatory testing and animal welfare: recommendations on best scientific practices for acute local skin and eye toxicity testing. ILAR J 43(Suppl):S105–S107

Broach JR, Thorner J (1996) High-throughput screening for drug discovery. Nature 384(6604 Suppl):14–16

Carnero A (2006) High throughput screening in drug discovery. Clin Transl Oncol 8:482–490

Caruthers JM, Lauterbach JA, Thomson KT, Venkatasubramanian V, Snively CM, Bhan A, Katare S, Oskarsdottir G (2003) Catalyst design: knowledge extraction from high-throughput experimentation. J Catal 216:98–109

Cocucci E, Kim JY, Bai Y, Pabla N (2017) Role of passive diffusion, transporters, and membrane trafficking-mediated processes in cellular drug transport. Clin Pharmacol Ther 101:121–129

Cramer RD, Patterson DE, Bunce JD (1988) Comparative molecular field analysis (CoMFA). 1. Effect of shape on binding of steroids to carrier proteins. J Am Chem Soc 110:5959–5967

Czarnik AW, Mei HY (2007) How and why to apply the latest technology. In: Comprehensive medicinal chemistry II. Elsevier, Amsterdam, pp 289–557

Daina A, Zoete V (2016) A BOILED-egg to predict gastrointestinal absorption and brain penetration of small molecules. ChemMedChem 11:1117–1121

Daina A, Michielin O, Zoete V (2017) SwissADME: a free web tool to evaluate pharmacokinetics, drug-likeness and medicinal chemistry friendliness of small molecules. Sci Rep 7:42717

Damale MG, Harke SN, Kalam Khan FA, Shinde DB, Sangshetti JN (2014) Recent advances in multidimensional QSAR (4D-6D): a critical review. Mini Rev Med Chem 14:35–55

Du G, Fang Q, den Toonder JMJ (2016) Microfluidics for cell-based high throughput screening platforms - a review. Anal Chim Acta 903:36–50

Ducki S, Mackenzie G, Lawrence NJ, Snyder JP (2005) Quantitative structure-activity relationship (5D-QSAR) study of combretastatin-like analogues as inhibitors of tubulin assembly. J Med Chem 48:457–465

Egan WJ, Merz KM, Baldwin JJ (2000) Prediction of drug absorption using multivariate statistics. J Med Chem 43:3867–3877

Everitt BS, Dunn G (1992) Applied multivariate data analysis. Oxford University Press, London

Fisher RA (1936) The use of multiple measurements in taxonomic problems. Ann Eugenics 7:179–188

Fritsche E, Crofton KM, Hernandez AF, Hougaard Bennekou S, Leist M, Bal-Price A, Reaves E et al (2017) OECD/EFSA workshop on developmental neurotoxicity (DNT): the use of non-animal test methods for regulatory purposes. ALTEX 34:311–315

Ghose AK, Viswanadhan VN, Wendoloski JJ (1999) A knowledge-based approach in designing combinatorial or medicinal chemistry libraries for drug discovery. 1. A qualitative and quantitative characterization of known drug databases. J Comb Chem 1:55–68

Glickman JF, Schmid A, Ferrand S (2008) Scintillation proximity assays in high-throughput screening. Assay Drug Dev Technol 6:433–455

Gramatica P (2013) On the development and validation of QSAR models. Methods Mol Biol 930:499–526

Guengerich FP (2001) Common and uncommon cytochrome P450 reactions related to metabolism and chemical toxicity. Chem Res Toxicol 14:611–650

Gupta PP, Bastikar VA, Kuciauskas D, Kothari SL, Cicenas J, Valius M (2017) Molecular modeling and structure-based drug discovery approach reveals protein kinases as off-targets for novel anticancer drug RH1. Med Oncol 34:176

Hann MM, Oprea TI (2004) Pursuing the leadlikeness concept in pharmaceutical research. Curr Opin Chem Biol 8:255–263

Hansch C (1969) Quantitative approach to biochemical structure-activity relationships. Acc Chem Res 2:232–239

Huang SM, Strong JM, Zhang L, Reynolds KS, Nallani S, Temple R, Abraham S et al (2008) New era in drug interaction evaluation: US Food and Drug Administration update on CYP enzymes, transporters, and the guidance process. J Clin Pharmacol 48:662–670

Jenkinson DH (1991) Principles of drug action, the basis of pharmacology (3rd edn). Trends Pharmacol Sci 12:77–78

Kirchmair J, Göller AH, Lang D, Kunze J, Testa B, Wilson ID, Glen RC, Schneider G (2015) Predicting drug metabolism: experiment and/or computation? Nat Rev Drug Discov 14:387–404

Kriegel H, Kröger P, Sander J, Zimek A (2011) Density-based clustering. WIREs Data Min Knowl Discovery 1:231–240

Kubinyi H (1988) Free Wilson analysis. Theory, applications and its relationship to Hansch analysis. Quant Struct-Act Relat 7:121–133

Li Pira G, Ivaldi F, Moretti P, Manca F (2010) High throughput T epitope mapping and vaccine development. J Biomed Biotechnol 2010:325720

Lipinski CA, Lombardo F, Dominy BW, Feeney PJ (2001) Experimental and computational approaches to estimate solubility and permeability in drug discovery and development settings. Adv Drug Deliv Rev 46:3–26

Liu B, Li S, Hu J (2004) Technological advances in high-throughput screening. Am J Pharmacogenomics 4:263–276

Liu L, Yang H, Lou Y, Wu JY, Miao J, Lu XY, Gao JQ (2019) Enhancement of oral bioavailability of salmon calcitonin through chitosan-modified, dual drug-loaded nanoparticles. Int J Pharm 557:170–177

Mahfouz NM, Omar FA, Aboul-Fadl T (1999) Cyclic amide derivatives as potential prodrugs II: N-hydroxymethylsuccinimide-/isatin esters of some NSAIDs as prodrugs with an improved therapeutic index. Eur J Med Chem 34:551–562

Mannhold R, Poda GI, Ostermann C, Tetko IV (2009) Calculation of molecular lipophilicity: state-of-the-art and comparison of log P methods on more than 96,000 compounds. J Pharm Sci 98:861–893

Mayr LM, Bojanic D (2009) Novel trends in high-throughput screening. Curr Opin Pharmacol 9:580–588

McNaughton R, Huet G, Shakir S (2014) An investigation into drug products withdrawn from the EU market between 2002 and 2011 for safety reasons and the evidence used to support the decision-making. BMJ Open 4:e004221

Michael S, Auld D, Klumpp C, Jadhav A, Zheng W, Thorne N, Austin CP, Inglese J, Simeonov A (2008) A robotic platform for quantitative high-throughput screening. Assay Drug Dev Technol 6:637–657

Milstein R, Schreyoegg J (2016) Pay for performance in the inpatient sector: a review of 34 P4P programs in 14 OECD countries. Health Policy 120:1125–1140

Montanari F, Ecker GF (2015) Prediction of drug-ABC-transporter interaction—recent advances and future challenges. Adv Drug Deliv Rev 86:17–26

Morales AH, Pérez MAC, Combes RD, González MP (2006) Quantitative structure activity relationship for the computational prediction of nitrocompounds carcinogenicity. Toxicology 220:51–62

Morya VK, Kumari S, Kim EK (2012) Virtual screening and evaluation of Ketol-Acid Reducto-Isomerase (KARI) as a putative drug target for Aspergillosis. Clin Proteomics 9:1

Muegge I, Heald SL, Brittelli D (2001) Simple selection criteria for drug-like chemical matter. J Med Chem 44:1841–1846

Ottaviani G, Gosling DJ, Patissier C, Rodde S, Zhou L, Faller B (2010) What is modulating solubility in simulated intestinal fluids? Eur J Pharm Sci 41:452–457

Parkinson EI, Bair JS, Cismesia M, Hergenrother PJ (2013) Efficient NQO1 substrates are potent and selective anticancer agents. ACS Chem Biol 8:2173–2183

Pavan M (2013) Factors affecting drug distribution. http://www.authorstream.com/Presentation/murari33pavan-971247-factors-affecting-d-murari/. Accessed 10 Jan 2019

Payne MP, Walsh PT (1994) Structure-activity relationships for skin sensitization potential: development of structural alerts for use in knowledge-based toxicity prediction systems. J Chem Inf Comput Sci 34:154–161

Plošnik A, Vračko M, Dolenc MS (2016) Mutagenic and carcinogenic structural alerts and their mechanisms of action. Arh Hig Rada Toksikol 67:169–182

Pratt WB, Taylor P, Goldstein A (1990) In: Pratt WB, Taylor P (eds) Principles of drug action: the basis of pharmacology, 3rd edn. Churchill Livingstone, New York

Prentis R, Lis Y, Walker S (1988) Pharmaceutical innovation by the seven UK-owned pharmaceutical companies (1964-1985). Br J Clin Pharmacol 25:387–396

Raies AB, Bajic VB (2016) In silico toxicology: computational methods for the prediction of chemical toxicity. Wiley Interdiscip Rev Comput Mol Sci 6:147–172

Richard AM, Yang C, Judson RS (2008) Toxicity data informatics: supporting a new paradigm for toxicity prediction. Toxicol Mech Methods 18:103–118

Ritchie TJ, Macdonald SJF, Peace S, Pickett SD, Luscombe CN (2013) Increasing small molecule drug developability in sub-optimal chemical space. Med Chem Commun 4:673

Roy K, Das RN (2014) A review on principles, theory and practices of 2D-QSAR. Curr Drug Metab 15:346–379

Roy K, Kar S, Das RN (2015) A primer on QSAR/QSPR modeling. Springer, Cham

Sakuratani Y, Horie M, Leinala E (2018) Integrated approaches to testing and assessment: OECD activities on the development and use of adverse outcome pathways and case studies. Basic Clin Pharmacol Toxicol 123(Suppl):20–28

Sanches BMA, Ferreira EI (2019) Is prodrug design an approach to increase water solubility? Int J Pharm 568:118498

Savjani KT, Gajjar AK, Savjani JK (2012) Drug solubility: importance and enhancement techniques. ISRN Pharm 2012:1–10

Schlichting I, Berendzen J, Chu K, Stock AM, Maves SA, Benson DE, Sweet RM, Ringe D, Petsko GA, Sligar SG (2000) The catalytic pathway of cytochrome p450cam at atomic resolution. Science 287:1615–1622

Schroeder R (2018) Where are lipids located in the body? https://sciencing.com/lipids-located-body-5387939.html. Accessed 27 Dec 2018

Schuster D, Laggner C, Langer T (2005) Why drugs fail - a study on side effects in new chemical entities. Curr Pharm Des 11:3545–3559

Sharma HL, Sharma KK (2011) What is ion trapping? (Pharmacology). http://prosciencepharma.blogspot.com/2011/10/what-is-ion-trapping-pharmacology.html. Accessed 29 Jan 2019

Sharom FJ (2008) ABC multidrug transporters: structure, function and role in chemoresistance. Pharmacogenomics 9:105–127

Shaw PJA (2003) Multivariate statistics for the environmental sciences. Oxford University Press, London

Shore PA, Brodie BB, Hobgen CA (1957) The gastric secretion of drugs: a pH partition hypothesis. J Pharmacol Exp Ther 119:361–369

Singh DB (2018) Natural lead compounds and strategies for optimization. In: Ul-Haq Z, Wilson AK (eds) Frontiers in computational chemistry. Bentham Science, Sharjah, pp 1–47

Singh DB, Dwivedi S (2019) Computational screening and ADMET-based study for targeting Plasmodium S-adenosyl-L-homocysteine hydrolase: top scoring inhibitors. Netw Model Anal Health Inform Bioinform 8:4

Singh S, Malik BK, Sharma DK (2006) Molecular drug targets and structure based drug design: a holistic approach. Bioinformation 1:314–320

Sliwoski G, Kothiwale S, Meiler J, Lowe EW (2014) Computational methods in drug discovery. Pharmacol Rev 66:334–395

Stephens ML (2010) An animal protection perspective on 21st century toxicology. J Toxicol Environ Health B Crit Rev 13:291–298

Stoner CL, Troutman MD, Laverty CE (2008) Pharmacokinetics and ADME optimization in drug discovery. In: Cancer drug design and discovery. Academic Press, New York, pp 131–153

Szakács G, Váradi A, Ozvegy-Laczka C, Sarkadi B (2008) The role of ABC transporters in drug absorption, distribution, metabolism, excretion and toxicity (ADME-Tox). Drug Discov Today 13:379–393

Tate J, Ward G (2004) Interferences in immunoassay. Clin Biochem Rev 25:105–120

Testa B, Krämer SD (2007) The biochemistry of drug metabolism – an introduction. Chem Biodivers 4:2031–2122

Thomas G (2007) CH5 combinatorial chemistry. In: Medicinal chemistry: an introduction. Wiley, Chichester, p 170

Thomas G (2008) Medicinal chemistry: an introduction, 2nd edn. Wiley, Chichester

Tropsha A (2010) Best practices for QSAR model development, validation, and exploitation. Mol Inf 29:476–488

Vallianatou T, Giaginis C, Tsantili-Kakoulidou A (2015) The impact of physicochemical and molecular properties in drug design: navigation in the "drug-like" chemical space. Adv Exp Med Biol 822:187–194

Van de Waterbeemd H, Testa B (2007) The why and how of absorption, distribution, metabolism, excretion, and toxicity research. In: Comprehensive medicinal chemistry II. Elsevier, Amsterdam, pp 1–9

van Waterschoot RAB, Schinkel AH (2011) A critical analysis of the interplay between cytochrome P450 3A and P-glycoprotein: recent insights from knockout and transgenic mice. Pharmacol Rev 63:390–410

Vasavid P, Chaiwatanarat T, Pusuwan P, Sritara C, Roysri K, Namwongprom S, Kuanrakcharoen P et al (2014) Normal solid gastric emptying values measured by scintigraphy using Asian-style meal: a multicenter study in healthy volunteers. J Neurogastroenterol Motil 20:371–378

Veber DF, Johnson SR, Cheng HY, Smith BR, Ward KW, Kopple KD (2002) Molecular properties that influence the oral bioavailability of drug candidates. J Med Chem 45:2615–2623

Veith H, Southall N, Huang R, James T, Fayne D, Artemenko N, Shen M et al (2009) Comprehensive characterization of cytochrome P450 isozyme selectivity across chemical libraries. Nat Biotechnol 27:1050–1055

Verma J, Khedkar VM, Coutinho EC (2010) 3D-QSAR in drug design—a review. Curr Top Med Chem 10:95–115

Virupaksha B, Alpana G (2012) CoMFA QSAR models of camptothecin analogues based on the distinctive SAR features of combined ABC, CD and E ring substitutions. Comput Biol Med 42:890–897

Wildey MJ, Haunso A, Tudor M, Webb M, Connick JH (2017) High-throughput screening. In: Annual reports in medicinal chemistry. Elsevier, Amsterdam, pp 149–195

Wolf CR, Smith G, Smith RL (2000) Science, medicine, and the future: pharmacogenetics. BMJ 320:987–990

Yang H, Lou C, Sun L, Li J, Cai Y, Wang Z, Li W, Liu G, Tang Y (2018) admetSAR 2.0: web-service for prediction and optimization of chemical ADMET properties. Bioinformatics 35 (6):1067–1069

Zhong Y, Guo Z, Che J (2015) Data quality assurance and statistical analysis of high throughput screenings for drug discovery. In: Frontiers in computational chemistry. Bentham Science Publishers, Sharjah, pp 389–425

Advances in Pharmacophore Modeling and Its Role in Drug Designing

10

Priya Swaminathan

Abstract

Pharmacophore modeling is a central method in the ligand-based drug designing module. Its basis lies in developing a scaffold or an empirical molecule based on a group of known inhibitors to a target. The empirical molecule will contain features that are common to the known inhibitors and specified as donors, acceptors, rings, positively charged, or negatively charged. These five features or a combination of some of these features at specific distances make a pharmacophore. This pharmacophore facilitates the identification of other novel compounds that are specific and sensitive as well as effective inhibitors to a receptor. This method is particularly effective when the structural annotations are unavailable for the target. Thus pharmacophore modeling is a tool in drug discovery where screening of the pharmacophore built leads to the discovery of novel compounds against the target. Using these techniques as well as variations of these techniques, millions of compounds can be screened in a matter of hours to shortlist actives. Variations might be based on building a pharmacophore by the energy contribution of features in a single molecule against a specific target. Otherwise, based on only the geometric features of the active site in a target, a pharmacophore can be designed. Thus a designed pharmacophore can be used to screen novel agonists and antagonists that are specific to targets, to screen toxicants, to identify unknown targets, and to screen out best molecular docking results.

Keywords

Pharmacophore · Features · Antagonist · Actives · Drug · Structural annotations

P. Swaminathan (✉)
Department of Biotechnology, SRM Institute of Science and Technology, Kattankulathur, Tamilnadu, India
e-mail: priyas@srmist.edu.in

D. B. Singh (ed.), *Computer-Aided Drug Design*,
https://doi.org/10.1007/978-981-15-6815-2_10

10.1 Introduction

Conventional drug discovery is a very tedious process with higher expenditure and lower success. The meaning of conventional is synthesizing novel compounds in a chemical lab and then developing an assay to check its efficacy against a said target. Under such high stakes, the pharma industries, as well as the disease specialist, were forced to find other cost-effective means to find new cures. Computer-aided drug designing (CADD) deals with finding new drugs against a target to be less random and more efficient (Dalkas et al. 2013). The rationale is based on the structure of the target active site or the structure of the molecules active against the target structure. Both these approaches tend to seek diverse or similar compounds that might be active against the target. A pharmacophore is a group of features that spell specificity against a target. Its advent was propelled when structure activity based analysis gained momentum in CADD (Böhm 1993). A definition that is widely used by medicinal chemists states that a pharmacophore is the ensemble of steric and electronic features that are necessary for the optimal supramolecular interactions with a specific biological target structure and to generate a biological response (Testa 2012).

Pharmacophore modeling is popular in the industry as it chalks a roadmap of preferred as well as unsought functional groups that are to yield good results. The first pharmacophore was built in as early as, 1940 where a 2D pharmacophore was developed to inhibit tetrahydrofolic acid. The basis of the pharmacophore was p-aminobenzoic acid (PABA), which was substituted with a sulphonamide at a distance of 2.4 Å from the para amine group (Woods 1940). This and many other studies led to the development of pharmacophores. The realization of the empirical features of a group of compounds is called pharmacophore mapping. Pharmacophore mapping is a prime method under ligand-based drug design (LBDD) where a group of active compounds is superimposed together to gather 3-dimensional (3D) empirical feature coordinates that spells out features that are active in the original group of compounds. The pharmacophore helps to fetch other diverse compounds with similar distanced features.

10.2 Features in a Pharmacophore

Pharmacophore is not chemical molecules, but just specification of the physico-chemical features desired in a structure. A typical pharmacophore will be a 3D coordinate system specified feature table. The pharmacophore can be considered as the largest common substructure shared by a set of active molecules.

The most common features in the feature table are shown in Fig. 10.1.

- Hydrogen bond donor (HBD).
- Hydrogen bond acceptor (HBA).
- Aromatic Rings(R).
- Hydrophobic groups(H).

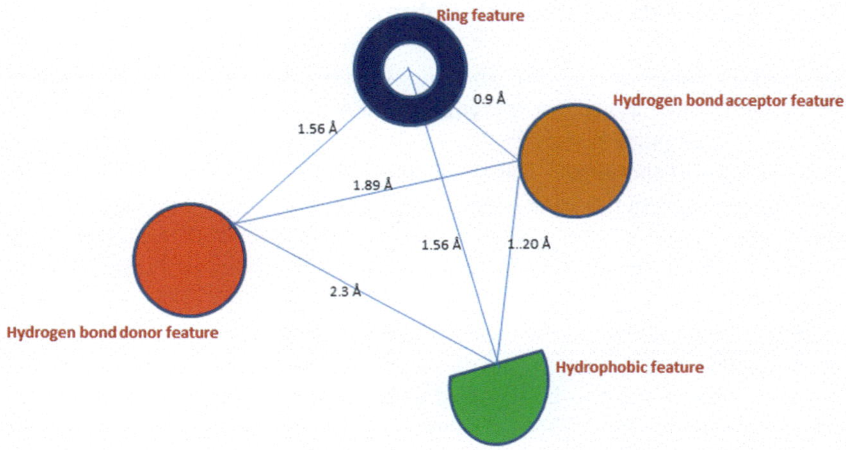

	-	1.56	1.89	2.3
	1.56	-	0.9	1.56
	1.89	0.9	-	1.20
	2.3	1.56	1.20	-

Fig. 10.1 A typical pharmacophore (cartoon representation)

- Positively charged groups(P).
- Negatively charged groups (N).

Other features like chiral centres, bulky groups, metal ions, and solvation penalty areas can be indicated.

These features are essential for understanding the active site of the protein if the protein structure is unavailable. The pharmacophore obtained is a mirror image of the active site (Soliman 2013). As the features help to identify the amino acids and the specific type of interaction it could have with a small molecule. A hydrogen bond acceptor in the pharmacophore specifies a hydrogen bond donor in the active site like serine or threonine. A ring feature specifies a pi-pi type of interacting group in the active site from an aromatic amino acid.

10.3 Pharmacophore Modeling

The two distinct types of pharmacophore models are ligand-based and structure-based pharmacophore models (Kaserer et al. 2015).

10.3.1 Ligand-Based Pharmacophore

This method is used specifically when the structure of the protein macromolecule is absent. Under such a scenario, a group of a known active small molecule is aligned together to find common features that would help to find other molecules that could be active against the target. The small molecules are conformationally rotated to get alternate forms of the same ligand. This is then stored into a database of ligands, which is superimposed to get features common among the set. The alternate forms of the same ligand are created to take into account the flexibility of a small molecule in a reaction.

The resultant model is a 4–7 featured pharmacophore that specifies the required properties that a novel new small molecule should have against the target. This pharmacophore is valid to be searched against another ligand database to identify new active molecules. The limitation of this method is finding additional conformations for each ligand (Yang 2010). Different algorithms such as Monte Carlo and genetic algorithms are applied to generate the additional conformations of the proteins. Another big limitation is that the conformation used for the pharmacophore might not be a free energy-based active form of the ligand. This risk is carried by all ligand-based pharmacophore.

10.3.2 Building a Pharmacophore

Pharmacophore modeling uses certain structured steps to build a rational scaffold that is useful to find other chemical moieties that have the same property against the target of interest in the disease (Fig. 10.2).

The steps include:

- Ligand preparation.
- Pharmacophore feature mapping.
- Searching for common pharmacophore.
- Scoring the common pharmacophore.

10.3.2.1 Ligand Preparation

Ligand preparation is a prerequisite for building a pharmacophore. This step provides a group of compounds in its active conformations. The catch in this step is to predict the conformationally active one. This knowledge is generally not known for most chemical compounds. The torsion angles of actives that are taken for pharmacophore development are rotated to get different conformations of each active, and among them, the thermo stable conformations are retained (Merz et al. 2010). This preparation yields plausible low-energy minima conformations of the actives that are near real to the actual conformation of the ligands. Thus any pharmacophore tool will incorporate an algorithm to generate conformers and a force field to predict the energy minima of each conformer.

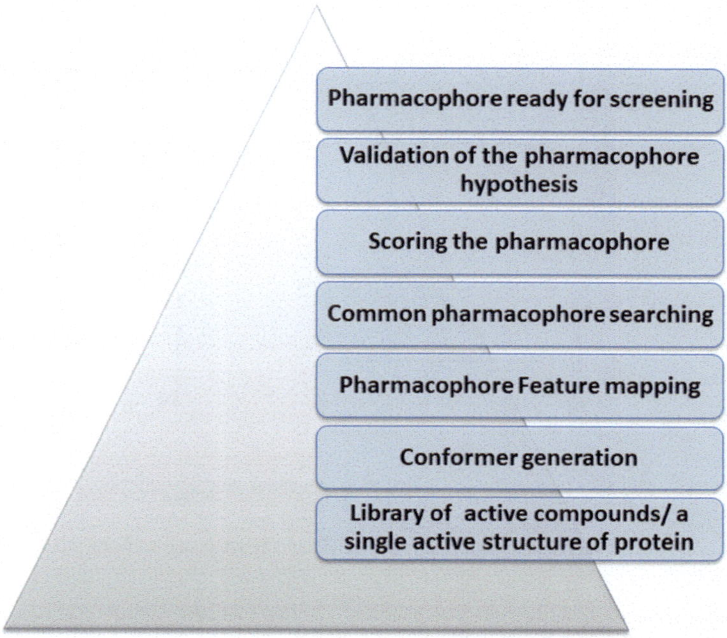

Fig. 10.2 Steps in pharmacophore modeling

In PHASE, the different conformers are generated by using a Monte Carlo multiple minimum (MCMM) sampling methods to explore actives for the torsion angles that are possibly rotatable. Then using the optimized potential for liquid simulations (OPLS) force field, the energy minima is obtained.

A typical structure with more than 10 rotatable bonds can generate tens of thousands of meaningful structures. But this can be computationally taxing, employing multiple days to months of CPU time. Thus, under such cases where databases of million compounds have to be handled, an empirical torsion sampling algorithm is employed to give not only the most accurate energy minima but, in the near range enabling lesser CPU time consumption (Dixon et al. 2006).

10.3.2.2 Pharmacophore Feature Mapping

In this step of the pharmacophore development, the tool is having a set of a prepared conformationally active dataset in hand. Now the compounds have to be individually screened for ligand binding features such as non-bonded interactions such as hydrogen bonding and hydrophobic bonding groups. Each structure has to be individually screen to map out its groups that might participate in the bond formation in the active site of the protein. The most common features, as mention earlier, are hydrogen bond donors and acceptors, ring atoms, atoms that can take part in hydrophobic interactions, as well as charged atoms that are useful for electrostatic interactions.

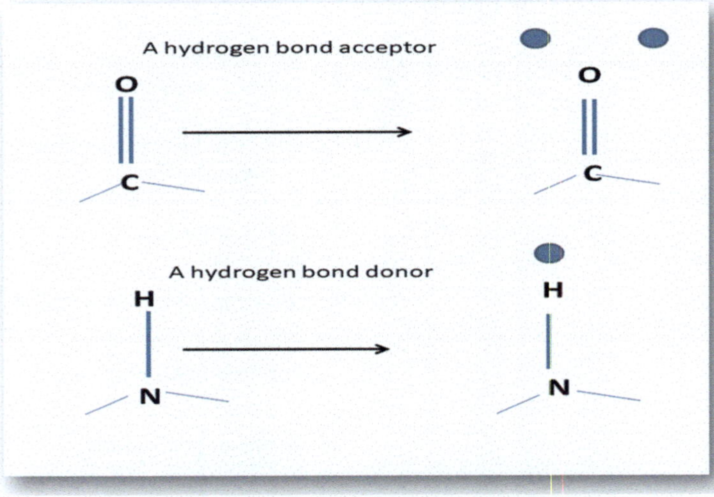

Fig. 10.3 The vector representation of a typical HBD and HBA in the form of a coloured sphere and its position

The mapping will result in identifying interacting groups in a conformer as well as a vector to indicate the directionality of the interaction.

In a tool like PHASE, the rules needed to identify features are given as SMARTS to an algorithm called feature definition. These feature definitions can be customized to the user's viewpoint, and only those features will be recognized for the pharmacophore. SMARTS (Daylight systems) is a format to describe feature patterns in computational chemistry. It is a simple set of rules that help to perceive and develop new pharmacophore patterns to search by the user. The pharmacophore feature sites, as well as vectors, are represented uniquely in PHASE with points indicating their interaction directionality. Figure 10.3 shows how a typical hydrogen bonding will be represented after a mapping step. Further, the mapped pharmacophore for a complete molecule is given in Fig. 10.4 taken from Levit et al. (2011), who used a ligand-based pharmacophore approach to elucidate binders for prokineticin receptors.

10.3.2.3 Searching Common Pharmacophore

This step involves searching the features exhaustively in the other actives to perceive a common feature list with a similar distance vector. These common features have to be searched in all the conformers of the dataset and then some features are shortlisted for being at a similar cut off distance from other features.

In PHASE, the algorithm starts with a k-point feature search in all the actives of the dataset. Bear in mind the k-points are zeroed in based on a set of features that are common in a small subset of active conformations. This is an exhaustive and recursive search that employs heuristic algorithms to search faster. The tolerance of the match can be user-defined. Thus, now this K-point feature list that matches a

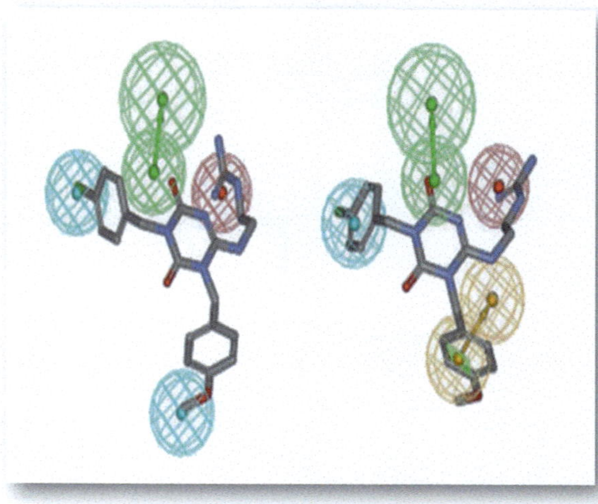

Fig. 10.4 The two conformations of a chemical structure and the pharmacophoric features in it (Levit et al. 2011). Green spheres represent hydrogen bond acceptors, red—positive ionizable, light blue—hydrophobic, and orange—aromatic ring

minimum subset of actives is passed through a tree-based partitioning algorithm where the tree is divided based on the intersite distance of the features. This tree partitioning checks whether every feature shares a common distance with other features (Lin 2000).

Suppose a partitioning tree is used to check the distance between an HBD and R feature for an intersite distance range of 1–6 Å in a subset of 30 active conformers, the tree would look something like in Fig. 10.5.

Here, those actives that fall in the same partition box of the tree or the neighbouring box of the tree can be accountably treated as a feature of a common pharmacophore. So supposedly, if all the 30 actives in the subset have an intersite distance between 2.6 Å and 2.9 Å, then in the partition tree, they would arrive in the box node $2 < d < 3$. Since this pair of the feature is contained in one box or its neighbouring box (sometimes due to overlapping conditions), thus these features are perceived as a common feature in the pharmacophore.

This partitioning is continued for all pairs of features in the k-site pharmacophore, and those that pass the partition tree are gathered as common pharmacophore. At this point, more than one common pharmacophore can be obtained.

10.3.2.4 Scoring the Common Pharmacophore

The surviving nodes in the partition tree are known to contribute as a common feature in the pharmacophore. But this cannot be taken on the face value as false positives may arise. For example, the mirror image of an active also gives rise to a surviving node in the tree. Thus these false positives have to be removed by comparing it to a reference score.

The reference score in the PHASE tool is a score based on RMSD and the average cosine of the angles formed by the corresponding features. Each pharmacophore from the surviving node in the tree is called as reference pharmacophore and is

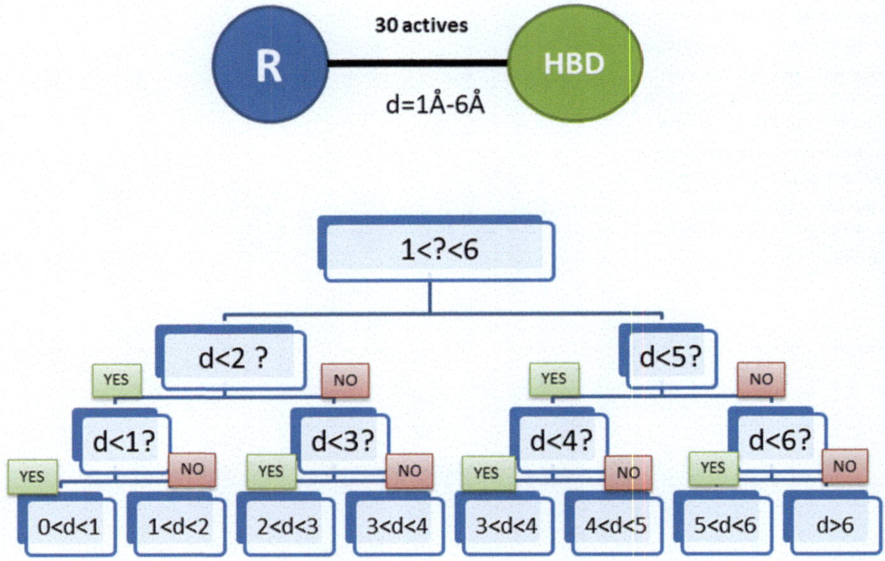

Fig. 10.5 A partition tree for an intersite distance of 1–6 Å between two hypothetical features of ring atom and hydrogen bond donor in 30 actives subset

compared with all other pharmacophore pairs in the surviving node of the partition tree (called as non-reference pharmacophore). The RMSD is calculated, and a cut-off is set (default is 1.2 Å), which helps to identify good reference pharmacophore as well as avoid false positives. The calculations are done to find a reference score. Equation (10.1) is the site vector score where the parameters wsite, wvector, and cut-off RMSD are used, and these parameters are user-adjustable. Propref in Eq. (10.2) is the value of the conformationally independent property for the ligand contributing to the reference pharmacophore.

$$\text{Site\_Vector\_Score}_i = w_{\text{site}}\text{Site\_Score}_i + \text{w}_{\text{vector}}\text{Vector\_Score}_i \tag{10.1}$$

Where

$$\text{Site\_Score}_i = 1 - \text{RMSD}_i/\text{cutoff}_{\text{RMSD}}$$

$$\text{Vector\_Score}_i = \frac{1}{n_v}\sum_{j=1}^{n_v}\cos\theta_{ij}$$

$$\text{Reference\_Score}_i = \text{Site\_Vector\_Score}_i + \text{w}_{\text{prop}}\text{Prop}_{\text{ref}} \tag{10.2}$$

The reference score for each of the 30 ligands in the surviving node is calculated, and the one that gives the highest becomes the reference ligand. Thus all partition trees are scored, and the reference score is refined iteratively. At this point, multiple reference ligands with equivalent reference scores might occur, where the user has to use some selectivity scores or the number of total actives matched to choose a good pharmacophore. The pharmacophoric sites in the chosen reference ligand are now called the pharmacophore hypothesis.

All of the above steps will be carried out by the pharmacophoric tools with slight variations like different algorithms and extra steps to improve the sensitivity of the hypothesis.

10.3.3 Algorithms Used to Build a Pharmacophore

The algorithms used to build ligand-based pharmacophore are

- Phase.
- Catalyst.
- Dantes.
- Disco.
- Galahad.

Phase algorithm uses a binary decision tree to cluster the 3D Cartesian space of ligands. It also produces a scorecard with geometric and heuristic scores to allow the user to choose the features in the pharmacophore model. The structure-based module in phase allows the generation of an energy-based pharmacophore that calculates the free energy contributions of each feature in the ligand and allows the user to choose the features for the pharmacophore (Van Drie 2012).

Catalyst uses two algorithms, Hypogen and HipHop algorithms, which develop the pharmacophore based on features with or without related IC_{50} values of the ligands, respectively. It gives an empirical value called as a cost, which allows the user to correlate structure to activity. Dantes uses a special algorithm to identify regions in the coordinate space of ligands that are intersecting with the coordinate space of all other ligands. Disco uses clique detection to identify all possible candidate pharmacophores as well as allows the user to sort the best pharmacophore.

Galahad, on the other hand, finds 3D similarity-based scores to manually overlay the molecules, and build a hyper molecule which has a less cost function to it. The hyper molecule with the least cost is the pharmacophore (Van Drie 2012).

10.3.4 Structure-Based Pharmacophore

The structure of the macromolecular target is used to locate the features for the pharmacophore. The features are based on the single X-ray crystallized target-ligand structure. The single ligand and its interactions with the protein are used to build the pharmacophore features. The main other difference between the former and the latter methods are the numbers of ligands used to build the pharmacophore. The ligand-based method requires a minimum of 30 actives, while structure-based can be done with a single ligand and its interaction map with the receptor. Also, the pharmacophore in the latter method is from an active conformation of the ligand (Qing et al. 2014).

Another way to build a structure-based pharmacophore is by using an APO target structure where the active site amino acids are identified, and based on their interaction property a feature list is generated that is suitable to be used in the pharmacophore (Thangapandian et al. 2011). The only limitation is where too many features are predicted in the list (greater than 7 features). Under such cases, the features have to be selected based on weight-based analysis or knowledge-based analysis using additional algorithms.

The structure-based method is unique, with its pre-set of the X-ray structure of a ligand–protein complex. This structure availability cuts down the step of finding active conformers of the ligand in the ligand-based method. Thus the steps involved are:

- Redocking of the co-crystal ligand.
- Scoring for pharmacophoric sites.
- Building a pharmacophoric hypothesis.

10.3.4.1 Redocking of Co-Crystal Ligand

The PDB protein structure with the bound ligand is used as the starting material. The bound state of the ligand gives an idea on the active conformer of the ligand, thus the prediction of the correct conformer is unwarranted in this type of method. However, redocking helps to calculate the energy related to each group in the ligand. Also, the refinement in the interactions of the ligand with the protein can be accurately measured. Structure-based pharmacophore is also known as energy-based pharmacophore (E-Pharmacophore). The name stands for building a hypothesis based on the energy contribution by each of the functional groups in the ligand. This is calculated by a docking program. In the E-Pharmacophore module of Schrodinger, the docking is performed using the Glide tool, and then the docked output is imported into the E-Pharmacophore module to predict the hypothesis (Pirhadi et al. 2013).

10.3.4.2 Scoring for Pharmacophoric Sites

The E-Pharmacophore module imports the docked file and assigns the energy function to each pharmacophore feature. The rule in this module of Schrodinger is that any functional group that has less than half of its heavy atom contributing to the glide score is not considered to be pharmacophoric features. For example, a ring structure with only two of its 5/6 atoms is contributing to the Glide score can be disregarded as a pharmacophoric feature. Thus, only the functional groups with a maximum contribution of energy to the glide score are considered for the pharmacophore hypothesis (Loving et al. 2009).

10.3.4.3 Building a Pharmacophoric Hypothesis

The identified features are ranked according to their glide contributions. The features that have a maximum contribution to the glide score are ranked first, and the features that are low contributors are ranked last. In this way, at least a list of a minimum of 3 or a maximum of 7 features will be generated. All the features predicted in the list can be used to build a hypothesis, or the user can use his/her discretion to leave out some features in the hypothesis. This will generate a single hypothesis along with its intersite distance that marks as the scaffold for a chemical structure that might be a suitable ligand for the given target (Langer and Wolber 2004).

10.4 Tools for Pharmacophore Building

There are many tools available for pharmacophore modeling or the mapping of common features. Each of the tools has its unique features in the way the pharmacophore is represented or in the way the algorithm is used to find common features (Sanders et al. 2012; Fei et al. 2013).

The pharmacophore generated by these tools can be cited for further studies that help to search new novel ligands for receptors or targets.

10.5 Validation of a Pharmacophore Hypothesis

Once the features have been selected, the pharmacophore is called a hypothesis. This hypothesis has to be tested for its robustness. The pharmacophore is validated for the features that were chosen.

The most common method is to take the actives from which the pharmacophore was built and mix it with a set of decoy molecules. The rationale is like a bootstrapping method where thousands of decoy molecules will be made to screen along with real active molecules to check the accuracy of the pharmacophore in screening out the actives among the decoys (Mysinger et al. 2012).

Decoys are computed based on similar physical properties but different chemical structures. For each active, 50 decoys with similar 1-D physicochemical properties to remove bias (molecular weight and calculated LogP) but dissimilar 2D topology to be likely non-binders are included (Li et al. 2015). This type of decoys can be obtained from the decoy database called directory of useful decoys and enhanced (DUD-E) (Mysinger et al. 2012). The compounds from the database can be downloaded, and a virtual screening protocol can be used to obtain hits using these compounds along with the actives. The ratio of actives in the top 1% of the total hits indicates the enrichment factor (EF) of the pharmacophore hypothesis. EF indicated in percentage where a high percentage means maximum actives were recovered in the top 1% of the hits, and then the pharmacophore is validated as efficient.

Another method called Güner–Henry (GH) scoring method is also used to indicate the efficiency of the pharmacophore in screening actives only.

$$\text{GH SCORE} = \left(\left(\frac{Ha}{4HtA}\right)(3A + Ht)\right) \times \left(1 - \frac{(Ht - Ha)}{(D - A)}\right)$$

where D is the total number of molecules in the database. A is the total number of actives in the database. Ht represents the total number of hit molecules from the database, and Ha represents the total number of active molecules in the hit list. An ideal score will be closer to 1 and ideally above 0.7. A value closer to zero indicates a null hypothesis.

The other measure of a robust pharmacophore is a receiver operating characteristic (ROC), which predicts a curve that can discriminate true positives and false positives, i.e. (sensitivity) and (1-specificity) in the X- and Y-axis, respectively. The curve formed that is closer to the Y-axis indicates 0 or near-zero false positives and higher true positives (Wang and Chen 2013). If the curve was farther away from the y-axis, then it would suffice to construe that the hypothesis is not efficient enough to screen true positives. Also, a term called area under the curve (AUC) helps to quantify the ROC results. AUC, when closer to 1 indicates a sensitive pharmacophore.

$$\text{AUC} = \sum i \left[\frac{(Se_{i+1})(Sp_{i+1} - Sp_i)}{2}\right]$$

where Sp_i and Se_i are specificity and sensitivity values of the ith data. In other words, the area under the curve is a sum of areas of all rectangles formed below the curve (Triballeau et al. 2005).

10.6 A Case Study of Structure and Ligand-Based Pharmacophore

Matrix metalloprotease is a zinc metal based enzyme that degrades collagen and implicates itself in pannus formation in rheumatoid arthritis. The authors use a combination of structure-based and ligand-based techniques to design and screen potent non-zinc binding inhibitors of MMP8 (Kalva et al. 2014).

Step I—Generation of Selective Pharmacophore Hypothesis

In the Phase tool, using the two ligands of MMP8 which are non-zinc binding, a seven point structure-based pharmacophore was built (AADDRR). This seven point pharmacophore was refined using ligand-based pharmacophore knowledge like an acceptor and a donor of the seven featured pharmacophore shared by a conserved residue in all MMPs were excluded to avoid non-specific binding. Another ring feature which did not form any specific interaction in the S1 loop of MMP8 was removed. Further excluded volumes were added to avoid ligands that might bind to zinc metal, as metal based docking analysis is highly unreliable. Finally, a hydrophobic group was added to the pharmacophore based on already known ligand studies of MMP8. This leads to a refined five featured pharmacophore with excluded volumes included (AADRH).

Step II—Validation of the Pharmacophore Hypothesis
Already known zinc metal binding and non-zinc metal binding ligands were pooled to make a 181 compound database and then using the AADRH pharmacophore to screen them. The pharmacophore screened out 15 compounds out of which the two known non-zinc binding inhibitors showed top fitness scores. This proves the efficacy of the pharmacophore to screen specific non-zinc binding compounds.

Step III—Database Screening, ADME, Docking, and Dynamics
The validated pharmacophore was screened with the ZINC database and 1000 compounds were obtained as hits. Using these ADME properties, 81 compounds were shortlisted and further these compounds were docked and validated with other MMP proteins. It was seen that 6 ZINC hits passed the tests among which ZINC 00673680 showed stable binding through validation and molecular dynamics. Thus the paper indicates a probable inhibitor for MMP8, and also sets a common pharmacophore that is specific to MMP8, non-zinc metal binding, and has only useful features using Pharmacophore mapping.

10.7 Uses of Pharmacophore

The importance of pharmacophore lies in finding new novel actives with the same physicochemical groups in its structure as in the hypothesis. Figure 10.6 gives the various avenues of searching the hypothesis to obtain novel compounds that are active against a receptor. The other way to phrase this is that a pharmacophore gives an insight into important interactions in the active site of the target. This is helpful to find new or novel compounds against the target. Based on this insight, applications of pharmacophore are:

10.7.1 Virtual Screening

The pharmacophore hypothesis is screened against a 3D database of chemical structures to find hits that match the template of the pharmacophore. It is used as a screening tool, which virtually clusters compounds that are similar in activity to the known actives. The hits receive a fitness score, which gives the extent of matching with the pharmacophore (Horvath 2011). The fitness score is a linear combination of site and vector alignment scores and the volume scores. A good hit to the hypothesis means, a fitness score nearest to 3 while a bad hit would have a score lesser than 1. The general norm is to take a cut-off of 1.5 and above as good hits (Kalva et al. 2016). One of the main factors that are unique in pharmacophore screening is that the molecules screened out are not necessarily belonging to the same class or analogs of

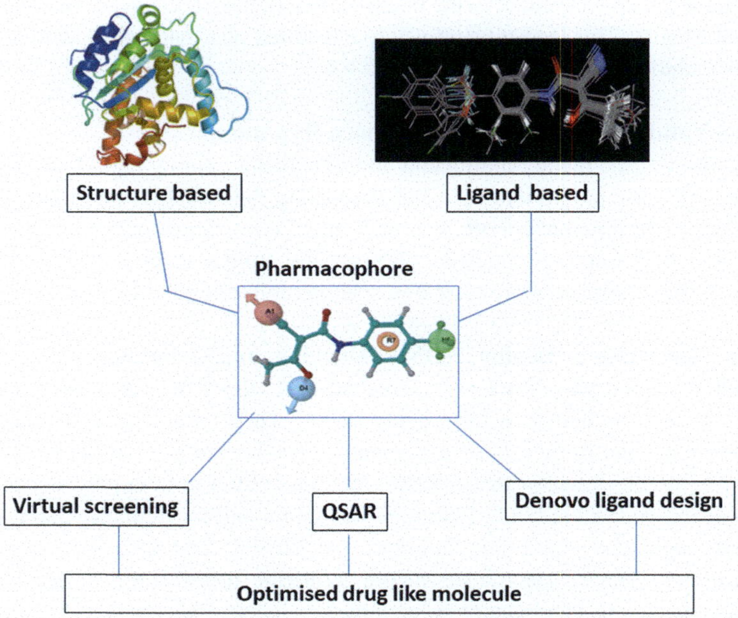

Fig. 10.6 Application of pharmacophore modeling

Table 10.1 Pharmacophore mapping tools

S. No.	Tool	Company	Availability
1	Phase	Schrödinger	Paid
2	Catalyst	Accelrys-Biovia	Paid
3	Sybyl	Tripos-Certara	Paid
4	OSPPREYS	MOE	Paid
5	PharmaGist	Tel Aviv University	Web server
6	LigandScout	Inte:ligand	Paid
7	Align-It	Silicos-it	Free

each other, hence not structurally related to the original actives. This leads to the clustering of a diverse group of compounds having an affinity towards the same active site. Another challenge in virtual screening is the conformers of the database structures. Depending upon the size of the database, each structure has to be energetically minimized by generating conformers. This task is no less complicated than the conformer generation step in ligand-based pharmacophore.

The tools in Table 10.1, along with building a pharmacophore hypothesis, also perform virtual screening. While the tools Pharmer (Koes and Camacho 2011) and Molsign (VLifeMDS 2010) allow only virtual screening to be performed on an already existing pharmacophore hypothesis.

10.7.1.1 An Instance of Virtual Screening and Its Workflow

In this work, a structure-based pharmacophore was built based on a known inhibitor of dihydroorotate dehydrogenase, a target for rheumatoid arthritis. This pharmacophore hypothesis had four features described in it. The hypothesis was validated using decoys, and further screened in the KEGG phytochemical subset. Eighteen hits were further filtered through docking protocols to obtain four diverse compounds from natural origins that have features important to inhibit dihydroorotate dehydrogenase (Swaminathan et al. 2014).

10.7.2 Pharmacophore Fingerprint

In this day and age, where technology gives unbound power to virtually screen and shortlist compounds, and move from lead to drug development in a short period, there is a need to quickly search for similarity between ligands (Choudhari et al. 2012). The computational expense of virtual screening is very high therefore, the natural language of bits and bytes are used to store the pharmacophoric features and then used for similarities searches. This method saves time and memory for the computational process. Initially, it starts with a triplet based hypothesis, where P1, P2, and P3 can be any of the pharmacophoric features like HBD, HBA, R, P, and N. The set of 3 edges are given different distance ranges between 6 and 14 ranges based on current knowledge. Now, using the different edges, the pharmacophore is converted to a bit map for each structure in the database. The bitmaps are weighted sometimes based on the hydrophobic capacity. Triplet fingerprints generated with the parameter set of 3 vertices of 5 types and 3 edges of 14 possible lengths imply 53 × 143 distinct combinations, including the geometrically impossible ones, which will be 275,674 bits in length. These bits can be shortened to bitmaps where the number of zeroes followed after a 1 can be summed up like this 0,100; 1,1; 0,200; 1; 0, 1000;. . .where "," followed by a number represents the number of times the number is repeated. This bitmap can then be matched with other bitmaps in the database to get a Tanimoto coefficient indicating similarity between the fingerprints (Juan Alvarez 2005). This bit similarity by a Tanimoto score is always between the ranges 0 and 1, and the 1 indicates the highest similarity. Tanimoto coefficient for two molecules A and B (Salim and Kinghorn 2008),

$$\mathrm{SIM}_{A,B} = c/a + b - c$$

c: bits set common in the two fingerprints.a and b: bits set in the fingerprints for A and B respectively.$\mathrm{SIM}_{A,B}$: Tanimoto coefficient for the similarity between molecules A and B.

The same can be applied to molecules in the training set of QSAR to obtain a partial least equation (McGregor and Muskal 1999). An algorithm like PharmPrint or a tool like ChemAxon is efficient to perform pharmacophore fingerprint screening (McGregor and Muskal 2000). These fingerprints can also be used as a descriptor in comparative molecular field analysis (COMFA) and comparative molecular similarity index analysis (COMSIA) analysis of QSAR (Verma et al. 2010).

10.7.2.1 An Instance of Pharmacophore Fingerprint Searching

In this 2013 work, the authors have compared manual 2D fingerprints and 3D pharmacophore fingerprints to search and cluster the most active molecules for a rational pharmacophore hypothesis. 5-HT1A receptor ligands (serotonin receptor) were used to cluster the actives into different groups. The workflow was applied in such a way that each clustering method tried to cover a different aspect in grouping the actives. One method among them was based on 3D fingerprints using the Canvas tool of Schrodinger. The Canvas tool used the pharmacophoric fingerprints of each 5-HT1A receptor to group them into 28 clusters. The segregation into clusters is based on the Tanimoto coefficient for any two pharmacophoric fingerprints. Each cluster had a chosen representative molecule selected as a pharmacophore hypothesis. These approaches lead to the discovery of new novel binders of serotonin receptors (Sharma et al. 2016).

10.7.3 De Novo Ligand Design

De novo ligand designing is building a chemical structure from scratch. Pharmacophoric features will help in the development of a new molecule that does not coincide with any patented molecules or already known toxic molecules (Warszycki et al. 2013). A set of disconnected molecular fragments that are specified by a pharmacophore hypothesis is joined by linkers such as chains, rings, or atom moieties. The pharmacophore acts as a feature guide to build novel molecules by taking fragments to match every feature of the pharmacophore and then adding linkers to make whole molecules (Hartenfeller and Schneider 2011). The problem here lies in finding the steric imbalance regions in the active site as well as synthesizing the molecule built by the de novo program. The knowledge of the receptor-binding region is a must to design new compounds. Also, abstract pharmacophoric features will not be useful for ligand construction. LUDI is a well-known tool for de novo synthesis based on pharmacophore (Böhm 1993).

10.7.3.1 An Instance of De Novo Ligand Design

In 2008, a few researchers employed fragment-based drug design using pharmacophore to build a novel ligand against cannabinoid receptors to combat obesity. Fragments were created from known cannabinoid receptor ligands, and then these ligands were pharmacophorically classified into different regions in the receptor site. These were then linked together to give novel ligands that are antagonists or inverse agonists to the constitutive drug designing (Alig et al. 2008).

10.8 Success Stories in Pharmacophore-Based Drug Designing

Pharmacophore-based drug designing is commonly seen in many drug designing projects. The pharmaceutical companies use this methodology either to discover new targets for already available drugs or to find new candidates for already known targets.

Renin is a good target in the renin-angiotensin pathway for hypertension. This rational drug design target was used by Novartis pharmaceuticals to develop a renin inhibitor. This inhibitor had to be a non-peptidic source as peptide inhibitors have less pharmacokinetic properties. Using the structure of the S3-S1 pocket in renin, a pharmacophore was designed, and a dipeptide like transition state mimetic was developed. Further optimizations in the structure of the dipeptide with the change of bulky groups to smaller alkyl ester groups resulted in better pharmacokinetic values. Further, the X-ray structure of renin revealed the presence of Tyr14 and Arg74 in the S3 pocket of renin and was important to make hydrogen bonds as hydrophobic interactions (Talele et al. 2010). Thus, all these structural and functional analysis lead to the development of Aliskiren (Norvatis) that passed the clinical trials in 2007 to become a pharmacophore-based drug for hypertension (Cohen 2007).

For drug target HIV 1 Integrase, inhibitors were screened out by building DKA pharmacophore. DKA pharmacophore was transferred to a naphthyridine carboxamide core, a class of n-alkyl hydroxypyrimidinone carboxylic acids, which was the result of the success with the DKA structural analog, and the drug named raltegravir became the first integrase inhibitor approved by the FDA (Summa et al. 2006).

To cite another example, an FDA approved drug against Ebola was taken to build a common pharmacophore and compared to a structure-based pharmacophore of VP35 protein. The pharmacophore, when compared were similar with one hydrogen bond acceptor feature and 1–4 hydrophobic features. This confirms that the FDA approved drugs of the Ebola virus might have VP35 as a primary target of action (Ekins et al. 2014).

10.9 Significance of Pharmacophore

Pharmacophores can be used to design rational drug candidates. It can also be used for optimizing already discovered drug candidates for better pharmacodynamics and lesser toxicity (Liu et al. 2013). It can be used to predict other receptors that the pharmacophore can react with, thus build more specific ligands to the target (Thai et al. 2013). Further, the pharmacophore can be built to specify and check ADMET properties of drugs, where a scaffold specifying the likable moieties for drug absorption, drug toxicity, and other properties can be built to predict the same in other chemical molecules (Guner and Bowen 2013). Another Study in recent times uses 3D pharmacophores to screen the safety of drugs against a specific receptor (Fan et al. 2019). Pharmacophores of virus epitopes can be reverse predicted to find drugs that might antagonize a virus receptor (Wadood et al. 2017). Techniques like pharmacophore mapping based on molecular dynamics could be the way to pharmacophore design, which predicts a tailored pharmacophore (Choudhury et al. 2015; Machaba et al. 2017). The pharmacophores can also be used to distinguish between the different active conformations of a protein active site (Shiri et al. 2019). In a recent paper, the pharmacophore and QSAR data were combined to a genetic algorithm to have a 4D-QSAR that can be used to predict the potential of a compound as anti-cancer leads (Sahin and Saripinar 2020).

Thus all these studies and examples reinstate that pharmacophores are important for increasing the accuracy of a prediction model in drug discovery as it takes into consideration the features or pairs of features and its contribution to the activity rather than the functional groups. This also serves to train a prediction workflow with diverse compounds that produce better results in combination with other in silico methods like QSAR and Docking.

10.10 Downside of Pharmacophore Modeling

Unlike docking or virtual screening, pharmacophore queries do not have a reliable general scoring metric, which indicates a good match or effective match. The identification of the actives depends on a pharmacophore-based screening on conformational databases. These databases only contain a limited number of low-energy conformations per molecule where an active molecule is missed due to conformation limitations. No clarity on the construction of a pharmacophore query is available, where two similar pharmacophores to the same target will give different molecules as hits when screened in the same database (Qing et al. 2014).

10.11 Conclusion

Pharmacophore analysis tools are powerful approaches in drug designing that are used as a template to find desirable drug features for a target. If used judiciously, it will group or help build a dataset of new novel drug-like molecules. The hypothesis can also be applied to steps like lead optimization, core hopping, and active site structural analysis. Thus, the use of ligand-based and structure-based pharmacophore design can yield important hits, but the level of false positives is also high in this technique. Thus rationale and a previous knowledge base can certainly guide to shortlist novel lead compounds. Pharmacophore modeling plays an important role in CADD, and any medicinal chemist should focus on pharmacophore during lead optimization.

Acknowledgement The author thanks the Editor, Dr. Dev Bukhsh Singh, for his critical comments and encouragement to write this chapter. The author also thanks SRMIST and the Department of Biotechnology for their support.

Competing Interest The author declares that there are no competing interests.

References

Alig L, Alsenz J, Andjelkovic M, Bendels S, Benardeau A, Bleicher K, Bourson A, David-Pierson P, Guba W, Hildbrand S, Kube D, Lubbers T, Mayweg AV, Narquizian R, Neidhart W, Nettekoven M, Plancher JM, Rocha C, Rogers-Evans M, Rover S, Schneider G, Taylor S, Waldmeier P (2008) Benzodioxoles: novel cannabinoid-1 receptor inverse agonists for the treatment of obesity. J Med Chem 51:2115–2127

Böhm HJ (1993) A novel computational tool for automated structure-based drug design. J Mol Recognit 6:131–137

Choudhari PB, Bhatia MS, Jadhav SD (2012) Pharmacophore identification and QSAR studies on substituted benzoxazinone as antiplatelet agents: KNN-MFA approach. Sci Pharm 80:283–294

Choudhury C, Priyakumar UD, Sastry GN (2015) Dynamics based pharmacophore models for screening potential inhibitors of mycobacterial cyclopropane synthase. J Chem Inf Model 55:848–860

Cohen NC (2007) Structure-based drug design and the discovery of aliskiren (tekturna): perseverance and creativity to overcome a R&D pipeline challenge. Chem Biol Drug Des 70:557–565

Dalkas GA, Vlachakis D, Tsagkrasoulis D, Kastania A, Kossida S (2013) State-of-the-art technology in modern computer-aided drug design. Brief Bioinform 14:745–752

Dixon SL, Smondyrev AM, Knoll EH, Rao SN, Shaw DE, Friesner RA (2006) PHASE: a new engine for pharmacophore perception, 3D QSAR model development, and 3D database screening: 1. Methodology and preliminary results. J Comput Aided Mol Des 20(10–11):647–671

Ekins S, Freundlich JS, Coffee M (2014) A common feature pharmacophore for FDA-approved drugs inhibiting the Ebola virus. F1000Res 3:277

Fan F, Warshaviak D, Hamadeh HK, Dunn RT II (2019) The integration of pharmacophore-based 3D QSAR modeling and virtual screening in safety profiling: a case study to identify antagonistic activities against adenosine receptor, A2A, using 1,897 known drugs. PLoS One 14(1): e0204378

Fei J, Zhou L, Liu T, Tang XY (2013) Pharmacophore modeling, virtual screening, and molecular docking studies for discovery of novel akt2 inhibitors. Int J Med Sci 10:265–275

Guner OF, Bowen JP (2013) Pharmacophore modeling for ADME. Curr Top Med Chem 13:1327–1342

Hartenfeller M, Schneider G (2011) De novo drug design. Methods Mol Biol 672:299–323

Horvath D (2011) Pharmacophore-based virtual screening. Methods Mol Biol 672:261–298

Juan Alvarez BS (2005) Virtual screening in drug discovery. CRC Press, Boca Raton

Kalva S, Vinod D, Saleena LM (2014) Combined structure- and ligand-based pharmacophore modeling and molecular dynamics simulation studies to identify selective inhibitors of MMP-8. J Mol Model 20:2191

Kalva S, Agrawal N, Skelton A, Saleena LM (2016) Identification of novel selective MMP-9 inhibitors as potential anti-metastatic lead using structure-based hierarchical virtual screening and molecular dynamics simulation. Mol BioSyst 12:2519–2531

Kaserer T, Beck KR, Akram M, Odermatt A, Schuster D (2015) Pharmacophore models and pharmacophore-based virtual screening: concepts and applications exemplified on hydroxysteroid dehydrogenases. Molecules 20:22799–22832

Koes DR, Camacho CJ (2011) Pharmer: efficient and exact pharmacophore search. J Chem Inf Model 51:1307–1314

Langer T, Wolber G (2004) Pharmacophore definition and 3D searches. Drug Discov Today Technol 1(3):203–207

Levit A, Yarnitzky T, Wiener A, Meidan R, Niv MY (2011) Modeling of human prokineticin receptors: interactions with novel small-molecule binders and potential off-target drugs. PLoS One 6(11):e27990

Li R-J, Wang Y-L, Wang QH, Wang J, Cheng MS (2015) In silico design of human IMPDH inhibitors using pharmacophore mapping and molecular docking approaches. Comput Math Methods Med 2015:1–11

Lin SK (2000) Pharmacophore perception, development and use in drug design. Edited by Osman F Guner. Molecules 5(7):987–989

Liu X, Zhu F, Ma XH, Shi Z, Yang SY, Wei YQ, ChenY Z (2013) Predicting targeted polypharmacology for drug repositioning and multi-target drug discovery. Curr Med Chem 20:1646–1661

Loving K, Salam NK, Sherman W (2009) Energetic analysis of fragment docking and application to structure-based pharmacophore hypothesis generation. J Comput Aided Mol Des 23 (8):541–554

Machaba KE, Mhlongo NN, Dokurugu YM, Soliman ME (2017) Tailored-pharmacophore model to enhance virtual screening and drug discovery: a case study on the identification of potential inhibitors against drug-resistant *Mycobacterium tuberculosis* (3r)-hydroxyacyl-ACP dehydratases. Future Med Chem 9:1055–1071

Mcgregor MJ, Muskal SM (1999) Pharmacophore fingerprinting. 1. Application to QSAR and focused library design. J Chem Inf Comput Sci 39:569–574

Mcgregor MJ, Muskal SM (2000) Pharmacophore fingerprinting. 2. Application to primary library design. J Chem Inf Comput Sci 40:117–125

Merz K Jr, Ringe D, Reynolds C (2010) Drug design: structure- and ligand-based approaches. Cambridge University Press, Cambridge

Mysinger MM, Carchia M, Irwin JJ, Shoichet BK (2012) Directory of useful decoys, enhanced (DUD-E): better ligands and decoys for better benchmarking. J Med Chem 55:6582–6594

Pirhadi S, Shiri FJ, Ghasemi B (2013) Methods and applications of structure based pharmacophores in drug discovery. Curr Top Med Chem 13(9):1036–1047

Qing X, Yin Lee X, De Raeymaecker J, Tame J, Zhang K, De Maeyer M, Voet A (2014) Pharmacophore modeling: advances, limitations, and current utility in drug discovery. J Recept Lig Channel Res 7:81–92

Sahin K, Saripinar E (2020) A novel hybrid method named electron conformational genetic algorithm as a 4D QSAR investigation to calculate the biological activity of the tetrahydrodibenzazosines. J Comput Chem 41:1091–1104

Salim AA, Kinghorn AD (2008) Drug discovery from plants. In: Ramawat KG, Mérillon JM (eds) Bioactive molecules and medicinal plants. Springer, Heidelberg, pp 1–24

Sanders MP, Barbosa AJ, Zarzycka B, Nicolaes GA, Klomp JP, De Vlieg J, Del Rio A (2012) Comparative analysis of pharmacophore screening tools. J Chem Inf Model 52:1607–1620

Sharma R, Dhingra N, Patil S (2016) COMFA, COMSIA, HQSAR and molecular docking analysis of ionone-based chalcone derivatives as antiprostate cancer activity. Indian J Pharm Sci 78:54–64

Shiri F, Pirhadi S, Ghasemi B (2019) Dynamic structure based pharmacophore modeling of the Acetylcholinesterase reveals several potential inhibitors. J Biomol Struct Dyn 37(7):1800–1812

Soliman MES (2013) A hybrid structure/pharmacophore-based virtual screening approach to design potential leads: a computer-aided design of South African HIV-1 subtype C protease inhibitors. Drug Dev Res 74:283–295

Summa V, Petrocchi A, Matassa VG, Gardelli C, Muraglia E, Rowley M, Paz OG, Laufer R, Monteagudo E, Pace P (2006) 4,5-dihydroxypyrimidine carboxamides and N-alkyl-5-hydroxypyrimidinone carboxamides are potent, selective HIV integrase inhibitors with good pharmacokinetic profiles in preclinical species. J Med Chem 49:6646–6649

Swaminathan P, Kalva S, Saleena LM (2014) E-pharmacophore and molecular dynamics study of flavonols and dihydroflavonols as inhibitors against dihydroorotate dehydrogenase. Comb Chem High Throughput Screen 17:663–673

Talele TT, Khedkar SA, Rigby AC (2010) Successful applications of computer aided drug discovery: moving drugs from concept to the clinic. Curr Top Med Chem 10:127–141

Testa B (2012) To Monty kier, a friendly tribute. Curr Comput Aided Drug Des 8:85–86

Thai KM, Ngo TD, Tran TD, Le MT (2013) Pharmacophore modeling for antitargets. Curr Top Med Chem 13:1002–1014

Thangapandian S, John S, Lee Y, Kim S, Lee KW (2011) Dynamic structure-based pharmacophore model development: a new and effective addition in the histone deacetylase 8 (hdac8) inhibitor discovery. Int J Mol Sci 12:9440–9462

Triballeau N, Acher F, Brabet I, Pin JP, Bertrand HO (2005) Virtual screening workflow development guided by the "receiver operating characteristic" curve approach. Application to high-throughput docking on metabotropic glutamate receptor subtype 4. J Med Chem 48:2534–2547

Van Drie JH (2012) Generation of three-dimensional pharmacophore models. WIREs Comput Mol Sci 3:449–464

Verma J, Khedkar VM, Coutinho EC (2010) 3D-QSAR in drug design—a review. Curr Top Med Chem 10:95–115

VLifeMDS (2010) Molecular design suite. VLife Sciences Technologies, Pune. www.vlifesciences.com

Wadood A, Mehmood A, Khan H, Ilyas M, Ahmad A, Alarjah M, Abu-Izneid T (2017) Epitopes based drug design for dengue virus envelope protein: a computational approach. Comput Biol Chem 71:52–160

Wang F, Chen Y (2013) Pharmacophore models generation by catalyst and phase consensus-based virtual screening protocol against Pi3kα inhibitors. Mol Simul 39:529–544

Warszycki D, Mordalski S, Kristiansen K, Kafel R, Sylte I, Chilmonczyk Z, Bojarski AJ (2013) A linear combination of pharmacophore hypotheses as a new tool in search of new active compounds—an application for 5-HT1A receptor ligands. PLoS One 8:E84510

Woods DD (1940) The anti-sulphanilamide activity (*in vitro*) of P-aminobenzoic acid and related compounds. Chem Ind 59:133–134

Yang SY (2010) Pharmacophore modeling and applications in drug discovery: challenges and recent advances. Drug Discov Today 15:444–450

In Silico Designing of Vaccines: Methods, Tools, and Their Limitations

11

Parvez Singh Slathia and Preeti Sharma

Abstract

In the post genomic era, the finding of new therapeutic targets has hugely been accelerated by the use of bioinformatics tools. The availability of genome sequences of pathogenic microbes has led to an increased finding of genes and proteins that could be potential targets for drug or vaccine design. The tools made available by bioinformatics have played a central role in the analysis of the genome and protein sequences for finding immunogenic proteins among the repertoire possessed by the organisms. The methods for prediction of immunogenicity are automated, and the whole proteome can be analyzed to find the top candidates that could have immunity inducing properties. Not only finding of immunogenic proteins has been achieved, but the mapping of the individual epitopes is also being done. The availability of methods for finding T and B cell epitopes can lead to the design of epitope-based vaccines. The description of different bioinformatics tools that are available for determining the immunogenic properties, finding of T and B cell epitopes, and in silico tools that are used in vaccine design is given in here. An account of epitope-based vaccine design employing bioinformatics methods reported in the literature is discussed. There are many shortcomings associated with these methods, which are discussed in the chapter. As is the case with other bioinformatics methods, there exist issues of prediction accuracy. Achievement of higher accuracy in predictions and their translation into in vivo/in vitro conditions still requires improvement. The chapter intends to provide the list of freely accessible software for epitope prediction and vaccine design with their merits/demerits and also throwing light on their applicability in vaccine research.

P. S. Slathia (✉) · P. Sharma
School of Biotechnology, Shri Mata Vaishno Devi University, Katra, Jammu and Kashmir, India

D. B. Singh (ed.), *Computer-Aided Drug Design*,
https://doi.org/10.1007/978-981-15-6815-2_11

Keywords

Epitope prediction · T cell epitopes · B cell epitopes · Antibody · Bioinformatics tools

11.1 Introduction

Treatment for infectious diseases is one of the most important aspects to improve the quality of human existence on the planet. However, the fight between the host and pathogen is not static but of a highly dynamic nature. The major pathogens threatening mankind are viruses, bacteria, and parasites. The co-evolution of pathogens and the emergence of new diseases have made the battle against pathogens continuous in nature. The issue has been further complicated by pathogens crossing the generic barriers, and pathogens from other animals have been showing presence in humans causing diseases. The most common pathogens that migrate from animals to humans are viruses and events of such transfer are reported. Animal viruses have infected humans, and caused diseases like SARS, MERS, SARS-CoV2 (coronaviruses), H1N1 (swine flu), and H5N1 (avian flu). SARS coronavirus originated most probably in bats though many argue its origin as uncertain (Chen et al. 2013) whereas MERS coronavirus was transmitted to humans from dromedary camels and its probable origin may be bats (Mohd et al. 2016). The most recent outbreak of pandemic novel coronavirus (SARS-CoV2) is still prevailing throughout the world and the reports to date suggest its bat or pangolin origin (Andersen et al. 2020; Zhang et al. 2020). The origins of H1N1 and H5N1 viruses are believed to be pigs and birds, respectively (Mena et al. 2016; Sims et al. 2005). All these viral outbreaks have happened in the first 20 years of this century.

Following the discovery of penicillin, the treatment for bacterial diseases has grown by leaps and bounds. With a battery of antibiotics available for therapy, the mortality and morbidity caused by bacterial diseases have been controlled. However, the drug resistance is increasing in bacteria and strains of multidrug resistant bacteria have been found, particularly in *Mycobacterium tuberculosis*. The therapeutic measures for parasitic disease treatment are limited by the availability of a few drugs. For example, nifurtimox and benznidazole are the only two drugs available for the treatment of Chagas disease. Furthermore, parasites have also started developing resistance against the current drugs used for treatment. Leishmania strains are showing increasing resistance to antimonates used for treatment. In addition, many of the drugs used for parasitic disease treatment are toxic including those mentioned in preceding sentences. This necessitates the development of better methods for finding a cure for pathogenic diseases and among these prophylactic measures like vaccines find an important place. Vaccines can help in reducing the disease burden by priming the immune system and inducing protective immunity against diseases. The success of vaccines has been well proven in eradicating smallpox and near elimination of polio from the world.

The modern era of vaccines started with the observation of cross-protection of smallpox through cowpox infection. It was Edward Jenner who observed that dairymaids who contracted cowpox subsequently never suffered from smallpox. He opined that cowpox somehow protected against smallpox and validated his theory experimentally by inoculating a boy first with cowpox pustule followed by smallpox pustule and the boy did not contract smallpox. This study published in 1798 was met with mixed reactions at that time (Riedel 2005). Louis Pasteur in 1879 accidentally discovered the "attenuation" while working on chicken cholera. His observations that chicken injected with old cultures of disease causing bacteria developed protection against subsequent injection of virulent cultures laid the foundation for a vaccination with attenuated organisms. Attenuation may thus be defined as the decrease in pathogenicity of a microbe without comprising its immune response generating properties. Later on, he used the same principle to develop protection against anthrax bacteria (Schwartz 2001). These discoveries towards the end of the nineteenth century paved the way for advances in immunology and vaccine development. In the forthcoming years, several new principles of developing vaccines were illustrated. Today vaccines based on various design platforms are being used commercially in the immunization regimens all over the world.

11.1.1 Live Attenuated Vaccine

The vaccines developed on the attenuation principle include the BCG vaccine for tuberculosis, Sabin polio vaccine (oral polio vaccine), measles vaccine, rotavirus vaccine, mumps vaccine, and varicella zoster (chickenpox) vaccine. Attenuation is generally achieved by growing the pathogen in abnormal conditions for long durations. In the case of Pasteur's chicken cholera vaccine, it was found subsequently that aerobic culture conditions were responsible for attenuation. These vaccines though efficient, yet require considerable time for development. BCG is an attenuated strain of *Mycobacterium bovis*, which took 13 long years for development. The attenuation was achieved by growing *Mycobacterium bovis* in increasing concentrations of bile salts by Albert Calmette and Camille Guerin (Luca and Mihaescu 2013). Sabin polio vaccine was developed by culturing poliovirus in monkey kidney epithelial cells. The reversion of attenuated organisms into virulent forms can occur thereby causing disease rather than providing immunity. Sometimes the administration of these vaccines has led to conditions like natural disease in a small percentage of recipient population like in the measles vaccine (Kindt et al. 2007).

11.1.2 Inactivated Vaccine

Inactivated vaccines contain the killed pathogen and hence are also called killed vaccines. This class of vaccines includes hepatitis A vaccine, Salk polio vaccine, rabies vaccine, etc. Inactivation is generally mediated by chemicals like formaldehyde that was used for the inactivation of poliovirus to produce the Salk vaccine. In the case of killed vaccines, the process involves killing or inactivation of pathogens, thus the workers involved in the process are exposed to pathogenic microbes posing a serious health challenge. Further, these individuals, if infected, can serve as a reservoir for other populations and can lead to the spread of disease. Sometimes, there can be a failure in the inactivation or killing of the pathogenic organism, which leads to disease outbreak upon vaccination. This has happened with the first Salk polio vaccine where the virus was not killed by formaldehyde, and a high number of recipients of the vaccine developed paralytic polio (Fitzpatrick 2006).

11.1.3 Subunit Vaccine

The dangers associated with killed and inactivated vaccines have led to the development of vaccines that do not use the whole organism but the parts of the organism, which are sufficient to generate immunity. The subunit vaccines have been developed, which use macromolecules like protein (Hepatitis B vaccine) or carbohydrates (Pneumococcal vaccine) for inducing protective immunity. Hepatitis B virus surface antigen (HBsAg) gene has been cloned into yeast and mammalian cells and this recombinant protein is used as a licensed vaccine. There is no handling of the virus involved during vaccine production (WHO Data n.d.). However, in the case of subunit vaccines comprised of carbohydrate moieties like a pneumococcal vaccine, the bacteria *Streptococcus pneumonia* is cultured and the polysaccharides are purified for use in vaccine formulations (Morais et al. 2018). Thus, handling is involved during the production process, which can make workers involved in production exposed to the pathogen. The subunit vaccines involving the use of immunogenic proteins are preferable as genes for proteins can easily be cloned in high expression vectors, and the production of such vaccines can be carried out with ease. Toxoid vaccines are produced by inactivating the exotoxin produced by bacteria. Tetanus and diphtheria toxoid vaccines were developed by inactivating the exotoxin with formaldehyde.

11.1.4 Recombinant Vector and DNA Vaccines

The knowledge that the proteins rather than the whole organism can provide immunity has led to the development of recombinant vector and DNA vaccines. The genes for immunogenic proteins can be cloned into attenuated viral or bacterial strains and are expressed for longer duration as the vector used replicates in the host. Adenoviruses, vaccinia virus, attenuated strains of *Salmonella*, BCG strain of *Mycobacterium bovis* are some examples of the vectors that can be used. The vaccine for SARS-CoV-2 being developed by Prof. Sarah Gilbert at the University of Oxford contains a gene sequence of spike glycoprotein cloned into the chimpanzee adenovirus vector. This vaccine is undergoing accelerated clinical trials for the remedy of the current prevailing COVID-19 pandemic (https://www.ovg.ox.ac.uk/news/covid-19-vaccine-development). The development of DNA vaccines involves the cloning of a gene for an antigenic protein in a plasmid that can be directly injected into a muscle. The muscle cells take up the DNA and express the protein to induce protective immunity by priming the immune system. Though there are no licensed vaccines based on these approaches yet they are very promising for future vaccine applications (Kindt et al. 2007).

11.1.5 Epitope-Based Vaccines

From the preceding two sections (Sects. 11.1.3 and 11.1.4) it becomes clear that bio-macromolecules particularly proteins alone are capable of generating protective immunity provided they are good antigens. Antigens may be defined as those molecules that can be recognized by B cell receptors (antibodies/immunoglobulins) or T cell receptors. The antigen-antibody binding is direct without the mediation of any other molecule. However, the recognition of the antigen by the T cell receptor requires that the antigen is presented by MHC (Major histocompatibility complex) protein molecule. The antigen loaded in the MHC molecule cleft interacts with the T cell receptor present on T cells. Immune cells, both B and T cells do not interact with the whole antigen molecule but on certain discrete sites present on the antigen called epitopes. Epitopes may be defined as antigenic determinants present in the antigen that directly interact with the antigen-specific receptors present on B and T cells. Epitopes are of immense importance as they can be potentially used in epitope-based vaccine design. Epitopes are regions of immune specificity within a protein and can elicit a protective immune response. Epitope-based vaccines comprise immuno-dominant epitopes of a pathogen. Epitope-based vaccines are considered to be safer than traditional vaccines and focus on the most crucial antigenic elements of the pathogen to generate protective immunity (De Groot et al. 2009). Furthermore, epitope-based vaccines have provided the opportunity to design multi-epitopic immunogens that contain epitopes from different proteins. Such chimeric vaccines generated can have a combined protective effect, which otherwise would have required all the proteins whose epitopes are incorporated in the vaccine, which is a difficult process. This approach derives the benefit of using epitopes derived from multiple proteins rather than focusing on a single protein molecule. The use of bioinformatics has been extensively made in designing such vaccines. There are no commercially available vaccines based on this strategy yet epitope-based vaccines hold a great promise for the future.

11.2 B and T Cell Epitopes

The prerequisite for epitope-based vaccines is the availability of epitopes. The nature of epitopes present in an antigen needs to be understood for such vaccine design. There is a difference between the recognition of epitopes by B and T cells. B cell receptors can bind to epitopes in antigen present either in soluble form or on the surface of pathogen and there is no requirement of mediation by any other molecule for this binding. However, the binding mechanism for T cell epitopes is different, as they require an epitope to be presented by MHC molecules for binding to the T cell receptor. The nature of B and T cells of epitopes and their interactions are detailed in the next sections.

11.2.1 B Cell Epitopes

B cell epitopes are located on the native protein and are both continuous and conformational. The continuous epitopes are also known as linear, or sequential epitopes comprise amino acids present sequentially in the protein. The conformational epitopes also called structural or discontinuous epitopes can comprise amino acids that are located distantly in sequence, but because of protein folding come close together to form a particular protein structure. B cell epitopes are mostly surface accessible, hydrophilic, polar regions of the antigens that can readily bind to the respective antibody molecule (Zobayer et al. 2019). The epitope and the antibody binding site are complementary and the epitope fits into the complementarity determining region (CDR) of the antibody molecule. The interactions between them are stabilized by weak forces like electrostatic interactions, hydrogen bonds, van der Waals forces, and hydrophobic interactions.

11.2.2 T Cell Epitopes and Their Processing

Unlike B cell epitopes that can be recognized directly, T cell epitopes require presentation of epitope with MHC molecules. T cell epitopes are only linear or sequential and the antigens need to undergo processing before being recognized by their receptors. The protein is first degraded into small peptides; these peptides bind to MHC molecule and subsequently form a trimolecular complex with T cell receptors. There are two types of T cells *viz* Tc cells or cytotoxic T cells that display CD8 protein molecule on their surface and Th cells or helper T cells displaying CD4 surface protein. The epitopes that are presented to Tc cells are displayed by Class I MHC molecules whereas Th cell epitopes are displayed by Class II MHC molecules. The pathways of processing and presenting epitopes to both types of T cells are different.

Tc cells recognize epitopes arising from proteins processed by the cytosolic pathway, which involves processing through proteasome and subsequent binding of the cleaved peptides to class I MHC molecule before presentation and recognition. Concisely, the proteasome (a multimeric protein complex) cleaves the protein into small peptides; these peptides are transported by TAP proteins (transporters associated with antigen processing) into the ER (endoplasmic reticulum) lumen. Class I MHC molecules are undergoing folding in the ER lumen where they bind to these transported peptides with the help of tapasin. The MHC-peptide complex is

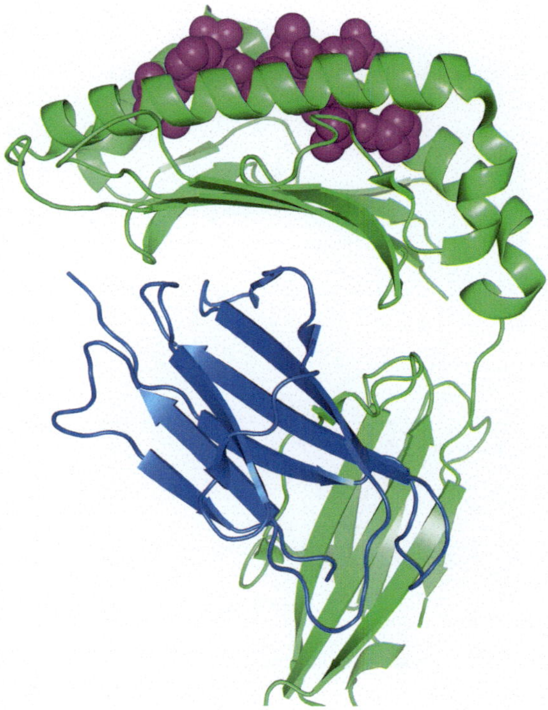

Fig. 11.1 Class I MHC molecule with peptide bound in the cleft (α chain: green, β2 microglobulin: blue and the peptide: purple color)

then transported by the secretory pathway to the cell surface (Hewitt 2003). Class I MHC glycoproteins are expressed by all nucleated cells and present antigen to cytotoxic T (Tc) cells. The peptide binding cleft of Class I MHC molecule is closed at both the ends and can bind peptides with 8–10 amino acids in length with nonamers showing best binding. The structure of Class I MHC with peptide bound in its cleft is shown in Fig. 11.1, whereas its antigen processing pathway is shown in Fig. 11.2.

Antigen processing for epitopes binding to Th cells takes place by the endocytic pathway involving phagocytosis and lysosomal cleavage of protein followed by binding to the Class II MHC molecule for presentation and recognition. Briefly, antigens are internalized into the cell by phagocytosis and it proceeds sequentially through early endosomes, late endosomes, and finally to lysosomes. In these increasingly acidic compartments, antigen gets cut into small peptides by the inherent proteases present there. Class II MHC molecules are transported from the Golgi complex to the endocytic pathway by an invariant chain. As the MHC II molecule moves through the endocytic pathway invariant chain gets cleaved leaving CLIP (class II-associated invariant chain peptide) occupying the peptide binding cleft of MHC II. HLA-DM catalyzes the exchange of CLIP with antigenic peptide and finally, Class II MHC molecule moves to the cell surface (Kindt et al. 2007). Class II MHC glycoproteins expressed on the surface of antigen presenting cells (dendritic cells, macrophages, and B cells) present antigen to helper T cells (Th). The

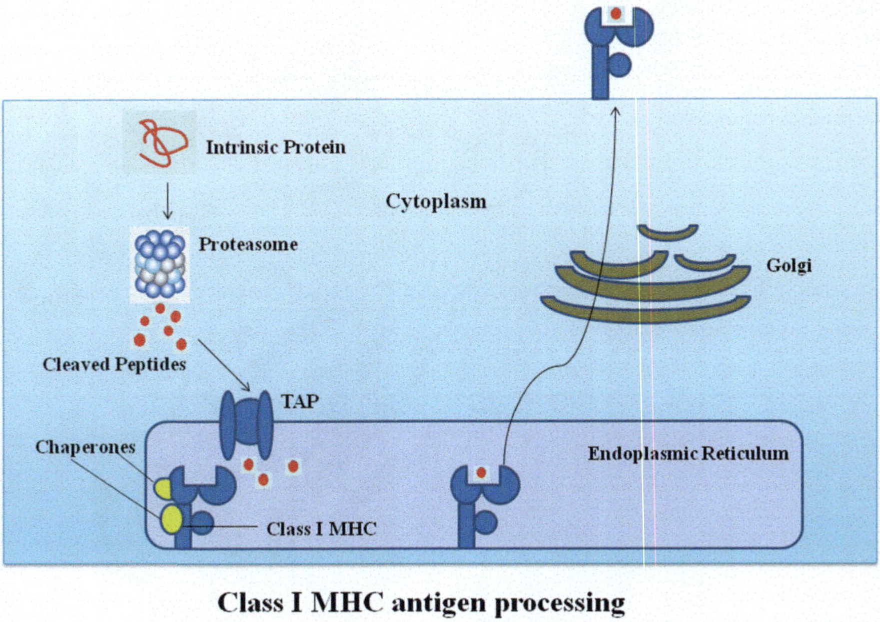

Fig. 11.2 Processing of antigen, binding of epitope to Class I MHC, and its display on cell surface

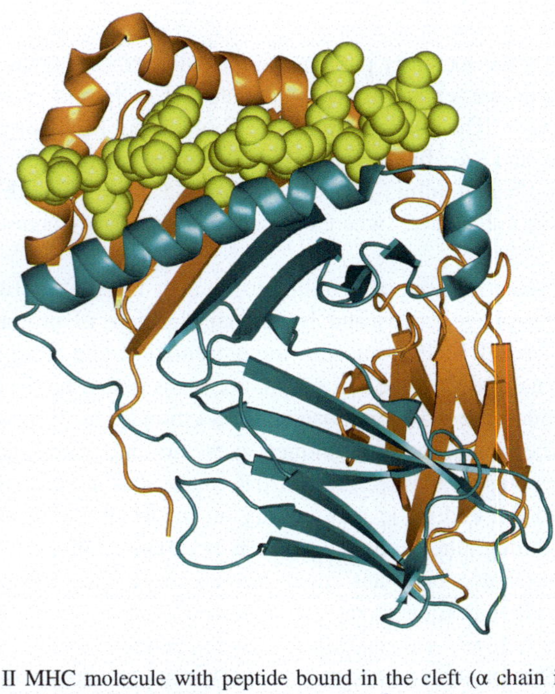

Fig. 11.3 Class II MHC molecule with peptide bound in the cleft (α chain in blue, β chain in orange, and the peptide in yellow color)

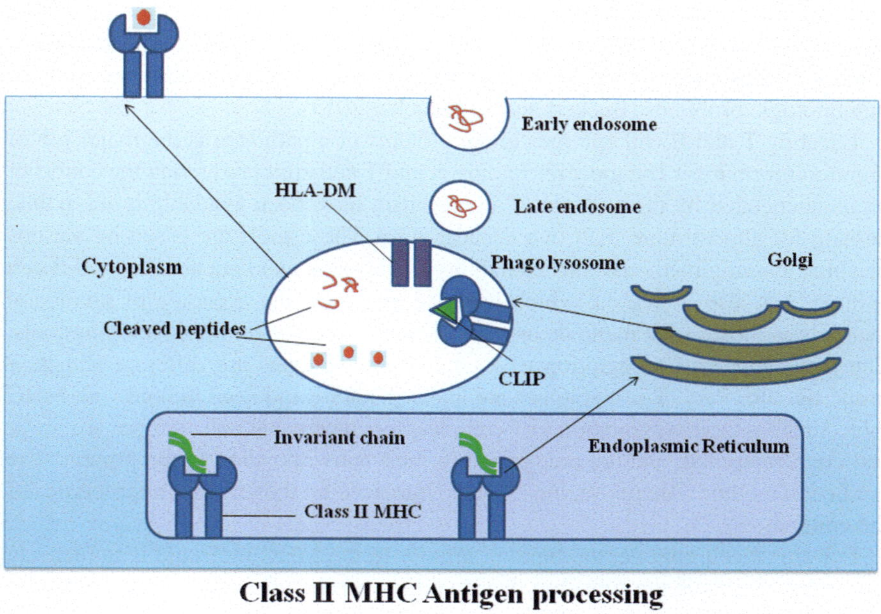

Fig. 11.4 Processing of antigen, binding of epitope to Class II MHC, and its display on cell surface

peptide binding site of Class II MHC is open at both ends and can bind peptides of 13–18 amino acids length. Figure 11.3 depicts the Class II MHC molecule with a bound peptide in its cleft and its antigen processing pathway is depicted in Fig. 11.4. Thus, the prerequisite for any protein which can be a possible T cell antigen is that it should be comprised of peptides that show binding affinity to MHC molecules. The proteins that upon passing through the antigen processing pathway generate peptides having an affinity for binding to the cleft of MHC molecules can be classified as T cell antigen proteins.

11.3 Bioinformatics in Vaccine Design

With the advance in genomic technologies in the recent past, the genomes of organisms are being sequenced at an unprecedented pace. The amount of the data available is immense and can provide insights into finding unexplored genome regions in search of novel targets for the treatment of diseases. The wealth of available genomic data has to be analyzed for deciphering the encoded proteins, and for vaccinology purposes. The total proteins encoded by the genome can be screened for finding out immunogenic proteins using bioinformatics tools. These antigenic proteins can further be used to find out epitopes located in them. Many of the genome databases have constructed proteomes of the sequenced genomes by automated methods. The repertoire of proteins encoded by the genome can be analyzed by bioinformatics servers to find antigenic proteins. The filtered antigenic proteins can be used to find specific B and T cell epitopes in these proteins by the available epitope prediction methods. A new branch called immunoinformatics has come into existence, which deals with the application of computational tools to immunologic problems (Backert and Kohlbacher 2015).

Locating T and B cell epitopes in the proteins of a pathogen is the major job of immunoinformatics. The tools for finding B and T cell epitopes among the cohort of proteins encoded by the genome of an organism have been available in the public domain for almost more than two decades now. These tools are based on various machine learning methods. The availability of experimental data about T and B cell epitopes has also increased, which has also enhanced the accuracy of prediction methods as most of the methods use this data as a training set for developing tools. The mechanisms of recognition of B and T cell epitopes are different and their properties also vary. T cell epitopes are linear in nature and need to bind with MHC molecules for their presentation to T cell receptors whereas B cell epitopes are linear and conformational, and are recognized in their native position in the protein. The prediction methods, therefore, have to take into account these different properties of the epitopes.

For any T cell epitope, the binding affinity to the MHC molecule is immensely important, as this is the first step that qualifies it to be an epitope. The prediction methods for finding such an affinity of peptides first progressively break antigenic protein into peptides and analyze their affinity for a particular MHC molecule. The diversity of MHC molecules further complicates the situation as the affinity for peptides changes with change in the molecule. The studies on the peptides eluted from MHC molecules reveal that there are differences in the properties of peptides bound in the cleft of different MHC proteins. The alleles for MHC are designated as HLA alleles; for class I these alleles are HLA- A, B, and C and for class II HLA-DP, DQ, and DR. In the human population, the number of HLA class I alleles is 14,800, and that of HLA class II alleles is 5288 (Statistics of HLA alleleshttps://www.ebi.ac.uk/ipd/imgt/hla/stats.html). Further, the distribution of HLA alleles differs among different population groups of the world. Thus, any software tool that is developed for the T cell epitope determination needs to consider these points. The epitope prediction tools used for B and T cell epitopes are discussed in subsequent sections. The B and T cell epitope prediction process is shown in Fig. 11.5.

11.4 Prediction Tools for Class I and II MHC Binding

A comprehensive list of the freely accessible tools available for determining the binding affinity of peptides in a protein to different MHC molecules is listed in Table 11.1. These tools are based on different machine learning methods like support vector machine (SVM), artificial neural networks (ANN), hidden Markov models (HMM), and position-specific scoring matrices (PSSM). Some tools can carry out the peptide binding predictions for both class I and II MHC molecules, whereas some of the tools are exclusive. Tools like NetMHC, NetMHCPan, ProPred-I, EpiJen, and nHLAPred carry out the binding affinity prediction of peptides to

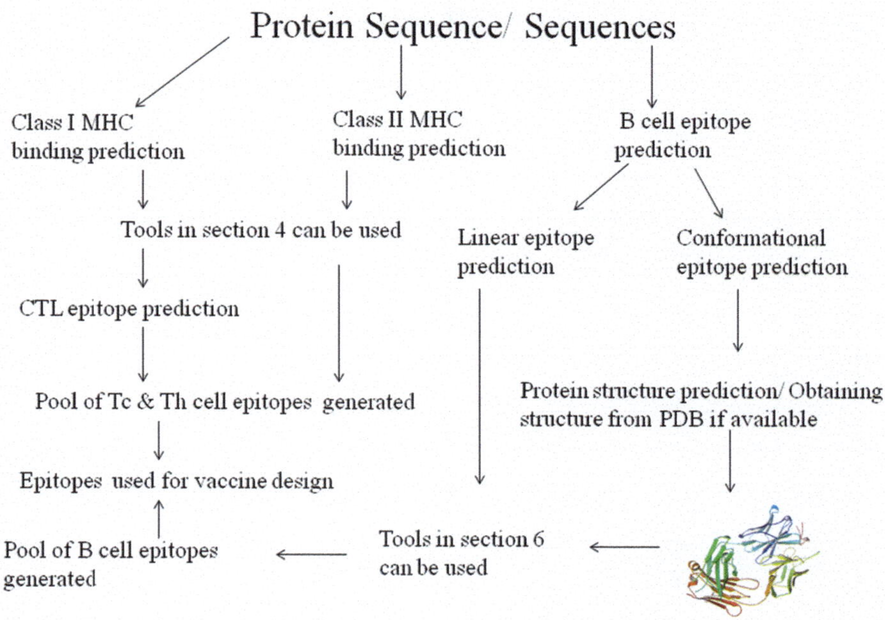

Fig. 11.5 Schematic process flow of B and T cell epitope prediction for epitope-based vaccine design

class I MHC molecules, and NetMHCII, NetMHCIIPan, and ProPred are exclusively used for class II MHC binding predictions. Most of the other tools have the capability of carrying out a prediction for both classes of MHC molecules. The number of alleles available for running predictions is different in each tool.

11.4.1 NetMHC

NetMHC utilizes the ANN approach to predict the binding affinity of a peptide for different class I MHC molecules. This predictive model has been trained for 81 different MHC alleles of humans, including HLA-A, HLA-B, HLA-C, and HLA-E (Andreatta and Nielsen 2016).

11.4.2 NetMHCPan

It predicts the binding affinity of peptides to any MHC of the known sequence. This ANN-based method is trained by more than 180,000 binding data, and MHC eluted ligands. The binding affinity data covers 172 MHC molecules from human, mouse (H-2), Cattle (BoLA), primates, and swine (SLA). It provides information about the likelihood of a peptide to be a natural ligand or the binding affinity (Jurtz et al. 2017).

Table 11.1 Tools for prediction of MHCI and MHCII binding peptides from the protein sequences

Prediction for MHC class	Prediction method	Tool	References
Class I MHC	ANN	NetMHC	Andreatta and Nielsen (2016)
Class I MHC	ANN	NetMHCPan	Jurtz et al. (2017)
Both class I and II MHC	Published motifs	SYFPEITHI	Rammensee et al. (1999)
Class I MHC	Addition/multiplication matrices	ProPred-I	Singh and Raghava (2003)
Both class I and II MHC	PSSM	RANKPEP	Reche et al. (2002)
Class I MHC	Additive method	EpiJen	Doytchinova et al. (2006)
Both class I and II MHC	Additive method	MHCPred	Guan et al. (2003)
Class I MHC	ANN and QM	nHLAPred	Bhasin and Raghava (2007)
Both class I and II MHC	SVR	SVRMHC	Liu et al. (2007)
Both class I and II MHC	SVM	SVMHC	Dönnes and Kohlbacher (2006)
Both class I and II MHC	Multiple methods	IEDB analysis	Zhang et al. (2008)
Both class I and II MHC	ANN	MULTIPRED2	Zhang et al. (2011)
Class II MHC	ANN	NetMHCII	Jensen et al. (2018)
Class II MHC	ANN	NetMHCIIPan	Jensen et al. (2018)
Class II MHC	QM	ProPred	Singh and Raghava (2001)
Class II MHC	SVM	MHC2Pred	Lata et al. (2007)

11.4.3 SYFPEITHI

This database contains MHC class I and class II ligands, peptide motifs of humans and other species, natural ligands, and T cell epitopes. It also provides connectivity to resources available at EMBL and PubMed databases (Rammensee et al. 1999).

11.4.4 ProPred-I

ProPred-I is used to identify the MHC class-I binding regions in antigens. It also helps the researcher to identify the promiscuous regions (Singh and Raghava 2003).

11.4.5 RANKPEP

It predicts the peptide binders to MHCI and MHCII from protein sequence information. It also identifies the MHCI ligands, whose C terminal end is likely to be the result of proteasomal cleavage (Reche et al. 2002).

11.4.6 MHCPred

This method assumes that each substituent present in a molecule has an additive and independent contribution to the biological activity. It considers the interaction between individual amino acids and the binding site, the interaction between adjacent and every second amino acids, and their effects on binding (Guan et al. 2003).

11.4.7 EpiJen

This method considers proteasome cleavage and TAP binding and can mimic the MHC binding mechanism in a real way (Doytchinova et al. 2006).

11.4.8 SVMHC

This tool is based on the SVM approach and used to predict both class I and class II MHC binding epitopes. This server is based on (Dönnes and Kohlbacher 2006).

11.4.9 MULTIPRED2

It is used to screen peptide that binds to multiple alleles belonging to HLA class I and class II DR super types. It performs binding predictions on 1077 alleles related to 26 HLA super types (Zhang et al. 2011).

11.4.10 ProPred

ProPred predicts class II MHC binding regions in the antigenic sequence. It assists in locating promiscuous binding regions which are useful in screening vaccine candidates (Singh and Raghava 2001).

11.4.11 MHC2Pred

This tool is used to predict promiscuous class II MHC binding peptides. For algorithm designing, the information of binders and non-binders for different alleles were taken from the MHCBN and JenPep database. The average accuracy of this method is ~80% (Lata et al. 2007).

11.5 CTL Epitope Prediction

Though multiple tools are available for prediction of binding affinity of peptides to different class I MHC molecules, yet only binding to a particular MHC is not sufficient to qualify a peptide to be a Tc cell epitope. In other words, not all class I MHC binders are Tc cell epitopes, whereas all Tc cell epitopes are good MHC binders. Also, the peptide should be amenable to the antigen processing pathway of class I MHC i.e. cytosolic pathway. Proteasomal cleavage and transport of peptides into the rough endoplasmic reticulum (RER) by TAP are other important steps involved in the cytosolic pathway of antigen processing and presentation (Hewitt 2003).

All intracellular proteins after spending a fixed time in the cell are marked for degradation by a small protein called ubiquitin. The marked proteins are then cleaved by proteasome into small peptides within its central hollow. The immune system modifies proteasome by the addition of extra protein molecules called LMP7, LMP2, and LMP10 to generate peptides having a preferential affinity for class I MHC molecules. Thus, for any peptide to act as a Tc cell epitope, it should be processed by the proteasome. The transport of peptides generated by the proteasome to RER is carried out by the transport by TAP. TAP also shows preference to transport peptides of 8–13 amino acid residues in length (Kindt et al. 2007). The peptide should have these properties to get transported from the cytosol to RER. These requirements are not as specific as binding to class I MHC molecule yet play an important role in making a peptide a Tc cell epitope.

There are bioinformatics tools that carry out proteasomal cleavage and TAP transport prediction and are listed in Table 11.2. Some of the class I MHC binding prediction tools have these two functions inbuilt in them. EpiJen server, in addition to MHC binding also uses proteasomal cleavage and TAP transport for predicting Tc cell epitopes (Doytchinova et al. 2006). nHLAPred also uses proteasomal cleavage matrices to refine the results of epitope prediction. ProPred-I uses the proteasomal model and immunoproteasome models for finding the epitopes. RANKPEP predicts class I MHC binding peptides whose C terminal end is likely to be the result of proteasomal cleavage. The description of tools that exclusively serve the purpose of proteasomal cleavage and TAP transport prediction is provided below.

11.5.1 NetCTL

It is used to predict peptide MHC class I binding, proteasomal C terminal cleavage, and efficiency of TAP transport. MHC class I binding and proteasomal cleavage is based on the ANN approach while the efficiency of TAP transport uses a weight matrix (Larsen et al. 2007).

Table 11.2 Prediction tools for accessing amenability to antigen processing pathway

Server	Application	References
NetCTL	Integrated	Larsen et al. (2007)
CTLPred	CTL prediction	Bhasin and Raghava (2004a)
NetChop	Proteasomal cleavage	Keşmir et al. (2002) and Nielsen et al. (2005)
MAPPP	Integrated	Hakenberg et al. (2003)
Pcleavage	Proteasomal cleavage	Bhasin and Raghava (2005)
PAProC	Proteasomal cleavage	Nussbaum et al. (2001)
TAPPred	Binding affinity for TAP transporter	Bhasin and Raghava (2004b)
EpiJen	Integrated	Doytchinova et al. (2006)

11.5.2 CTLPred

This tool uses a quantitative matrix, SVM, and ANN approach for prediction. It has been developed by training and testing the results from the dataset of T cell epitopes and non-epitopes (Bhasin and Raghava 2004a).

11.5.3 NetChop

NetChop is based on the ANN method to predict the cleavage sites of the human proteasome. Since the method is trained using human data, therefore, it shows better performance in predicting sites of proteasomal cleavage for humans. The method is used by NetCTL for predicting proteasomal cleavage sites (Keşmir et al. 2002; Nielsen et al. 2005).

11.5.4 MAPPP

MAPPP is used to predict antigenic epitopes on the cell surface by class I MHC to CD8 positive T lymphocytes. It also predicts the proteasomal cleavage with peptide anchoring to MHC I molecules (Hakenberg et al. 2003).

11.5.5 Pcleavage

It is an SVM based method used to predict constitutive and immunoproteasome cleavage sites in the antigenic molecule. The method only predicts proteasomal cleavage sites, but no prediction of TAP transport is available (Bhasin and Raghava 2005).

11.6 B Cell Epitope Prediction

B cell epitopes are recognized by the B cell receptors i.e. antibodies without the process of processing and presentation, unlike T cell epitopes. Linear B cell epitopes, in principle, can be predicted by the same methods as used for T cell epitopes. However, the prediction of conformational or structural epitopes is a challenging job. The requirement of structural data of a protein is absolute for finding discontinuous epitopes. There are methods available for prediction of both continuous and discontinuous B cell epitopes (Table 11.3), but the efficiency of these methods is less when compared to T cell epitope prediction methods. The methods use many different approaches like ANN, SVM, HMMs for predictions.

11.6.1 BCPred

In BCPred server, the user can select the method such as amino acids pair scaling (AAP), BCPred, and FBCPred for prediction. AAP has good accuracy in the prediction of antigenicity, hydrophilicity, and flexibility (Chen et al. 2007; EL-Manzalawy et al. 2008).

Table 11.3 B cell epitope prediction tools

Server	Type of epitope	References
BCPRED	Linear/continuous	Chen et al. (2007) and EL-Manzalawy et al. (2008)
LBtope	Linear/continuous	Singh et al. (2013)
ABCpred	Linear/continuous	Saha and Raghava (2006a)
BepiPred	Linear/continuous	Jespersen et al. (2017)
Bcepred	Linear/continuous	Saha and Raghava (2004)
SVMTriP	Linear/continuous	Yao et al. (2012)
Discotope	Conformational/ discontinuous	Kringelum et al. (2012)
BEpro	Conformational/ discontinuous	Sweredoski and Baldi (2008)
ElliPro	Conformational/ discontinuous	Ponomarenko et al. (2008)
Epitopia	Both linear and conformational	Rubinstein et al. (2009)
CBTOPE	Conformational/ discontinuous	Ansari and Raghava (2010)
PEASE	Conformational/ discontinuous	Sela-Culang et al. (2014a, b)
EpiPred	Conformational/ discontinuous	Krawczyk et al. (2014)
EPSVR and EPMeta	Conformational/ discontinuous	Liang et al. (2010)

11.6.2 LBtope

LBtope is based on data of B cell epitopes and non-B cell epitopes from the immune epitope database. Models like SVM and K-nearest neighbor are used in discriminating epitopes and non-epitopes. The features like binary profile, dipeptide composition, AAP (amino acid pair) profile have been used in design of the method, and the accuracy of prediction ranges from 54% to 86% (Singh et al. 2013).

11.6.3 ABCPred

It is an ANN-based approach used to predict continuous B cell epitopes using a fixed length pattern. This tool is developed using the dataset of epitopes from parasites, viruses, bacteria, and fungi from the BciPep database, and it has a prediction accuracy of 65.9% (Saha and Raghava 2006a).

11.6.4 BepiPred 2.0

It is based on the random forest algorithm and developed from a dataset of epitopes annotated from the antibody-antigen structure from PDB. This tool requires a FASTA format of the protein as input (Jespersen et al. 2017).

11.6.5 Bcepred

Bcepred predicts the linear B cell epitopes using physicochemical properties, such as hydrophilicity, accessibility, flexibility, polarity, exposed surface, etc. The accuracy of this server is 58.7% (Saha and Raghava 2004).

11.6.6 DiscoTope

This server is used for the prediction of discontinuous B cell epitopes from protein 3D structures using surface accessibility and a novel epitope propensity score of residues (Kringelum et al. 2012).

11.6.7 ElliPro

ElliPro is used for prediction and analysis of antibody epitopes in a protein structure. Here, PDB ID or a PDB file of a protein is used as input. It has been designed using the information of discontinuous epitopes present in antibody-protein complexes (Ponomarenko et al. 2008).

11.6.8 PEASE

This server predicts antibody-specific epitopes using sequence information of the antibody. The epitopes related information is provided at the residue level and also on the structure of antigen (Sela-Culang et al. 2014a, b).

11.7 Methods for In Silico Designing of Epitope-Based Vaccines

Vaccines designing using immunoinformatics tools have come a long way, and many strategies have been employed for this purpose. Before the advent of such tools and precedent to the availability of genome, the classical vaccinology approaches were used which required more time and labor. The prime requirement in the case of subunit vaccines is a biomolecule, mostly proteins that have the potential to induce immunity. In the case of epitope-based vaccines, epitopes from more than one protein can be amalgamated in a single construct for enhancing immunity. The general account of approaches used is given below. The process and tools used for epitope analysis and selection of epitopes for vaccine design are displayed in Fig. 11.6.

11.7.1 Selection of Proteins

As mentioned earlier, the requirement of immunogenic proteins is prime for epitope-based vaccine designing. The databases like NCBI Protein and UniProt can serve as the source of proteins for analysis. NCBI Protein database is a collection of protein sequences from SwissProt, PIR (Protein Information Resource), PRF (Protein Research Foundation), and PDB (Protein Data Bank) in addition to translated sequences obtained from annotated coding regions of GenBank sequences. UniProt contains protein sequences obtained from SwissProt and translated EMBL (trEMBL) database. The finding of immunogenic proteins from the genome can be achieved by using various criteria. The total proteins encoded by the genome of a pathogen i.e. its proteome can be analyzed for immunogenic proteins by using servers like VaxiJen (Doytchinova and Flower 2007). This server can take as input multiple protein sequences from bacteria, viruses, fungi, parasites, and the threshold value can be controlled by the user. The total proteome of an organism can be provided as input and depending upon the threshold, antigenic proteins can be selected. Proteomics approaches to find the stage-specific expression of proteins can also aid in vaccine development (Soria-Guerra et al. 2015). The other approach for fishing proteins from the proteome is to find the surface proteins. The surface proteins are easily accessible to immune effector molecules particularly to antibodies and can suffice the purpose. The servers like CELLO (Yu et al. 2006), Cell-PLoc (Chou and Shen 2008) that predict the subcellular localization of proteins can help in finding the surface

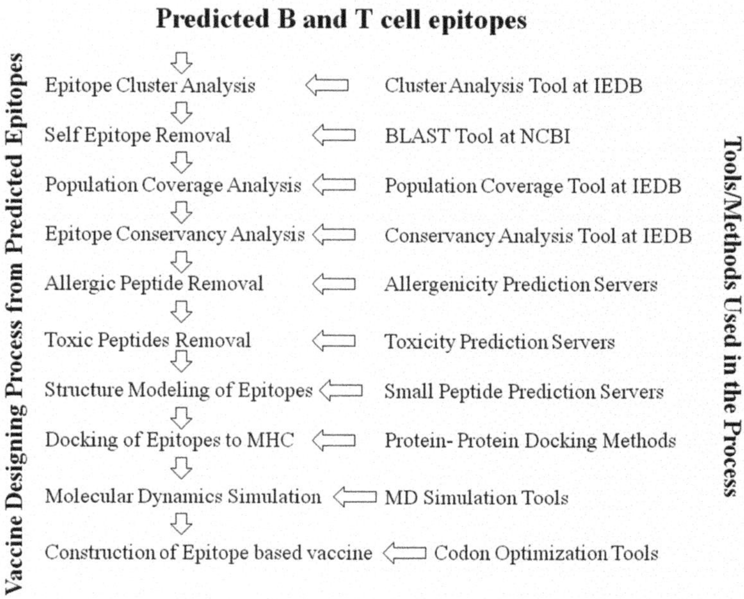

Fig. 11.6 Graphic representation of steps involved in the analysis of predicted B and T cell epitopes for designing of in silico vaccines

proteins from the proteome. The combined approach in which first surface proteins are predicted from the proteome and these proteins are then subjected to immunogenicity prediction by VaxiJen has also been employed (Pritam et al. 2019). There is another server by the name of Vaxign which provides two modes of usage; one in which pre-computed results of more than 350 genomes are available and can be used for finding immunogenic proteins, and the second involves the protein input to be provided by the user and results are computed by the server (He et al. 2010). This server can also assist in protein selection. Literature studies are also a good source for finding immunogenic proteins. The previously reported proteins capable of generating immunity can also be used for epitope prediction and a vaccine can be designed from the epitopes derived from multiple proteins.

In some cases, several variant sequences exist for a single immunogenic protein. This could be due to the protein sequences arising from different strains of a pathogen or the variability induced by the pathogen itself in its surface proteins for evading the immune response. This variability provides an advantage to the pathogen and poses a major hindrance in vaccine development. The conserved regions in such a protein are deciphered by multiple sequence alignment of the different variant sequences. Tools like Clustal Omega, TCOFFEE, etc. can be employed to carry out multiple sequence alignment. These conserved regions can then serve as the source for the prediction of epitopes.

11.7.2 Epitope Prediction and Analysis

Once the protein/proteins have been selected the subsequent step involves the prediction of B and T cell epitopes. The prediction of epitopes can be carried out using the tools mentioned in Sects. 11.4, 11.5, and 11.6. Tc cell epitope prediction involves predicting the affinity of peptides (by tools in Sect. 11.4) for the respective HLA allele (Class I MHC) followed by the proteasomal cleavage and TAP transport prediction (tools in Sect. 11.5). The epitopes qualifying these criteria are generally selected for the vaccine designing process. For helper T cells the epitope prediction involves ascertaining the binding affinity of peptides for HLA alleles (Class II MHC) and there are no methods available for determining the antigen processing and presentation prediction by endocytic pathway. The tools presented in Sect. 11.6 can predict continuous and discontinuous B cell epitopes. Thus, a pool of B and T cell epitopes can be generated which can be further analyzed.

Many a time the epitopes particularly, T cell epitopes predicted for different HLA alleles can share considerable sequence similarity. Such epitopes can be clustered together and a single representative of this cluster can be used in the final design. The process of clustering removes the unwanted repetitiveness of epitopes and prevents the vaccine construct from unnecessary elongation. Epitope cluster analysis tool (Dhanda et al. 2018) can be used to carry out this step as the epitopes are clustered together based on the sequence identity threshold set by the user. Another important aspect is to check the similarity of the epitopes with the host proteins and this can be achieved by using BLAST (Basic Local Alignment Search Tool) available at NCBI (National Center for Biotechnology Information). After BLAST analysis the epitopes sharing similarity with host proteins need to be omitted, as they may not generate any response. Population coverage tool helps in finding the predicted immune response to T cell epitopes in a population group based on HLA allele distribution (Bui et al. 2006). The Allele Frequency Net Database is the source of HLA allele distribution frequencies in different populations of the world used in the tool. The epitopes should be able to provide a higher percentage of population coverage to be used in vaccine design as this ensures immune response generation in most of the individuals in a population group. To check the conservancy of epitopes across the variants of a protein, Epitope conservancy analysis tool can be used which calculates the degree of the conservancy of a particular epitope in the cohort of protein sequences (Bui et al. 2007). This tool becomes an important asset when different variants of a protein exist as mentioned in Sect. 11.7.1.

The epitopes can also be checked for the presence of any allergenic and toxic peptides among them. The tools for allergenicity prediction like AlgPred (Saha and Raghava 2006b), AllergenFP (Dimitrov et al. 2014a), AllerCatPro (Maurer-Stroh et al. 2019), and AllerTop (Dimitrov et al. 2014b) are freely available and can be used to remove the epitopes possessing allergenic nature from the group to be used in vaccine design. The toxic peptides can be predicted by the ToxinPred server (Gupta et al. 2013) and any epitopes that are toxic in nature have to be omitted from the final construct. The epitopes after filtration by the above methods can be used further.

11.7.3 Molecular Docking and Molecular Dynamics Simulation

The use of molecular docking and simulation to find interactions between epitopes and immune effector molecules is an important aspect of in silico vaccine designing. T cell epitopes should bind to MHC molecules for presentation to T cell receptors. This binding can be studied by molecular docking, for which the structure of epitope needs to be determined. The servers like PEPFOLD (Lamiable et al. 2016), QUARK (Xu and Zhang 2012), etc. are freely available for modeling of small peptides and can provide models of T cell epitopes. The structures of MHC molecules (Class I and II) can either be obtained from Protein Databank (PDB) if available or models of these proteins be generated by homology modeling servers like SwissModel (Waterhouse et al. 2018) and many others. The docking of the epitopes with respective MHC molecule/HLA allele can be carried out by protein-protein docking servers like ZDOCK (Pierce et al. 2014), ClusPro (Kozakov et al. 2017), etc. The docking results can validate whether the epitope binds in the cleft of the MHC molecule. The docked complexes can further be analyzed by molecular dynamics simulation to explore the interaction between the epitope and MHC molecule in conformational space. These results, if positive can further strengthen the possibility of epitopes being presented by MHC molecules to T cells. Software suites like GR OMACS (van der Spoel et al. 2005) are freely available that can be used for simulation studies. T cell epitopes that dock into the peptide binding cleft of MHC molecules are selected for vaccine designing. The docking and simulation studies for B cell epitopes are not required as they bind directly to the antibody molecules and are not presented through MHC molecules. The complementarity determining regions (CDRs) of antibody molecules are highly diverse and thus binding studies cannot be carried out.

11.7.4 Construction of Vaccine

Many studies culminate at the finding of epitopes that qualify the processes mentioned in Sects. 11.7.1–11.7.3 and the resultant cohort of epitopes is left for the designing of vaccines in future studies followed by experimental validations. In some studies, the epitopes are used for cell culture based studies, and their ability to initiate an immune response is validated by the cytokine response generated by them in peripheral blood mononuclear cells (PBMCs). However, in silico vaccine designing based on the predicted epitopes is also widely carried out. The epitopes are joined in tandem with the insertion of specific linkers for efficient processing of epitopes such as AAY linker is generally used between two CTL epitopes. AAY linker possesses the cleavage site of proteasomes, which leads to the generation of natural epitopes and it can also reduce the unwanted joining of two neighboring epitopes in the vaccine construct. Similarly, the GPGPG linker is used for the separation of T helper cell epitopes as this linker is reported to facilitate immune processing and prevent the joining of two epitopes. In many of the vaccine constructs protein adjuvants like cytokines have also been fused with the epitopes and in these constructs linker like EAAAK has been widely used. The linker EAAAK causes the separation of fusion proteins (Arai et al. 2001) and can prevent the interaction between the vaccine and adjuvant domains of the vaccine protein construct. Thus, the final fusion protein obtained as vaccine consists of epitopes, linkers, and may be adjuvants in certain cases. The structure of this fusion protein can be deduced using web-based protein modeling severs like I-TASSER (Roy et al. 2010). In some instances, homology modeling servers can successfully model the vaccine protein structure, whereas for some structures ab initio modeling approaches have to be used. The protein sequence can be reverse translated into DNA and thus gene constructs of vaccine can be made. JCat tool can be used for reverse translation as well as for codon optimization for efficient translation in the host cells (Grote et al. 2005). The protein sequence of the vaccine construct can be reverse translated using codon bias for expression in eukaryotic or prokaryotic cells for heterologous production of the vaccine. Alternatively, the gene for human expression can be codon optimized for direct use as a DNA vaccine in humans. Thus, this section summarizes the methods that can be used for designing of in silico vaccines. Some examples of the development of vaccines using these approaches are depicted in Sect. 11.8.

11.8 Case Studies of Vaccine Designing

There have been various studies on vaccine designing using bioinformatics tools. Immunoinformatics has been widely used for in silico designing of vaccines for various pathogens like viruses, bacteria, and parasites. The details of these vaccine designing studies are given in the forthcoming sections.

11.8.1 Vaccine Designing for Viral Pathogens

Viruses are nucleoprotein particles, which have imposed a heavy disease burden throughout human history. Since, viruses use the host cell machinery for their replication and other functions, it limits the availability of drug targets in them. Vaccines have been the prime means for the treatment of viral diseases. Recently, vaccines for viruses have been designed using in silico methods. The vaccines have been designed based on epitopes derived from a single viral protein. The criteria used for the selection of protein are either immunogenicity or surface accessibility. Ebola virus vaccine was designed using predicted B and T cell epitopes present in the glycoprotein of the virus. VaxiJen server was used to find immunogenic protein followed by epitope predictions, which were further validated by molecular docking and molecular dynamics simulation approach (Dash et al. 2017). T and B cell epitopes (linear and discontinuous) were predicted in the Spike protein of MERS-COV using bioinformatics tools, which could be used in vaccine design (Ul Qamar et al. 2019). In some cases, epitopes have been predicted from more than one viral protein for use in vaccine design. The proteins E, prM, NS1, NS3, and NS5 of Japanese Encephalitis Virus (JEV) were used in a recent study for the prediction of T and B cell epitopes. Based on different parameters assessed four T cell and one B cell epitope were found to have potential in inducing immunity and could be used in vaccines against the virus (Chakraborty et al. 2020). B and T cell epitopes from five structural polyproteins (capsid, E2, 6K, E3, and E1) of the Mayaro virus were predicted using immunoinformatics tools. Multi-epitope vaccine was designed, molecular docking with TLR-3 was done, and finally in silico expression was carried out in *E. coli* (Khan et al. 2019). The phenomenon of cross reactivity found among related viruses has earlier led to the development of immunity as in the case of smallpox (details in Sect. 11.1). With this background, attempts have been made to find common epitopes present in two or more related viruses for designing vaccines that could generate immunity across these viruses.

A study on four antigenically important proteins (HA, NA, NP, and M2) of H1N1, H2N2, H3N2, and H5N1 viruses revealed the presence of 18 conserved epitopes across these viruses which have the potential for future vaccines (Muñoz-Medina et al. 2015). Hendra virus and Nipah virus proteins (F, G, and M), when subjected to B and T cell epitope prediction, showed common epitopes which could be used for vaccine design against both the viruses (Saha et al. 2017). In the envelope protein of the Japanese Encephalitis virus and West Nile virus, a common conserved epitope was detected which contained both B and T cell epitopes that could find use in designing epitope-based vaccines (Slathia and Sharma 2019).

11.8.2 Vaccine Designing for Bacteria

Since the discovery of penicillin, the therapeutic interventions for bacterial diseases have increased by leaps and bounds, and antibiotics remain the most important treatment for bacterial infections. Prophylactic vaccines like DTP (Diphtheria, Tetanus, Pertussis), Hib (*Haemophilus influenzae* type B), pneumococcal are included in immunization schedules throughout the world and have been helpful in reducing the disease burden considerably. Bacteria have a larger genome and proteome as compared to viruses, therefore finding immunogenic proteins is a little laborious job. The full proteome of bacteria has been studied to find immunogenic proteins, which can be used for vaccine design. The total proteome of *M. tuberculosis* H37Rv, when used for finding the best vaccine candidates by in silico methods, revealed six novel vaccine candidates, EsxL, PE26, PPE65, PE_PGRS49, PBP1, and Erp, which could be used to design new TB vaccines (Monterrubio-López and Ribas-Aparicio 2015). In another study proteomes of three serotypes of *Shigella: S. dysenteriae* type1 (sd197), *S. flexneri* 2a (str. 301 and str. 2457T), and *S. sonnei* (ss046) were investigated to determine the common proteins of these three bacteria. The epitope prediction for these common proteins was done and five peptides were used for in vivo animal and human serum studies. The peptides elicited antibody and cytokine (Th1 and Th2) response confirming that these cross protective and conserved peptides have the potential to be used in future vaccines (Pahil et al. 2017). Studies have also been focused on a group of proteins or even a single protein for epitope prediction and vaccine design. Essential hypothetical proteins of five *Salmonella* strains were studied to find out drug and vaccine targets. Out of 106 proteins, 4 proteins were found to be immunogenic for which conserved B and T cell epitopes were predicted which can be used for future vaccine design (Sah et al. 2020). Nine epitopes were predicted from 11 multidrug resistance (MDR) proteins of *Salmonella typhi* that had the potential to generate B and T cell response and can find use in vaccine design (Jebastin and Narayanan 2019). A DNA vaccine based on cytotoxic T cell epitopes predicted from a single protein Listeriolysin-O of *Listeria monocytogenes* was constructed using in silico methods. T cell epitopes were fused in tandem, human and mouse gene constructs were made in addition to determining posttranslational modifications like phosphorylation and glycosylation (Jahangiri et al. 2011). An outer membrane protein of *Vibrio cholera* was used for epitope prediction by different tools and one surface exposed peptide was found containing both B and T cell epitopes, which could have future vaccine design applications (Rauta et al. 2016).

11.8.3 Vaccine Designing for Other Parasites

Parasitic diseases caused by helminths and protozoans are difficult to treat, as these organisms are eukaryotic in nature and drug targets that are non-homologous with host tend to be less in number. The therapeutic measures for their treatment are limited and there are no licensed vaccines for use in humans. A vaccine against *Plasmodium falciparum* "RTS, S" has been introduced under the aegis of WHO in Ghana, Kenya, and Malawi and is undergoing pilot scale trials since 2019. The efforts, therefore, are required to develop vaccines against parasitic diseases. The use of bioinformatics tools has been made to design vaccines for these diseases. A multi-epitope peptide vaccine derived from epitopes obtained from six proteins of *Onchocerca volvulus* was designed using in silico methods. The epitopes used in the peptide vaccine showed varying degrees of conservation in related species *Onchocerca ochengi, Loa loa*, *Onchocerca flexuosa*, *Brugia malayi*, and *Wuchereria bancrofi* indicating its cross protective capability. The peptide vaccine was reverse translated, codon optimized, and conceptually cloned in the pET vector after carrying out other analysis like docking, immune simulation (Shey et al. 2019). The proteome of *Taenia solium* was used to find surface accessible immunogenic proteins for which B and T cell epitopes were predicted. A peptide construct based on the epitopes was made and the structure was determined by modeling and after that, it was docked with immune receptors and finally, a gene was constructed to express the peptide vaccine (Kaur et al. 2020). B and T cell epitopes were predicted from the enolase protein of *Echinococcus granulosus* and a multi-epitope vaccine was designed after analyzing its immune response generating properties (Pourseif et al. 2019). Triosephosphate isomerase from the same organism has also been used to predict epitopes for use in vaccines (Wang and Ye 2016).

From the total proteome of *Plasmodium falciparum*, five surface accessible antigenic proteins were selected for the prediction of T cell epitopes. These epitopes upon docking and population coverage revealed their efficiency to be used in epitope-based vaccines (Pritam et al. 2019). B and T cell epitopes from AMR1, a surface exposed protein of *Plasmodium falciparum* have been predicted in a study that has the potential for use in future subunit vaccines (Sanasam and Kumar 2019). An approach for developing epitope-based vaccine *Trypanosoma cruzi* involved epitope prediction from the proteome of the pathogen. mRNA construct and the structure of the peptide vaccine comprising epitopes were made (Michel-Todó et al. 2019). Conserved T cell epitopes were predicted from variants of an amastin protein of *Trypanosoma cruzi* for future vaccine designing (Slathia and Sharma 2018). In silico prediction of T cell epitopes from promastigote surface antigen (PSA), LmlRAB (*L. major* large RAB GTPase), and histone (H2B) proteins of *Leishmania* was done followed by testing of these epitopes for inducing different cytokines in peripheral blood mononuclear cells (PBMCs) isolated from cured and healthy individuals. The epitopes were able to induce specific cytokine producing helper and cytotoxic T cells and could be used in future vaccine design (Hamrouni et al. 2020).

11.9 Limitations and Challenges

The major step involved in epitope-based vaccine designing using bioinformatics tools is the prediction of epitopes. Therefore, the accuracy of epitope prediction methods is of prime importance, as this will govern the success of vaccines in the real world. More is the accuracy of the epitope prediction methods greater are the chances of success of inducing protective immunity by the vaccine. The methods available for epitope prediction have been benchmarked using the experimental steps in many studies. The limitations and their prediction efficiency have been studied. Many new methods have been redesigned as new data becomes available. Generally, it has been observed that modern machine learning methods like SVM and ANN perform better than linear methods like PSSM. The prediction efficiency achieved in class I MHC epitope prediction is better as compared to predictions for class II MHC and B cell epitope prediction. The benchmarking of automated servers for class I MHC prediction is carried out weekly, and the results are available on the immune epitope database (IEDB). These benchmarking results show that among the participating servers, NetMHCPan is the best performing server.

The next best performing methods are SMM and ANN. The ranking scores are indicative of the performance of methods among each other and do not indicate the absolute predictive performance. The ranks are concerning each other and not in the context of their prediction efficiency (Trolle et al. 2015). Many of the binding peptides are not immunogenic, and even if they are amenable to processing and presentation, they do not act as epitopes. There are still loopholes in the methods, and the binding stability of peptide and HLA molecule has also to be taken into account. The only tool available for this is NetMHCstab (Jørgensen et al. 2014) which is an ANN-based tool and has only been trained on 13 HLA alleles. With the increase in data about HLA alleles and their binding peptides, these tools are bound to increase their efficiency in the future.

The prediction methods for class II MHC are yet to achieve the efficiency of class I MHC predictors. In most of the methods, the prediction is limited to HLA-DR alleles, and few servers like NetMHCII 2.3, RANKPEP, NetMHCIIPan carry out predictions for HLA-DP and HLA-DQ as well. The nature of peptides binding to class II MHC is different from that showing binding to class I. Peptides binding to class II MHC molecules have a binding core rather than anchor residues seen in class I MHC binding peptides. Besides, the peptides binding to class II MHC are longer, and the position of the binding core is not fixed (Kindt et al. 2007). The peptide binding mode of class II is less specific than class I, and the genotype structure of class II allotypes is more complicated. This makes designing of class II MHC binding prediction methods more challenging. The tools need to address these issues, and the lack of data available makes these prediction tools less efficient.

The benchmarking of class II predictors is also done weekly, and among the different prediction tools, the NN-align method which is the basis of NetMHC2.0 (Nielsen and Lund 2009) outperforms the other methods. NetMHCIIPan is the next best performing method followed by Comblib matrices, (Sidney et al. 2008) a method available at IEDB analysis resource and SMM-align. Next-generation sequencing (NGS) has increased the inflow of genomic data in an unprecedented manner, and the data for HLA alleles is now being generated at a high pace. This data along with other high throughput experimental data about the class II MHC-peptide binding is required to increase the efficiency of prediction tools.

Prediction of continuous B cell epitopes follows the same principles, albeit the length of epitopes is not fixed. For discontinuous epitopes, the prediction requires different approaches as the classic machine learning methods need continuous sequence data (Backert and Kohlbacher 2015). There are fewer benchmarking studies for B cell epitope predictors, and most of them conclude that the efficiency of these methods is yet far from meeting the requirements in the biological context. Since there are no universal properties that are present in antigenic epitopes but absent in other protein surfaces, therefore, designing methods for prediction is a challenging job. The methods for linear epitope prediction are based on the hypothesis that certain amino acids occur more frequently in the epitopic regions. A benchmarking study for linear B cell epitope prediction concluded that these methods require improvement, and new approaches need to be taken into account for devising more efficient methods (Blythe and Flower 2005).

In a study on discontinuous epitope prediction tools, it was found out that DiscoTope and PEPITO have the highest predictive performance (Kringelum et al. 2012). The prediction efficiency of different discontinuous epitope predictors was done by Yao et al. (Yao et al. 2013), wherein they found out that the highest prediction accuracy obtained was only 25.6% by the EPMeta server. In the case of lowering the threshold for prediction, the prediction accuracy rose to 31.6%. There is a huge scope of improvement in the B cell epitope prediction methods to reach the accuracy levels of T cell epitope prediction methods. An important consideration for designing epitope-based vaccines is the prevalence of HLA alleles in the target populations. HLA alleles have a varied affinity towards the binding peptides and their distribution also varies in different population groups. The selection of epitopes for vaccine designing without taking this into account may fail vaccine to provide immunity (Oyarzun and Kobe 2015).

The challenge in vaccine design using only epitopes is that the peptides mostly fail to generate the immune response required for producing long lasting immunity. Because of their small size, the peptides are often weakly immunogenic and this thwarts the basic function of designing vaccines. Epitope-based peptide vaccines are mostly known to initiate antibodies (humoral response) and fail to induce T cell-mediated immunity. The generation of humoral immunity is not enough to protect against disease (Li et al. 2014). Since the molecular size is an important feature for immune response development such as small-sized peptides harboring epitopes need to be conjugated with carriers/adjuvants. The "RTS, S" vaccine for *Plasmodium falciparum* developed recently is based on truncated (C terminal end) of circumsporozoite protein (CSP) containing B and T cell epitopes. However, this 188 amino acid part of CSP has been fused with HBsAg protein to generate an immunogenic construct (Oyarzún and Kobe 2016). The CSP alone is weakly immunogenic predominantly generating antibody response but its fusion with HBsAg enhances its immunogenicity (Collins et al. 2017). Therefore, suitable carriers are required for vaccines based on epitopes as most of the times they are not enough immunogenic to induce both cell-mediated and humoral immunity.

The use of carriers/adjuvants becomes critical in designing epitope-based vaccines and many studies involving in silico designing of vaccines have taken this into account by the addition of adjuvant in the final vaccine construct (Shey et al. 2019; Khatoon et al. 2017). The usage of adjuvants like toxoids, Freund's incomplete adjuvant, and the most recent TLR (Toll-like receptor) agonists enhances the immunogenicity of vaccines. These adjuvants are an essential requirement for the success of epitope-based vaccines; however, in silico studies can only design a construct using protein-based adjuvants and for other adjuvants, lab studies need to be undertaken.

11.10 Conclusion

In silico methods can provide a huge impetus to vaccine design and development. The B and T cell epitope prediction methods form the core of in silico epitope-based vaccine designing. The prediction of epitopes reduces the huge cost and labor involved in experimentally finding out the epitopes. These methods ease out the efforts involved in deducing T and B cell epitopes. The methods for T cell epitope prediction are more advanced in terms of prediction accuracy when compared to B cell epitope prediction tools. These methods need to be improved so that prediction accuracy can be increased, and we may be able to design more efficient vaccines in the future. The tools and methods for the analysis of the predicted epitopes though appear to be subsidiary yet their importance cannot be ignored. The checking of epitope clusters to avoid undue repetition of epitopes, checking their conservancy, finding toxic and allergic epitopes are essentiality that cannot be done away with. Analyzing the population coverage that can be achieved by the epitopes has far reaching consequences for the success or failure of vaccines in different population groups. Molecular docking and dynamics simulation strengthen the chances of epitope binding to MHC molecules. Finally, the construction of vaccines using these rational approaches can strengthen the possibility of it being successful, which of course needs to be validated by laboratory studies.

Competing Interest The authors declare that there are no competing interests.

References

Andersen KG, Rambaut A, Lipkin WI, Holmes EC, Garry RF (2020) The proximal origin of SARS-CoV-2. Nat Med 26(4):450–452

Andreatta M, Nielsen M (2016) Gapped sequence alignment using artificial neural networks: application to the MHC class I system. Bioinformatics 32(4):511–517

Ansari HR, Raghava GPS (2010) Identification of conformational B-cell Epitopes in an antigen from its primary sequence. Immunome Res 6(1):6

Arai R, Ueda H, Kitayama A, Kamiya N, Nagamune T (2001) Design of the linkers which effectively separate domains of a bifunctional fusion protein. Protein Eng 14(8):529–532

Backert L, Kohlbacher O (2015) Immunoinformatics and epitope prediction in the age of genomic medicine. Genome Med 7(1):119

Bhasin M, Raghava GPS (2004a) Prediction of CTL epitopes using QM, SVM and ANN techniques. Vaccine 22:3195–3204

Bhasin M, Raghava GPS (2004b) Analysis and prediction of affinity of TAP binding peptides using cascade SVM. Protein Sci 13(3):596–607

Bhasin M, Raghava GPS (2005) Pcleavage: an SVM based method for prediction of constitutive proteasome and immunoproteasome cleavage sites in antigenic sequences. Nucleic Acids Res 33(2):W202–W207

Bhasin M, Raghava GPS (2007) A hybrid approach for predicting promiscuous MHC class I restricted T cell epitopes. J Biosci 32(1):31–42

Blythe MJ, Flower DR (2005) Benchmarking B cell epitope prediction: underperformance of existing methods. Protein Sci 14(1):246–248

Bui HH, Sidney J, Dinh K, Southwood S, Newman MJ, Sette A (2006) Predicting population coverage of T-cell epitope-based diagnostics and vaccines. BMC Bioinf 7:153

Bui HH, Sidney J, Li W, Fusseder N, Sette A (2007) Development of an epitope conservancy analysis tool to facilitate the design of epitope-based diagnostics and vaccines. BMC Bioinf 8:361

Chakraborty S, Barman A, Deb B (2020) Japanese encephalitis virus: a multi-epitope loaded peptide vaccine formulation using reverse vaccinology approach. Infect Genet Evol 78:104106

Chen J, Liu H, Yang J, Chou KC (2007) Prediction of linear B-cell epitopes using amino acid pair antigenicity scale. Amino Acids 33(3):423–428

Chen F, Cao S, Xin J, Luo X (2013) Ten years after SARS: where was the virus from? J Thorac Dis 5(2):S163–S167

Chou KC, Shen HB (2008) Cell-PLoc: a package of Web servers for predicting subcellular localization of proteins in various organisms. Nat Protoc 3(2):153–162

Collins KA, Snaith R, Cottingham MG, Gilbert SC, Hill AV (2017) Enhancing protective immunity to malaria with a highly immunogenic virus-like particle vaccine. Sci Rep 7:46621

Dash R, Das R, Junaid M, Akash MF, Islam A, Hosen SZ (2017) In silico-based vaccine design against Ebola virus glycoprotein. Adv Appl Bioinforma Chem 10:11–28

De Groot AS, Moise L, McMurry JA, Martin W (2009) Epitope-based Immunome-derived vaccines: a strategy for improved design and safety. In: Clinical applications of immunomics. Springer, New York, pp 39–69

Dhanda SK, Vaughan K, Schulten V, Grifoni A, Weiskopf D, Sidney J, Peters B, Sette A (2018) Development of a novel clustering tool for linear peptide sequences. Immunology 155 (3):331–345

Dimitrov I, Naneva L, Doytchinova I, Bangov I (2014a) AllergenFP: allergenicity prediction by descriptor fingerprints. Bioinformatics 30(6):846–851

Dimitrov I, Bangov I, Flower DR, Doytchinova I (2014b) AllerTOP v.2—a server for in silico prediction of allergens. J Mol Model 20(6):2278

Dönnes P, Kohlbacher O (2006) SVMHC: a server for prediction of MHC-binding peptides. Nucleic Acids Res 34(2):W194–W197

Doytchinova IA, Flower DR (2007) VaxiJen: a server for prediction of protective antigens, tumour antigens and subunit vaccines. BMC Bioinf 8(1):4

Doytchinova IA, Guan P, Flower DR (2006) EpiJen: a server for multistep T cell epitope prediction. BMC Bioinf 7(1):131

EL-Manzalawy Y, Dobbs D, Honavar V (2008) Predicting linear B cell epitopes using string kernels. J Mol Recognit 21(4):243–255

Fitzpatrick M (2006) The cutter incident: how America's first polio vaccine led to a growing vaccine crisis. J R Soc Med 99(3):156

Grote A, Hiller K, Scheer M, Münch R, Nörtemann B, Hempel DC, Jahn D (2005) JCat: a novel tool to adapt codon usage of a target gene to its potential expression host. Nucleic Acids Res 33 (2):W526–W531

Guan P, Doytchinova IA, Zygouri C, Flower DR (2003) MHCPred: bringing a quantitative dimension to the online prediction of MHC binding. Appl Bioinforma 2:63–66

Gupta S, Kapoor P, Chaudhary K, Gautam A, Kumar R, Raghava GP, Open Source Drug Discovery Consortium (2013) In silico approach for predicting toxicity of peptides and proteins. PLoS One 8(9):e73957

Hakenberg J, Nussbaum AK, Schild H, Rammensee HG, Kuttler C, Holzhütter HG, Kloetzel PM, Kaufmann SH, Mollenkopf HJ (2003) MAPPP: MHC class I antigenic peptide processing prediction. Appl Bioinforma 2(3):155–158

Hamrouni S, Bras-Gonçalves R, Kidar A, Aoun K, Chamakh-Ayari R, Petitdidier E, Messaoudi Y, Pagniez J, Lemesre JL, Meddeb-Garnaoui A (2020) Design of multi-epitope peptides containing HLA class-I and class-II-restricted epitopes derived from immunogenic Leishmania proteins, and evaluation of CD4+ and CD8+ T cell responses induced in cured cutaneous leishmaniasis subjects. PLoS Negl Trop Dis 14(3):e0008093

He Y, Xiang Z, Mobley HL (2010) Vaxign: the first web-based vaccine design program for reverse vaccinology and applications for vaccine development. J Biomed Biotechnol 2010:297505

Hewitt EW (2003) The MHC class I antigen presentation pathway: strategies for viral immune evasion. Immunology 110(2):163–169

Jahangiri A, Rasooli I, Gargari SL, Owlia P, Rahbar MR, Amani J, Khalili S (2011) An in silico DNA vaccine against Listeria monocytogenes. Vaccine 29(40):6948–6958

Jebastin T, Narayanan S (2019) In silico epitope identification of unique multidrug resistance proteins from Salmonella Typhi for vaccine development. Comput Biol Chem 78:74–80

Jensen KK, Andreatta M, Marcatili P, Buus S, Greenbaum JA, Yan Z, Sette A, Peters B, Nielsen M (2018) Improved methods for predicting peptide binding affinity to MHC class II molecules. Immunology 154(3):394–406

Jespersen MC, Peters B, Nielsen M, Marcatili P (2017) BepiPred-2.0: improving sequence-based B-cell epitope prediction using conformational epitopes. Nucleic Acids Res 45(W1):W24–W29

Jørgensen KW, Rasmussen M, Buus S, Nielsen M (2014) Net MHC stab–predicting stability of peptide–MHCI complexes; impacts for cytotoxic T lymphocyte epitope discovery. Immunology 141(1):18–26

Jurtz V, Paul S, Andreatta M, Marcatili P, Peters B, Nielsen M (2017) NetMHCpan-4.0: improved peptide–MHC class I interaction predictions integrating eluted ligand and peptide binding affinity data. J Immunol 199(9):3360–3368

Kaur R, Arora N, Jamakhani MA, Malik S, Kumar P, Anjum F, Tripathi S, Mishra A, Prasad A (2020) Development of multi-epitope chimeric vaccine against Taenia solium by exploring its proteome: an in silico approach. Expert Rev Vaccines 19(1):105–114

Keşmir C, Nussbaum AK, Schild H, Detours V, Brunak S (2002) Prediction of proteasome cleavage motifs by neural networks. Protein Eng 15(4):287–296

Khan S, Khan A, Rehman AU, Ahmad I, Ullah S, Khan AA, Ali SS, Afridi SG, Wei DQ (2019) Immunoinformatics and structural vaccinology driven prediction of multi-epitope vaccine against Mayaro virus and validation through in-silico expression. Infect Genet Evol 73:390–400

Khatoon N, Pandey RK, Prajapati VK (2017) Exploring Leishmania secretory proteins to design B and T cell multi-epitope subunit vaccine using immunoinformatics approach. Sci Rep 7(1):1–2

Kindt TJ, Goldsby RA, Osborne BA, Kuby J (2007) Kuby immunology. W. H. Freeman, New York

Kozakov D, Hall DR, Xia B, Porter KA, Padhorny D, Yueh C, Beglov D, Vajda S (2017) The ClusPro web server for protein–protein docking. Nat Protoc 12(2):255–278

Krawczyk K, Liu X, Baker T, Shi J, Deane CM (2014) Improving B-cell epitope prediction and its application to global antibody-antigen docking. Bioinformatics 30(16):2288–2294

Kringelum JV, Lundegaard C, Lund O, Nielsen M (2012) Reliable B cell epitope predictions: impacts of method development and improved benchmarking. PLoS Comput Biol 8(12): e1002829

Lamiable A, Thévenet P, Rey J, Vavrusa M, Derreumaux P, Tufféry P (2016) PEP-FOLD3: faster de novo structure prediction for linear peptides in solution and in complex. Nucleic Acids Res 44(W1):W449–W454

Larsen MV, Lundegaard C, Lamberth K, Buus S, Lund O, Nielsen M (2007) Large-scale validation of methods for cytotoxic T-lymphocyte epitope prediction. BMC Bioinf 8(1):424

Lata S, Bhasin M, Raghava GPS (2007) Application of machine learning techniques in predicting MHC binders. Methods Mol Biol 409:201–215

Li W, Joshi MD, Singhania S, Ramsey KH, Murthy AK (2014) Peptide vaccine: progress and challenges. Vaccine 2(3):515–536

Liang S, Zheng D, Standley DM, Yao B, Zacharias M, Zhang C (2010) EPSVR and EPMeta: prediction of antigenic epitopes using support vector regression and multiple server results. BMC Bioinf 11(1):381

Liu W, Wan J, Meng X, Flower DR, Li T (2007) In silico prediction of peptide-MHC binding affinity using SVRMHC. Methods Mol Biol 409:283–291

Luca S, Mihaescu T (2013) History of BCG vaccine. Maedica 8(1):53–58

Maurer-Stroh S, Krutz NL, Kern PS, Gunalan V, Nguyen MN, Limviphuvadh V, Eisenhaber F, Gerberick GF (2019) AllerCatPro—prediction of protein allergenicity potential from the protein sequence. Bioinformatics 35(17):3020–3027

Mena I, Nelson MI, Quezada-Monroy F, Dutta J, Cortes-Fernández R, Lara-Puente JH, Castro-Peralta F, Cunha LF, Trovão NS, Lozano-Dubernard B, Rambaut A (2016) Origins of the 2009 H1N1 influenza pandemic in swine in Mexico. elife 5:e16777

Michel-Todó L, Reche PA, Bigey P, Pinazo MJ, Gascón J, Alonso-Padilla J (2019) In silico design of an epitope-based vaccine ensemble for Chagas disease. Front Immunol 10:2698

Mohd HA, Al-Tawfiq JA, Memish ZA (2016) Middle East respiratory syndrome coronavirus (MERS-CoV) origin and animal reservoir. Virol J 13(1):87

Monterrubio-López GP, Ribas-Aparicio RM (2015) Identification of novel potential vaccine candidates against tuberculosis based on reverse vaccinology. Biomed Res Int 2015:483150

Morais V, Dee V, Suárez N (2018) Purification of capsular polysaccharides of Streptococcus pneumoniae: traditional and new methods. Front Bioeng Biotechnol 6:145

Muñoz-Medina JE, Sánchez-Vallejo CJ, Méndez-Tenorio A, Monroy-Muñoz IE, Angeles-Martínez J, Santos Coy-Arechavaleta A, Santacruz-Tinoco CE, González-Ibarra J, Anguiano-Hernández YM, González-Bonilla CR, Ramón-Gallegos E (2015) In silico identification of highly conserved epitopes of influenza A H1N1, H2N2, H3N2, and H5N1 with diagnostic and vaccination potential. Biomed Res Int 2015:813047

Nielsen M, Lund O (2009) NN-align. An artificial neural network-based alignment algorithm for MHC class II peptide binding prediction. BMC Bioinf 10(1):296

Nielsen M, Lundegaard C, Lund O, Keşmir C (2005) The role of the proteasome in generating cytotoxic T-cell epitopes: insights obtained from improved predictions of proteasomal cleavage. Immunogenetics 57(1–2):33–41

Nussbaum AK, Kuttler C, Hadeler KP, Rammensee HG, Schild H (2001) PAProC: a prediction algorithm for proteasomal cleavages available on the www. Immunogenetics 53(2):87–94

Oyarzun P, Kobe B (2015) Computer-aided design of T-cell epitope-based vaccines: addressing population coverage. Int J Immunogenet 42(5):313–321

Oyarzún P, Kobe B (2016) Recombinant and epitope-based vaccines on the road to the market and implications for vaccine design and production. Hum Vaccin Immunother 12 (3):763–767

Pahil S, Taneja N, Ansari HR, Raghava GPS (2017) In silico analysis to identify vaccine candidates common to multiple serotypes of Shigella and evaluation of their immunogenicity. PLoS One 12:8

Pierce BG, Wiehe K, Hwang H, Kim BH, Vreven T, Weng Z (2014) ZDOCK server: interactive docking prediction of protein–protein complexes and symmetric multimers. Bioinformatics 30 (12):1771–1773

Ponomarenko J, Bui HH, Li W, Fusseder N, Bourne PE, Sette A, Peters B (2008) ElliPro: a new structure-based tool for the prediction of antibody epitopes. BMC Bioinf 9(1):514

Pourseif MM, Yousefpour M, Aminianfar M, Moghaddam G, Nematollahi A (2019) A multi-method and structure-based in silico vaccine designing against Echinococcus granulosus through investigating enolase protein. Bioimpacts 9(3):131–144

Pritam M, Singh G, Swaroop S, Singh AK, Singh SP (2019) Exploitation of reverse vaccinology and immunoinformatics as promising platform for genome-wide screening of new effective vaccine candidates against Plasmodium falciparum. BMC Bioinf 19(13):468

Rammensee HG, Bachmann J, Emmerich NPN, Bachor OA, Stevanović S (1999) SYFPEITHI: database for MHC ligands and peptide motifs. Immunogenetics 50(3–4):213–219

Rauta PR, Ashe S, Nayak D, Nayak B (2016) In silico identification of outer membrane protein (Omp) and subunit vaccine design against pathogenic *Vibrio cholerae*. Comput Biol Chem 65:61–68

Reche PA, Glutting JP, Reinherz EL (2002) Prediction of MHC class I binding peptides using profile motifs. Hum Immunol 63(9):701–709

Riedel S (2005) Edward Jenner and the history of smallpox and vaccination. Proc Baylor Univ Med Cent 18(1):21–25

Roy A, Kucukural A, Zhang Y (2010) I-TASSER: a unified platform for automated protein structure and function prediction. Nat Protoc 5(4):725–738

Rubinstein ND, Mayrose I, Martz E, Pupko T (2009) Epitopia: a web-server for predicting B-cell epitopes. BMC Bioinf 10(1):287

Sah PP, Bhattacharya S, Banerjee A, Ray S (2020) Identification of novel therapeutic target and epitopes through proteome mining from essential hypothetical proteins in Salmonella strains: an in silico approach towards antivirulence therapy and vaccine development. Infect Genet Evol 83:104315

Saha S, Raghava GPS (2004) BcePred: prediction of continuous B-cell epitopes in antigenic sequences using physico-chemical properties. In: International conference on artificial immune systems. Springer, Berlin, pp 197–204

Saha S, Raghava GPS (2006a) Prediction of continuous B cell epitopes in an antigen using recurrent neural network. Proteins 65(1):40–48

Saha S, Raghava GPS (2006b) AlgPred: prediction of allergenic proteins and mapping of IgE epitopes. Nucleic Acids Res 34:W202–W209

Saha CK, Hasan MM, Hossain MS, Jahan MA, Azad AK (2017) In silico identification and characterization of common epitope-based peptide vaccine for Nipah and Hendra viruses. Asian Pac J Trop Med 10(6):529–538

Sanasam BD, Kumar S (2019) In-silico structural modeling and epitope prediction of highly conserved Plasmodium falciparum protein AMR1. Mol Immunol 116:131–139

Schwartz M (2001) The life and works of Louis Pasteur. J Appl Microbiol 91(4):597–601

Sela-Culang I, Ashkenazi S, Peters B, Ofran Y (2014a) PEASE: predicting B-cell epitopes utilizing antibody sequence. Bioinformatics 31(8):1313–1315

Sela-Culang I, Benhnia MRI, Matho MH, Kaever T, Maybeno M, Schlossman A, Nimrod G, Li S, Xiang Y, Zajonc D (2014b) Using a combined computational-experimental approach to predict antibody-specific B cell epitopes. Structure 22(4):646–657

Shey RA, Ghogomu SM, Esoh KK, Nebangwa ND, Shintouo CM, Nongley NF, Asa BF, Ngale FN, Vanhamme L, Souopgui J (2019) In-silico design of a multi-epitope vaccine candidate against onchocerciasis and related filarial diseases. Sci Rep 9(1):4409

Sidney J, Assarsson E, Moore C, Ngo C, Pinilla C, Sette A, Peters B (2008) Quantitative peptide binding motifs for 19 human and mouse MHC class I molecules derived using positional scanning combinatorial peptide libraries. Immunome Res 4(1):2

Sims LD, Domenech J, Benigno C, Kahn S, Kamata A, Lubroth J, Martin V, Roeder P (2005) Origin and evolution of highly pathogenic H5N1 avian influenza in Asia. Vet Rec 157 (6):159–164

Singh H, Raghava GPS (2001) ProPred: prediction of HLA-DR binding sites. Bioinformatics 17 (12):1236–1237

Singh H, Raghava GPS (2003) ProPred1: prediction of promiscuous MHC Class-I binding sites. Bioinformatics 19(8):1009–1014

Singh H, Ansari HR, Raghava GPS (2013) Improved method for linear B-cell epitope prediction using antigen's primary sequence. PLoS One 8(5):e62216

Slathia PS, Sharma P (2018) Conserved epitopes in variants of amastin protein of Trypanosoma cruzi for vaccine design: a bioinformatics approach. Microb Pathog 125:423–430

Slathia PS, Sharma P (2019) A common conserved peptide harboring predicted T and B cell epitopes in domain III of envelope protein of Japanese Encephalitis Virus and West Nile Virus for potential use in epitope based vaccines. Comp Immunol Microbiol Infect Dis 65:238–245

Soria-Guerra RE, Nieto-Gomez R, Govea-Alonso DO, Rosales-Mendoza S (2015) An overview of bioinformatics tools for epitope prediction: implications on vaccine development. J Biomed Inform 53:405–414

Sweredoski MJ, Baldi P (2008) PEPITO: improved discontinuous B-cell epitope prediction using multiple distance thresholds and half sphere exposure. Bioinformatics 24(12):1459–1460

Trolle T, Metushi IG, Greenbaum JA, Kim Y, Sidney J, Lund O, Sette A, Peters B, Nielsen M (2015) Automated benchmarking of peptide-MHC class I binding predictions. Bioinformatics 31(13):2174–2181

Ul Qamar MT, Saleem S, Ashfaq UA, Bari A, Anwar F, Alqahtani S (2019) Epitope-based peptide vaccine design and target site depiction against Middle East Respiratory Syndrome Coronavirus: an immune-informatics study. J Transl Med 17:362

Van Der Spoel D, Lindahl E, Hess B, Groenhof G, Mark AE, Berendsen HJ (2005) GROMACS: fast, flexible, and free. J Comput Chem 26(16):1701–1718

Wang F, Ye B (2016) In silico cloning and B/T cell epitope prediction of triosephosphate isomerase from Echinococcus granulosus. Parasitol Res 115(10):3991–3998

Waterhouse A, Bertoni M, Bienert S, Studer G, Tauriello G, Gumienny R, Heer FT, de Beer TA, Rempfer C, Bordoli L, Lepore R (2018) SWISS-MODEL: homology modelling of protein structures and complexes. Nucleic Acids Res 46(W1):W296–W303

WHO Data (n.d.) Hepatitis B. https://www.who.int/biologicals/vaccines/Hepatitis_B/en/

Xu D, Zhang Y (2012) Ab initio protein structure assembly using continuous structure fragments and optimized knowledge-based force field. Proteins 80(7):1715–1735

Yao B, Zhang L, Liang S, Zhang C (2012) SVMTriP: a method to predict antigenic epitopes using support vector machine to integrate tri-peptide similarity and propensity. PLoS One 7(9):e45152

Yao B, Zheng D, Liang S, Zhang C (2013) Conformational B-cell epitope prediction on antigen protein structures: a review of current algorithms and comparison with common binding site prediction methods. PLoS One 8(4):e62249

Yu CS, Chen YC, Lu CH, Hwang JK (2006) Prediction of protein subcellular localization. Proteins 64(3):643–651

Zhang Q, Wang P, Kim Y, Haste-Andersen P, Beaver J, Bourne PE, Bui HH, Buus S, Frankild S, Greenbaum J, Lund O (2008) Immune epitope database analysis resource (IEDB-AR). Nucleic Acids Res 36(2):W513–W518

Zhang GL, DeLuca DS, Keskin DB, Chitkushev L, Zlateva T, Lund O, Reinherz EL, Brusic V (2011) MULTIPRED2: a computational system for large-scale identification of peptides predicted to bind to HLA supertypes and alleles. J Immunol Methods 374(1–2):53–61

Zhang T, Wu Q, Zhang Z (2020) Probable pangolin origin of SARS-CoV-2 associated with the COVID-19 outbreak. Curr Biol 30(7):1346–1351

Zobayer N, Hossain AA, Rahman MA (2019) A combined view of B-cell epitope features in antigens. Bioinformation 15(7):530–534

Machine Learning Approaches to Rational Drug Design 12

Salman Akhtar, M. Kalim A. Khan, and Khwaja Osama

Abstract

Pharmaceutical industries are multibillionaire setups with a diligent team of scientists, researchers, technical manpower, and investors. A major concern of such industries is to always curtail the time and cost factor associated with them. Bioinformatics involving machine learning (ML) methods have come to the forefront to address this problem. The predictive and statistical efficacy of ML methodologies has even proven to propose better leads than a wet lab pipeline. This chapter aims to give a brief overview of underlying principles of mainly GAs and ANNs as popular ML algorithms and deeper insight into their robust applications in the field of modern day drug design. It also attempts to share the future prospects of such ML techniques and their limitations with possible solutions hereafter.

Keywords

Machine learning · Genetic algorithms · Artificial neural networks · Drug designing · Deep learning · Support vector machines

12.1 Drug Industry

In the current era, drug industries have expanded rapidly and have seen enormous growth in terms of infrastructure, scientific manpower, technical handling, and business product outputs. Health is a major concern nowadays of every livelihood and various health and pharmaceutical products have become a routine part of the daily diet of many individuals. More importantly, a rapid increase in the knowledge

S. Akhtar (✉) · M. K. A. Khan · K. Osama
Department of Bioengineering, Integral University, Lucknow, India

D. B. Singh (ed.), *Computer-Aided Drug Design*,
https://doi.org/10.1007/978-981-15-6815-2_12

and recognition of various diseases, improved diagnostics tools, government promotional health schemes for all and varied networking between scientific researchers around the world has facilitated the need of development and subsequent financial investment in more and more pharmaceutical industrial setups (Zhong et al. 2018; Leelananda and Lindert 2016; Fox and Kriegl 2006).

Usually, the development of a novel drug against a particular disease requires a timeline of 12–15 years and a hugely expensive experimental setup accounting to nearly 200–800 million$. However, with the development of novel and affordable bioinformatics in silico setups, the time and cost required in the development of new drug molecules has been significantly curtailed. Bioinformatics, an in silico science comprising of various computer based prediction, search and optimization methods, has not only opened newer dimensions and directions in novel drug development but has also been instrumental in reducing the cost and time required against the expensive wet lab setups. Normally US Food and Drug Administration (FDA) approves a drug as defined by "a substance intended for use in the diagnosis, cure, mitigation, treatment, or prevention of disease affecting the structure or any function of the body." In a simpler sense, we can say a drug molecule as a chemical compound or more recently organic peptides which has the potential to interact with a specific biological target. Target usually includes any of the biological macromolecules that may be proteins, nucleic acids, carbohydrates, and lipids. Proteins among these stand to be the most prominent target accounting for nearly 95% of drugs designed against them followed by nucleic acids (3–4%), lipids, and carbohydrates (Yosipof et al. 2018; Huang et al. 2017).

12.2 Drug Discovery Pipeline

The important steps involved in the process of drug discovery are as described below (Fig. 12.1).

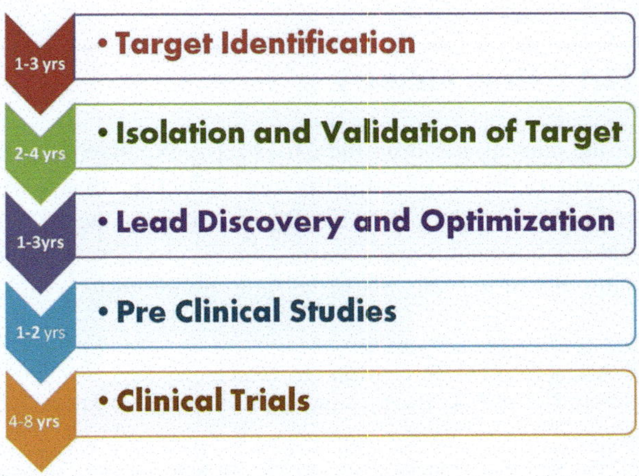

Fig. 12.1 Different phases of drug designing

12.2.1 Target Discovery

Target identification is the major step in any drug discovery process. The clinical relevance, therapeutics, and disease-causing potential of the selected target actually govern the efficacy of the designed drug molecule inside the body. The step may include a blend of diverse biological assays to high-throughput screening studies in the identification of a single target for a particular disease (Huang et al. 2018; Leelananda and Lindert 2016).

12.2.2 Target Validation

Target validation is carried out mainly by applying biotechnological innovational wet lab studies involving gene knock out in animal models, small molecule agonists/antagonists, aptamers, siRNA, ribozymes, and neutralizing antibodies. The basic aim of target validation studies is to validate the therapeutic efficacy of the selected target and to scientifically reveal that the selected target possesses significant disease-causing potential.

12.2.3 Lead Identification

Lead compounds are all those chemical compounds that possess some but not all properties of the drug molecule. These are the starting clinical agents from where potential drug candidates can be synthesized. The discovery of lead compounds is fastened up by employing virtual HTS studies, where millions of chemical compounds are screened using computer-generated models. Molecular docking along with pharmacophore modeling offers faster, effective, and cost-efficient screening solutions in large drug discovery projects (Schneider and Bohm 2002).

12.2.4 Lead Optimization

This process involves the customized structural and functional modification of lead compounds to enhance their inhibition potential against a particular target. Involving various structure-based and ligand-based drug designing strategies and quantitative structure–activity relationship (QSAR) studies it promotes structural and functional modification of leads for its increased efficacy. This is often the most critical and diligent step in drug discovery.

12.2.5 Preclinical Phase

It involves the studies on animal systems to access the dosage and adverse side effects associated with drug molecules.

12.2.6 Clinical Trials

The clinical lead development involves 4 phases:

12.2.6.1 Phase I

A small group of normal healthy volunteers is selected in this phase and the developed drug is tested on them to assess its safety, tolerability, dosage levels, pharmacokinetics, and pharmacodynamics inside the human body. 80% of drugs are reported to fail in Phase I clinical trial.

12.2.6.2 Phase II

Phase II involves the controlled clinical studies in a hospital conducted on a group of patients, to obtain average data on the efficacy and dosage level of the drug. The trial is carried out in an unbiased manner using a placebo and newer drug simultaneously, in a group of patients with particular indications and symptoms of the disease.

12.2.6.3 Phase III

Phase III studies involve randomized controlled trials on large patient groups in medical institutions and hospitals. Market regulatory affairs and commercial launch decisions of the drug are also accompanied in this step.

12.2.6.4 Phase IV

It involves long term monitoring process after the drug is launched in the market. Post-launch this may be mandated or initiated by the pharmaceutical company in assistance with a team of medical representatives and medicos. It is destined to detect long term rare or serious adverse effects of medicine over a large patient population.

12.3 Dimensions and Complexity of the Problem and Role of ML Techniques

With the ever-increasing number of potential lead compounds derived through virtual screening studies and its subsequent attempt to prioritize them based on their functional groups, physicochemical descriptors, and biological activities has compelled the researchers to use statistical function optimizers and prediction algorithms in handling such big data. The dimensionality in terms of thousands of molecular descriptors and complexity in terms of libraries of millions of compounds guarantees the application of computers as inevitable rather than any sort of manual calculations.

Chemoinformatics, a built-up control in which meaningful data are extricated, processed, and extrapolated from chemical structures plays an important role in solving such a problem (Lo et al. 2018). Chemoinformatics is the utilization of informatics strategies to fix chemical issues including chemical information retrieval and extraction, database exploration, and molecular chart mining (Varnek and Baskin 2011). The design of new drugs is an extremely difficult and complex issue (Schneider and Schneider 2016). The scalable search space for new and unfamiliar particles is one of the major barriers in drug design and configuration. Scientific experts need to choose and analyze particles from this extensive space to discover molecules that are dynamically active towards the respective target protein. Drug discovery is an enormously expensive and sluggish procedure representing various formidable difficulties. In this manner, technological advances in drug research and preclinical improvement could bring down the expenses of conveying another novel drug to the market (Pu et al. 2019).

In silico strategies try to quicken this procedure and diminish the expensive late-stage breakdown by utilizing the chemical information produced through computational techniques and high-performance assays to reveal hidden connections among the data. Screening vast virtual libraries of compounds with improved biological properties such as explicitness and selectivity toward the respective target, lower toxicity, or reduced cost help to discover new chemical or synthetic inhibitors. With the advancement of "big data" from HTS and combinatorial synthesis, ML has turned into a key apparatus for drug designers to extract synthetic information from substantial compound libraries to devise drugs with essential biological features (Lo et al. 2018). Other modules of chemoinformatics additionally incorporate computer-aided drug synthesis, chemical space investigation, pharmacophore, and scaffold testing, library preparation, etc. (Kapetanovic 2008). Various ML-based strategies thus have been created and ubiquitously used to identify and develop new drugs having superior biological activities to uncover complex relations between chemical and their biological targets. Mathematical mining of chemical graphs enables the inference of a group of 2D or 3D chemical descriptors bundled in a variety of ML models and probabilistic tasks as chemical fingerprints. Integrating numerous data types and sources or the so-called data fusion procedures which collate hereditary, structural, and pharmacological information in different level of organisms seems to be critical for the disclosure of more efficacious and safer drugs (Chen et al. 2018; Searls 2005).

In recent years, ML techniques have gained significant attention as a prominent research and development tool in rational drug designing approaches. Precisely genetic algorithms (GA) and artificial neural networks (ANN) have come to the forefront to uptake the diversified challenges faced in pharmaceutical industries (Zhong et al. 2018; Lavecchia 2015). GAs in the category of evolutionary algorithms and ANNs as a case of artificial intelligence (AI) have seen immense applications as ML techniques in searching and optimizing compounds, predicting active site and binding conformations, the establishment of quantitative structure–activity relationships, gene prediction, pharmacophore analysis, design of combinatorial libraries, and so forth (Goswami et al. 2018; Sayeed et al. 2016; Gupta et al. 2014; Arif et al. 2013). Furthermore, applications of NNs in drug design are concerned in areas like lead discovery; designation and estimation of biological activity; and A (absorption), D(distribution), M(metabolism), E(excretion)/Tox(toxicity) assets; multidimensional data processing; compound library analogy; combinatorial library striking similarity and diversity analysis; HTS data analysis (Terfloth and Gasteiger 2001).

Techniques of ML can be comprehensively named as supervised or unsupervised learning. Training data are assigned to labels for supervised learning, and once prepared; the model can foresee labels for specific data inputs. Regression analysis, random forests, naive Bayes, ANN, k-nearest neighbor (kNN), SVM algorithms are some popular instances of supervised ML models. While unsupervised ML methods gain directly from unlabeled data of molecular patterns, i.e. it needs only input data with no relating output factors. Here, the basic pattern or structure in the data is determined for the use of future analysis and prediction. Independent components analysis (ICA) and principal components analysis (PCA) are the common algorithms of unsupervised learning. These problems could be further categorized into clustering and association groups (Yang et al. 2018).

Capacity in handling the large amount of data employing immense computational power, these ML strategies are certainly proving a state-of-art in the field of rational drug design.

12.4 Genetic Algorithms

GA is a family of population-based computational models that are inspired by nature's natural process of evolution. GAs come under the category of search and optimization algorithms and are often viewed as function optimizers to solve a varied kind of problems. Initially given by John Holland in 1975, GAs have come a long way in providing solutions to the problems which have multiple inputs and multiple outputs (Goswami et al. 2018; Mandal et al. 2007).

A simple implementation of the normal GAs begins with a population of individuals which are encoded on a simple chromosome like data structures. One then evaluate these structures and allocate reproductive opportunities to these individuals based on their relative fitness. Selection is applied to this population so that the individuals with good solutions to the problems are given more chances to reproduce than the individuals with poorer solutions. The selection process generates an intermediate population after which recombination and mutation are applied to generate the next population (offspring). The process of evaluation, selection, recombination, and mutation from parent to offspring constitutes one generation of the GA. GAs likewise nature run for a number of generations and provide an optimal considerable solution to the problems, much better than their initial parent solution.

12.4.1 Working of Genetic Algorithms

Usually, there are two major components of GA:

1. Problem encoding.
2. Evaluation function.

The first assumption that is typically made is that the variable representing the parameters is represented in the form of binary strings (0, 1). This means the variable is discretized in the search space to some power of two (2^n). After creating the initial population, each binary string is then evaluated and assigned a fitness value. The fitness is defined by

$$\text{Relative fitness} = f_i / < f >$$

where f_i = evaluation associated with string i and $<f>$ = average evaluation of all the strings in the population.

The fitness value obtained normally translates the measure of performance of each individual into the number of copies passed onto the next generation. The obtained fitness function generates a population of current parent individuals, which are now subjected to various GA operators to create the next generation of offspring.

12.4.2 Genetic Algorithm Operators

12.4.2.1 Natural Selection Operator

Selection is normally applied to the current population to generate the intermediate population (F_1 generation) in such a way that the highly fit individuals with good solutions to problems are given more share or reproductive opportunities into the next generation against lesser fit individuals (Fig. 12.2). There are a number of ways to do the selection.

Parameter	Fitness
P1	15
P2	10
P3	8
P4	3
P5	2
P6	2
P7	1
P8	1

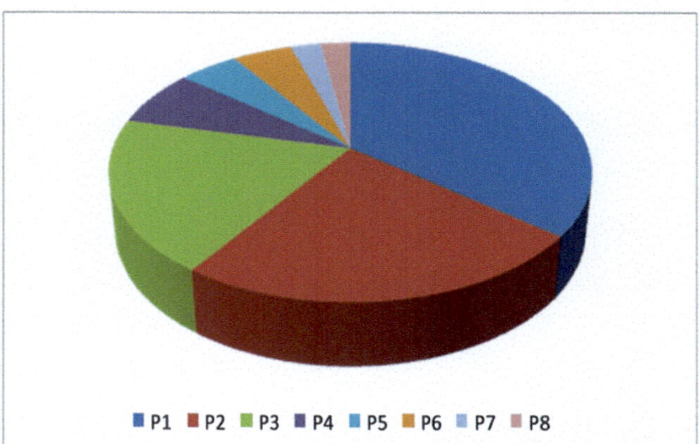

Parameters are assigned the space according to their fitness value

Fig. 12.2 Application of selection operator in GA

12.4.2.2 Stochastic Sampling with Replacement

We might visualize the population as mapping onto a Roulette wheel, where each individual is assigned its space which is directly proportional to its relative fitness. Iteratively spinning this roulette wheel, individuals are chosen and assigned to the intermediate generation in an unbiased manner.

12.4.2.3 Stochastic Universal Sampling

The population is laid down in random order on a simple pie (360°) graph, where again each individual is assigned its space, which is directly proportional to its relative fitness. Next, an outer Roulette wheel with N equally spaced pointers is placed over this pie graph. A single spin of this outer Roulette wheel unbiasedly picks all N individuals as members of the intermediate population.

12.4.2.4 Crossover/Recombination Operator

After the selection process has been carried out, the construction of the F_1 generation is complete and now crossover can be implemented. Crossover or recombination is a natural phenomenon that occurs during the late pachytene stage of meiosis and generally involves the exchange of chromosomal segments between the paternal and maternal chromosomes. The prime benefit of the crossover operator is

1. It introduces important genetic diversity in the current population.
2. It naturally preserves the critical information of the parents.

Crossover is normally applied in GAs by randomly recombining a pair of strings with a certain probability $P_{c.}$

Picking a pair of strings and recombining them by using 1-point crossover has the potential to generate newer solutions to the problems or may also subside as like its previous solutions.

1|100 & 1|001 [Same strings obtained after 1 – point crossover]

1100 & 1001

11|00 & 10|01 [Different strings obtained after 1- point crossover]

1101 & 1000

12.4.2.5 Mutation Operator

After recombination, we can apply a mutation operator. A mutation is simply interpreted to actually mean flipping of a bit to form a novel combination. For each bit in the population, mutate with some low probability *Pm*.

1100

↓

1000

The mutation is generally seen as a necessary evil in nature and therefore is typically applied as less than 1% in F_1 generation. The mutation has a capacity to introduce a novel solution to the problem which a normal recombination operator even after running for multiple generations fails to do so.

After the completion of selection, recombination, and mutation, the next population can be evaluated. The process keeps on running for multiple generations (>25,000 generations) to provide an optimal solution in the end in the successful compilation of GA.

12.5 Artificial Neural Networks

ANN is an information processing paradigm, which works in a way as the human biological nervous system works. Unlike GA, ANNs are purely a case of AI and are inspired by the human brain mechanism of information processing and exchange (Fig. 12.3). The key element of any ANN is its novel design of information processing structure, destined to solve a particular problem.

Likewise, the human nervous system, ANNs are composed of a large number of interconnected processing elements (artificial neurons) that work in unison to provide an optimal response to a specific stimulus/problem (Fig. 12.4). ANN works by learning from examples, by making adjustments in synaptic connections among its artificial processing elements/neurons. ANNs have been largely used in data recognition, pattern recognition, and prediction problems owing to their capacity for self-intelligence (Gupta et al. 2012). However, the results of ANNs are expected to be unpredictable and cannot be seen to perform miracles but if used sensibly in correct design ANNs have been seen to produce amazing results often.

Neural network simulations have witnessed its establishment even before the advent of computers but due to lack of funding and enthusiasm, this field suffered frustration and disrepute. The first artificial neuron was even developed by Warren McCulloch and Walter Pits in 1943 but later in a research paper published by Minsky and Papert in 1969, the concept of ANNs suffered a major setback and significant limitations.

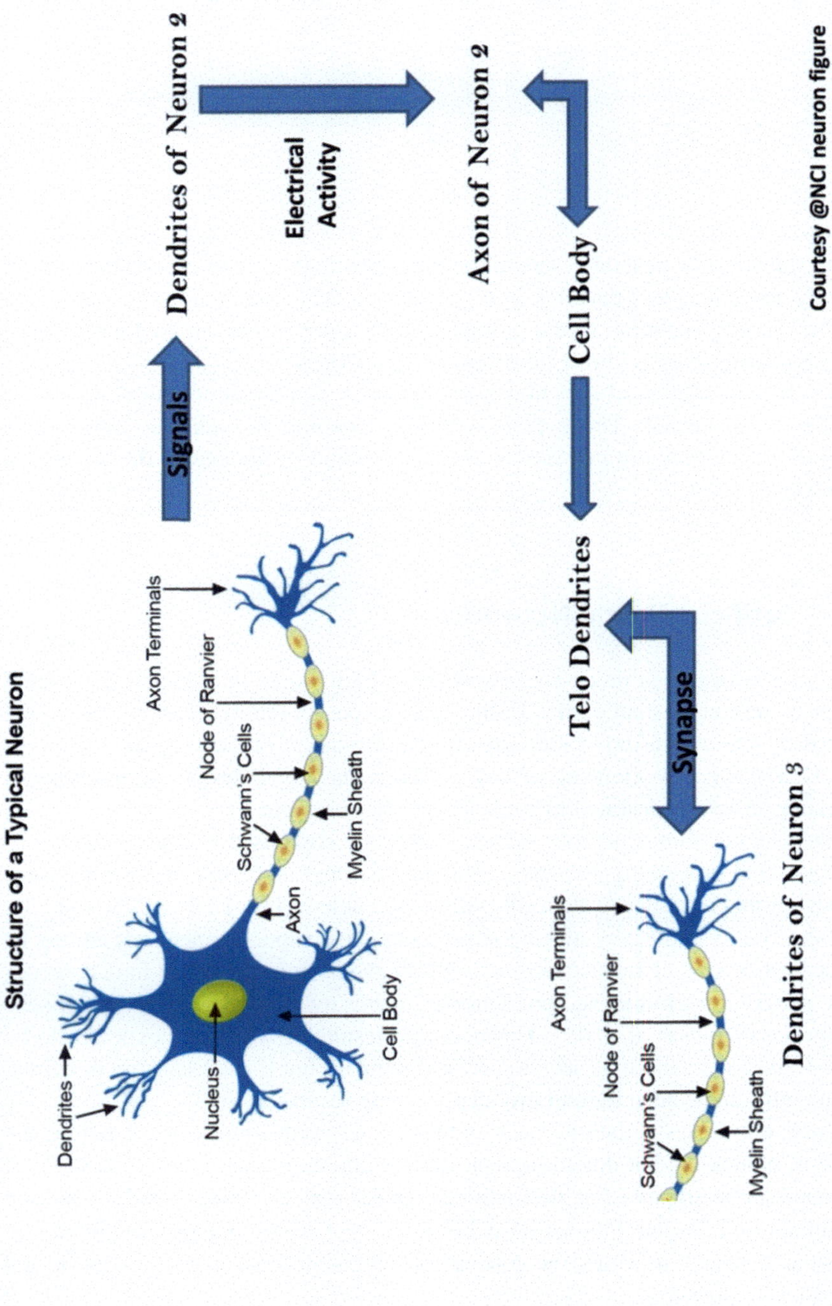

Fig. 12.3 Working of human neuronal cells

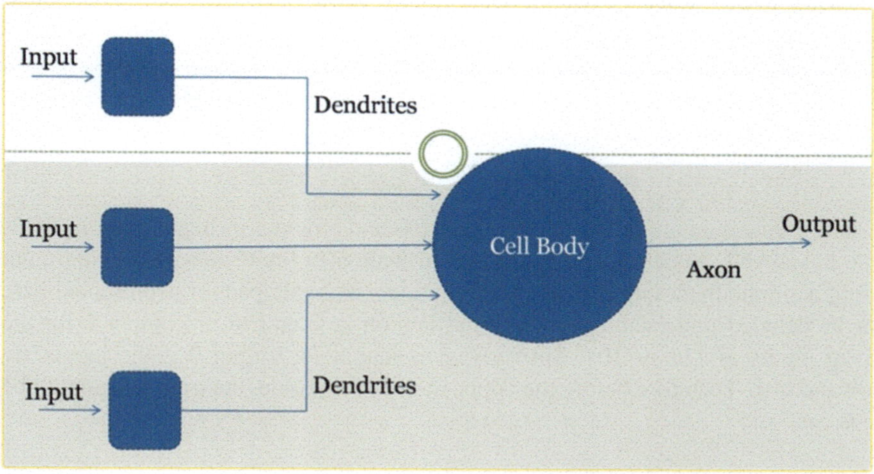

Fig. 12.4 Structure of an artificial neuron

In the current scenario, ANNs have seen a major comeback, and owing to its remarkable ability to derive meaning from complex and indefinite data, they are now been routinely used in data handling and data analysis problems in various sectors. The major advantages of ANNs include its self-organization capacity, adaptive learning mechanism, real-time operation, and fault tolerance attitude. ANN can learn from its training data and automatically create its own organization by making synaptic adjustments. In a state of its dynamic equilibrium, it can produce amazing results (Doyle et al. 2015; Oquendo et al. 2012).

Unlike computers, ANNs are not programmed and are unpredictable. They are better known to solve those problems which it does know how to solve. They can give any result based on the input and do not use a cognitive or algorithmic approach to problem-solving.

12.5.1 How the Human Brain Works?

Learning in the human brain occurs by changing the effectiveness of the synapses which is expressed in the form of change in the influence of one neuron on others. Synapses are the region of contact between two neurons or a neuron with a non-neuronal cell. The transmission of information between neurons may be electrical or chemical depending on the myelination of the nerve fibers.

12.5.2 A Simple Artificial Neuron

An artificial neuron is simply a device with multiple inputs but a single output. It works in basically two modes of operation

1. Using mode: When a taught input pattern is detected, its associated output becomes its natural output.
2. Training mode: In case when the input is not a part of a taught pattern, the neuron is trained to fire or not based on the learning from the taught input patterns. Learning from the taught input pattern is accomplished via the implementation of a firing rule.

Firing rule is an important concept of ANNs and provides its high flexibility. This rule helps ANN to calculate for a particular neuron to fire or not for a given input using a simple hamming distance technique. An example pattern is presented here for its detailed understanding. A 3-input neuron is taught to fire output 1 for the given inputs as 111 or 101. Similarly, it is taught for output 0 if the inputs are 000 and 001. Therefore before the application of firing rule, the truth table could be seen as:

X1	0	0	0	0	1	1	1	1
X2	0	0	1	1	0	0	1	1
X3	0	1	0	1	0	1	0	1
OUT	0	0	0/1	0/1	0/1	1	0/1	1

Implementing the firing rule, we take a pattern, for example, 010. It differs from 000 in 1 element and 001 in 2 elements. Similarly, it differs from 101 in 3 elements and 111 in 2 elements. Therefore, applying the hamming distance concept to the nearest input it belongs to 000. The output associated with 000 is 0, so after the learning process, the output associated with 010 also stands to be 0, instructing the neuron not to fire for this particular input.

By applying the firing rule to every column of this truth table, the new truth table generated can be represented as:

X1	0	0	0	0	1	1	1	1
X2	0	0	1	1	0	0	1	1
X3	0	1	0	1	0	1	0	1
OUT	0	0	0	0/1	0/1	1	1	1

In this way, the firing rule helps the neurons to understand the output for a given novel input based on the learning from the already present taught input patterns. A more complicated neuron model includes the application of weighted inputs. The weight is usually a number, which is multiplied with input to gives its total weighted input value, and if they exceed a certain threshold value, the neurons fire else not.

The commonest type of ANNs comprises of three layers of networking (Fig. 12.5):

1. Input layer: represents the raw data fed into the network.
2. Hidden layer: activity depends upon the input units and weights of connection between input and hidden layer.

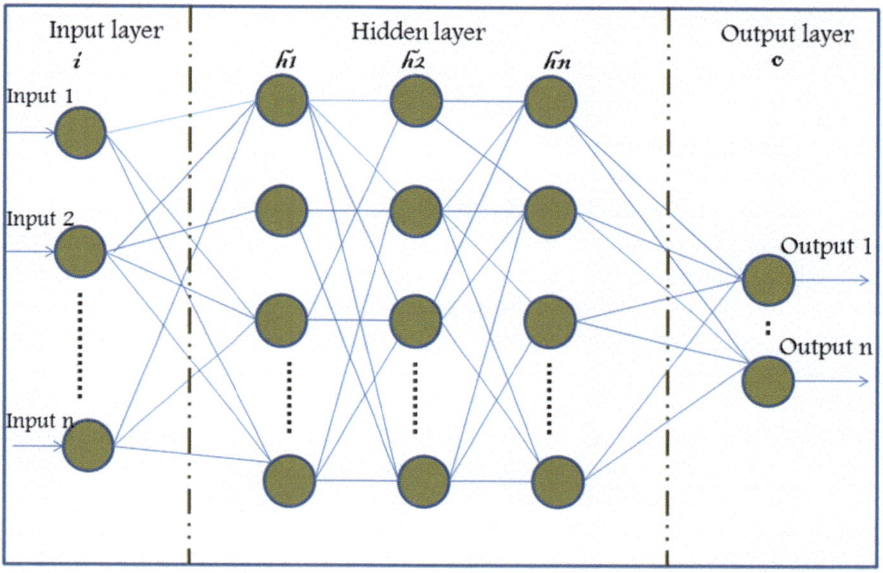

Fig. 12.5 Network layers in artificial neuron network

3. Output layer: activity depends upon the hidden layer and weights of connections between hidden and output units.

Hidden layers are just like a black box and are free to construct their own representation of data and reach a dynamic equilibrium stage (Patra et al. 2017).

12.5.3 Architecture of ANNs

There are two types of ANN architectures:

1. Feedforward networks: These networks allow the signals to travel in a unidirectional manner, i.e. from input to output only. They are straight forward networks and no feedback is provided to the back end layers or the layers on the same level. Also referred to as bottom up or top down networks, these feedforward networks are regularly used in data and pattern recognition problems.
2. Feedback networks: These networks allow the signals to travel in both directions by the introduction of loops in the network. They are a very powerful network and can get extremely complicated in their action. Their signal state keeps on changing continuously until it reaches a dynamic equilibrium. Also referred to as interactive or recurrent networks, they are well known to solve more complex problems.

ANNs are further known to work via supervised and unsupervised learning techniques.

12.6 Deep Learning(DL)

Deep learning or Deep neural networks (DNN), a class of ML algorithms uses ANNs with many layers of nonlinear processing units for learning data representations. DL is a highly adopted method to address the activity prediction problems in the first place. In drug discovery, compounds are presented by the same number of molecular descriptors, the straight forward method is to use fully connected DNNs to build models. Ma et al. 2015 applied a DNN on the Merck Kaggle challenge dataset using a large number of 2D topological descriptors, and the DNN showed better performance than the standard RF method. DNNs can handle thousands of descriptors without the need for feature selection; dropout can avoid the notorious overfitting problem faced by a traditional ANN. The study showed that multitask DNN models perform better than single-task models (Duvenaud et al. 2015). Convolutional neural networks (CNNs) are a type of neural network commonly used in image recognition (LeCun et al. 2015). CNNs are used for virtual screening in drug discovery effectively.

ANN has been applied to drug discovery in the context of ligand-based QSAR, and they have not traditionally been used in structure-based virtual screening methods. Currently, the enormous amount of biological data available for training, and the recent evolution of GPU-accelerated computation, and neural network-based techniques have very high potential to transform the in silico prediction of molecular recognition with maximum accuracy.

Durrant et al. created two fast and accurate neural network scoring functions for rescoring docked ligand poses (NNScore 1.0 and 2.0) (Durrant and McCammon 2010). Unlike traditional docking scoring functions, these nonparametric functions are not constrained to predetermined physical formulae or statistical analyses; rather, they "learn" directly from existing experimental data how best to predict binding and so can, in theory, better capture the nonlinear, synergistic relationships among binding determinants. These are the first neural network scoring functions that predict affinity by directly examining atomic resolution ligand–protein interactions. Recently, Abdul et al. developed a quadratic phenotypic optimization platform (QPOP), which provides new or best drug combinations for patients using the small dataset. This QPOP platform uses system-specific experimental data to determine the best drug combinations for a specific disease model or a patient sample rather than previous assumptions of molecular mechanisms of disease (Rashid and Chow 2019). DeepMind Technologies, a subsidiary of Google, collaborated with Royal Free London NHS Foundation Trust to assist in the management of acute kidney injury (Rashid and Hodson 2017). Atom wise (https://www.atomwise.com/) is a pioneer in healthcare AI and is the first DL technology for novel small molecule discovery and has assisted in the invention of new potential medicines for 27 disease targets and is working with top institutions such as Harvard University and Stanford

University, as well as pharmaceutical companies. Exscientia is an AI company that specializes in phenotypic drug discovery (Lee et al. 2017).

DL helps to resolve several prediction issues in bioinformatics. CNN and deep belief network (DBN) are used to unearth RNA-protein binding motifs DNA binding proteins, peptide-MHC binding, etc., stacked sparse autoencoder (SSAE) pooled with a Legendre moment (LM) has been utilized for predicting protein–protein interaction within cells (Huang et al. 2017). DL has also been effectively utilized in the QSAR studies in the form of 2D-QSAR, 3D-QSAR, and multidimensional (nD) QSAR. 2D fingerprint-based ANN (FANN)- QSAR, a novel ANN technique has been reported to calculate biological activities of structurally diverse chemical ligands efficiently. Successful implementation of FANN-QSAR is done to forecast the cannabinoid receptor (CB2) binding activity through data collection from structurally diverse sources (Myint et al. 2012).

DL has profoundly being used in computational biology for predicting protein disorder, enhancement of docked protein complexes, modeling structural properties of protein targets binding to RNA, etc. Compared to other techniques for predicting the mechanism of action with the help of high content image analysis data, it was observed to be advance to SVM with 87.62. The various framework of DL such as DNN, DBN, and RNN is currently being used for analyzing gene expression, genomic sequencing, and prediction of protein structure (Ekins 2016).

NiftyNet, a modular DL pipeline has emerged as the latest accelerated DL technique for solving problems related to medical imaging and computer-assisted intrusion. Gibson et al. utilized capability of NiftyNet to build analysis application like for layerwise separation of different organs that were obtained from computed tomography (CT), applied regression technique to foresee computed tomography attenuation maps from images of magnetic resonance of brain and creation of simulated images of ultrasound for defined poses of anatomy (Gibson et al. 2018).

12.7 Support Vector Machines

SVMs are by far the most ubiquitously used ML procedures in the design and discovery of drug case studies. SVM classifies various compound varieties to predict new molecules, biological activity (from regression models), and rank substances in virtual screening assays. SVM approaches involve several critical challenges, such as selecting kernel functions and optimizing parameters for specific issues. Integrating SVMs with other approaches has ended up being an excellent strategy for studying medicinal chemistry (Maltarollo et al. 2019).

Various ML algorithms, viz. PCA can help to identify various alveolar cells, BP—back error propagation algorithm can be used for predicting the secondary structure of the protein. CNN can assist in the detection of DNA sequence variation sites. In the field of data science, ML is a strategy to design complex models and algorithms for prediction (Niculescu 2003). A profound convolutional neural system can identify hereditary variations in aligned next-generation sequencing read data by learning factual relationships (probabilities) between pictures of read pileups framing putative variation points and ground-truth genotypes (Yang et al. 2018).

12.8 Artificial Intelligence and Drug Discovery

The AI pioneered in 1950 discussed a technique that could sense, reason, and think like people. Later on with the advancement in computer processing power and huge datasets now targeted AI has been developed which is more focused and accurate. AI has attracted pharmaceutical industries for de novo designing of peptides and chemical compounds against a particular target, along with its retrosynthesis. AI helps in predicting the physicochemical properties (i.e. ADMET) of candidate drug compounds among the large dataset. Several machine learning technologies like Random forests (RF), SVM, or Bayesian learning have been implied for optimizing the process of compound designing (Hessler and Baringhaus 2018). AI has transformed the methods of a pathway or target identification to treat diseases. In a study, it was shown that possibility of predicting therapeutic targets using a computational prediction application known as "open targets" a platform consisting of gene-disease association data, and it was reported that animal models exhibiting a disease-relevant phenotype with a neural network classifier of greater than 71% accuracy provided the most predictive power (Ferrero et al. 2017). IBM Watson, an AI platform for drug discovery has identified five new RNA-binding proteins (RBPs) linked to the pathogenesis of a neurodegenerative disease known as amyotrophic lateral sclerosis (ALS) (Bakkar et al. 2018). AI also contributes to the identification of target-specific virtual molecules and association of the molecules with their respective target while optimizing the safety and efficacy attributes. AI, with the ability to prioritize molecules based on the ease of synthesis also assist in development of tools that are effective for the optimal synthetic route (Segler et al. 2018). ML-based tools used in drug designing are listed in Table 12.1.

AI in the drug-like compound synthesis process has proven accurate in predicting the best sought-after reactions by filling the voids that cause high failure in expected organic synthesis (commonly known as "out of scope" compounds). The voids in organic synthesis are mainly the result of unpredictable steric and electronic effects and incomplete understanding of the reaction mechanism. Although several computers aided organic compound synthesis (CAOCS) systems are available to assist chemists in selecting the synthesis route a new AI platform named 3N-MCTS developed by Seglar et al. has proven to be much faster and better than that of traditional computer-assisted retrosynthesis systems (Segler et al. 2017). An innovative AI tool, SPiDER, has been developed (Rodrigues et al. 2018) as an alternative to chemoproteomics to advance natural products for drug discovery. The SPiDER was used to predict the molecular target of b-lapachone, a clinical-stage natural naphthoquinone with antitumor activity. The platform predicted b-lapachone as an allosteric and reversible modulator of 5-lipoxygenase (5-LO). The prediction is validated using a 5-LO functional assay. Read-across structure-activity relationship

Table 12.1 Common ML-based drug designing tools

Tool	Developer	Brief description and algorithm used	Website	Freely available
Docking				
AutoDock	The Scripps Research Institute	Simulated annealing, GA	http://autodock.scripps.edu	Yes
DOCK	University of California	Incremental construction, merged target structure ensemble	http://dock.compbio.ucsf.edu	Yes
FlexX	BioSolveIT GmbH	Incremental construction, merged target structure ensemble	https://www.biosolveit.de/FlexX	No
FRED	OpenEye Scientific Software	Exhaustive search algorithm	https://docs.eyesopen.com/oedocking/fred.html	Yes
Glide	Schrödinger, Inc.	Conformational expansion, Monte Carlo, torsional search	https://www.schrodinger.com/glide	No
GOLD	The Cambridge Crystallographic Data Centre	GA	https://www.ccdc.cam.ac.uk/solutions/csd-discovery/components/gold	No
ICM-Pro	Molsoft LLC.	Monte Carlo minimization procedure	http://www.molsoft.com	No
Surflex-Dock	Tripos Inc.	Hammerhead's empirical scoring function with morphological similarity	http://www.jainlab.org	No
Homology modeling				
Modeller	University of California	It performs comparative modeling using spatial restraints and performs optimization of protein structure models	https://salilab.org/modeller	Yes
MOE	Chemical Computing Group	It uses a precompiled antibody x-ray database (Fab database) to model	https://www.chemcomp.com	No

(continued)

Table 12.1 (continued)

Tool	Developer	Brief description and algorithm used	Website	Freely available
		F_v region of the immunoglobulin		
Prime	Schrödinger, Inc.	Prime performs comparative modeling using homology modeling and fold recognition	https://www.schrodinger.com/prime	No
SWISS-MODEL	Swiss Institute of Bioinformatics	Fully automated protein homology modeling server	https://swissmodel.expasy.org/	Yes
Molecular dynamics				
Amber	University of California, San Francisco, USA	*A*ssisted *m*odel *b*uilding with an *e*nergy *r*efinement suite is used for molecular dynamics simulations of biomolecules	http://ambermd.org/	No
CHARMM	Harvard University, USA	It simulates the peptides, proteins, prosthetic groups, ligands, nucleic acids, lipids, and carbohydrates in aqueous, crystals, and membrane environments	https://www.charmm.org	No
Desmond	D. E. Shaw Research	It uses novel parallel algorithms and numerical techniques to perform molecular dynamic simulations	https://www.deshawresearch.com	Yes
GROMACS	University of Groningen, The Netherlands	Performs molecular dynamic simulations of biomolecules like proteins, lipids, and nucleic acids	http://www.gromacs.org	Yes
NAMD	University of Illinois, USA	NAMD is used for the simulation of large biomolecules	http://www.ks.uiuc.edu/Research/namd/	Yes
Quantum mechanics				
GAMESS	Iowa State University, USA	The general atomic and molecular	https://www.msg.chem.	Yes

(continued)

Table 12.1 (continued)

Tool	Developer	Brief description and algorithm used	Website	Freely available
		electronic structure system (GAMESS) is a general ab initio quantum chemistry package	iastate.edu/gamess	
Gaussian	Gaussian Inc.	It is used for estimating analytic frequency, geometric optimizations and IR, Raman, VCD and ROA spectra of large biomolecules	https://gaussian.com	No
Jaguar	Schrödinger Inc.	Jaguar uses density-functional theory (DFT) and local second-order Møller–Plesset perturbation theory to compute a comprehensive array of molecular properties	https://www.schrodinger.com/jaguar	No
MOPAC	Stewart Computational Chemistry	It is used to calculate vibrational spectra, thermodynamic quantities, isotopic substitution effects, and force constants for molecules, radicals, ions, and polymers	http://openmopac.net	Yes
NWChem	Environmental Molecular Sciences Laboratory	Ground and excited solutions of many-electron Hamiltonian are obtained utilizing density-functional theory, many-body perturbation approach, and coupled cluster expansion. It is used to analyze potential energy surface and perform dynamical simulations	http://www.nwchem-sw.org	Yes

(continued)

Table 12.1 (continued)

Tool	Developer	Brief description and algorithm used	Website	Freely available
ADMET prediction				
ADMET Predictor	Simulations Plus, Inc.	ANN, SVM, Kernel partial least squares (KPLS), and multiple linear regression (MLR)	https://www.simulations-plus.com/software/membraneplus/admet-predictor/	No
StarDrop	Optibrium, Ltd	Pareto optimization, GA	https://www.optibrium.com/stardrop	No
Percepta Platform	Advanced Chemistry Development, Inc.	It provides two predictive algorithms—Classic (based on Hammett-type equation) and GALAS	https://www.acdlabs.com/products/percepta	No
ADMEWORKS Predictor	Fujitsu Kyushu Systems	It is a virtual screening system with a simultaneous evaluation of ADMET properties	https://www.fujitsu.com	No
Sarchitect	Syngene	Bayesian methods, ANN, SVM, decision trees and forests, and other algorithms are used for model building and prediction	https://www.syngeneintl.com	No
QikProp	Schrödinger, Inc.	Monte Carlo statistical mechanics simulation is used to correlate different descriptors to experimental properties	https://www.schrodinger.com/qikprop	No
Derek Nexus	Lhasa, Ltd	Uses the knowledge base of Lhasa for accurate toxicity predictions	https://www.lhasalimited.org/products/derek-nexus.htm	No
PASS	V. N. Orechovich Institute of Biomedical Chemistry under the aegis of the	Uses 20,000 principle compounds from MDDR databases	http://195.178.207.233/PASS/index.html	No

(continued)

Table 12.1 (continued)

Tool	Developer	Brief description and algorithm used	Website	Freely available
	Russian Foundation of Basic Research			
Hazard Expert Pro	CompuDrug, Ltd	A neural network-based approach is used to model the relationship between human cytotoxicity and atomic descriptors	https://www.compudrug.com/hazardexpertpro	No
VolSurf+	Molecular Discovery, Ltd.	It correlates128 molecular descriptors with the pharmacokinetic properties of the drugs	https://www.moldiscovery.com/software/volsurf	No
Bioclipse	Uppsala University, Sweden and European Bioinformatics Institute	It is a Java based workbench, which provides molecular editing and visualization as well as prediction of physio-chemical properties of the molecules	https://sourceforge.net/projects/bioclipse	Yes
MetaDrug	GeneGo, Inc.	It is an Oracle based software, which uses several network-building algorithms like Dijkstra's algorithm to predict ADME/Tox properties of drugs	https://omictools.com/metadrug-tool	No
TIMES	OASIS-LMC	It is a heuristic algorithm used to predict the toxicity of metabolites based on their metabolic maps	http://oasis-lmc.org	No
Molecular visualization				
UCSF Chimera	RBVI, University of California, San Francisco, USA	Provides visualization and analysis of molecular structures and related data.	https://www.cgl.ucsf.edu/chimera	Yes
Jmol	University of Notre Dame, USA	It is a free tool for academicians and	http://jmol.sourceforge.net	Yes

(continued)

Table 12.1 (continued)

Tool	Developer	Brief description and algorithm used	Website	Freely available
		researchers written in Java for 3D visualization of molecules		
PyMOL	Schrödinger Inc.	It is a molecular visualization tool for animating 3D structures	https://pymol.org	No
Swiss-Pdb Viewer (Deep View)	Swiss Institute of Bioinformatics	Active sites of the proteins can be compared and structural alignment can be performed. Several proteins can be analyzed at the same time	https://spdbv.vital-it.ch	Yes
VMD	University of Illinois, USA	It can model, analyze, and visualize biomolecules. It includes multiple sequence alignment and can be used for both sequence and structure data	https://www.ks.uiuc.edu/Research/vmd	Yes

(RASAR), an AI tool, links molecular structures and toxic properties by mining a large database of chemicals (Luechtefeld et al. 2018).

ANNs are further extensively used in the interpretation of analytical data, drug, and dosage form design through bio pharmacy to clinical pharmacy. In pharmaceutical, supervised associating networks are applied as an alternative to conventional response surface methodology (Bourquin et al. 1997; Rojas 2013).

12.9 Applications of ANNs, GAs, and Other ML Algorithms in Drug Discovery

The cost of discovering and developing a drug has since escalated from US$ 800 million in the year 2001 to the current estimated figure of US$ 3 billion (DiMasi et al. 2015). Finding successful new drugs is daunting and predominantly the most difficult part of drug development. ML and other computational technologies are playing a vital role in hunting new drugs quicker, cheaper, and more effective.

ML uses experimental data to optimize clustering or classification of samples or features to develop augment or verify models that can be used to predict the behavior or properties of systems (Cuperlovic-Culf 2018). In recent years bioinformatics and metabolism analyses have witnessed a variety of ML methods including self-organizing maps, SVM, the kernel machine, Bayesian networks, or fuzzy logic. ML has optimized metabolic network models and their analysis with the availability of enormous genomics and metabolomics data. ML also has been successfully applied for the determination of drug side effects using the data available for human diseases, drugs, and their associated phenotypes to determine metabolically associated side effect predictors (Shaked et al. 2016). Recently, as part of the consortium for metabonomics toxicology (COMET), relational and logic-based ML has successfully utilized to provide causal explanations of rat liver cell responses to toxins using high throughput NMR metabolomics data (Tamaddoni-Nezhad et al. 2006; Chen et al. 2008).

ML techniques prominently in the form of ANNs and GAs have been used in chemoinformatics, computational biology, and pharmaceutical research. The models built with these ML methods are been routinely used for robust external predictions.

Commonly the ANNs are employed in drug design to build predictive models involving:

(a) Hepatotoxicity, genotoxicity profiling.
(b) Pharmacokinetics; clearance, and permeability studies.
(c) Activity predictions against target proteins.
(d) ADME-Tox (absorption, distribution, metabolism, excretion, and toxicity) and biological activity prediction and classification studies.
(e) Lead discovery and similarity, diversity analysis of combinatorial libraries.
(f) HTS data analysis.

ML techniques also find their application in representation, classification, and categorization studies of small lead molecules in drug discovery. In this, the molecules are usually seen as fingerprints featuring its molecules substructures, molecular spaces, bonds, and interatomic distances, represented in the form of binary vectors. At the technical architectural level, these small molecular representations are further searched for their similar substructural and molecular spatial arrangements like compounds. The obtained results are ranked based on their biological activity by the application of various ML prediction filters (Panteleev et al. 2018; Terfloth and Gasteiger 2001).

GAs usually found such applications in the automated generation of small organic molecules. The compounds here are represented as SMILES strings. The electronic, lipophilicity, and shape parameters of these compounds are used to carry out virtual screening studies against the database. A specific GA based search and scoring function then generates a list of similar natural and synthetic compounds as potent leads (Douguet and Thoreau 2000).

There are two major benefits of the application of ML techniques in drug designing:

1. Generation of faster hypotheses overcoming the time and cost factor associated with lengthy wet lab studies;
2. Development of better hypothesis: A well planned and diligently designed ML models may propose better leads than conventional methods curtailing the drug discovery time to 8–10 years period.

GA and ANNs have been instrumental at various levels of lead discovery and lead optimization and are routinely used in various drug discovery tools as follows:

12.9.1 Molecular Docking

Docking is used for finding the preferred pose of a ligand with another molecule to form a stable complex. In the simplest sense, in silico interaction studies between any two molecules are referred to as molecular docking. GA is commonly used in docking algorithms as a favorable search and optimization function. It runs for several generations (>25,000) to generate an optimal binding conformation between a protein and ligand molecule. Some common programs involving GA enabled docking include genetic optimization for ligand docking (GOLD), family competition evolutionary approach (FCEA), and Auto dock (Yang and Kao 2000; Morris et al. 1998)

12.9.2 Pharmacophore Modeling

A pharmacophore may be defined as the essential geometric arrangement of atoms or functional groups necessary to produce a given biological response. A strict IUPAC definition of pharmacophore comes as: "A pharmacophore is the ensemble of steric and electronic features that is necessary to ensure the optimal supramolecular interactions with a specific biological target structure and to trigger (or to block) its biological response." (Choudhury and Narahari 2019; Seifert et al. 2007).

This method involves the generation of a quantitative pharmacophore model for a structurally diverse set of input compounds. A suitable hypothesis predicting the biological activity of compounds is developed using training set compounds which is further validated with test compounds. The correlation and regression analysis data obtained with training and test set compounds evaluate the predictive efficacy of developed pharmacophore models. A program named as GA for multiple molecule alignment (GAMMA) is a similar function allowing flexible alignment for multiple small molecules applying Newton optimizer based GA (Terfloth and Gasteiger 2001). GAs have also its relevance in the analogue preparation or automated creation of small molecules by altering the SMILES format along with maintaining the physicochemical descriptors and drug-like standards, i.e., electronic properties, lipophilicity, and conformational features for evaluating scoring function to show better interaction with the respective protein (Douguet and Thoreau 2000).

12.9.3 Quantitative Structure–Activity Relationship (QSAR)

QSAR studies involve the mathematical and statistical analysis of how the chemical structure of a compound is related to its biological activity. The chemical structure of the compound is usually defined as a function of its molecular descriptors such as molecular weight, functional groups, atom, and bond counts to more complex topological, geometric, connectivity, and physicochemical parameters. These descriptors can frequently be calculated by several ML-based popular programs such as MOE, DRAGON, Molconn-Z, ADMET predictor, CODESSA, and PowerMV. Once the descriptors are obtained, statistical modeling methods are employed to derive the correlation and regression between the activity and descriptors. In addition, ANNs are applied to these QSAR models for the prediction of physicochemical and pharmacokinetic properties (Terfloth and Gasteiger 2001).

In the recent era, several pharmaceutical companies and research institutions have come up with novel and intelligently developed ML programs to be used in drug discovery. Two such programs have been developed at Merck pharmaceuticals, namely classification & regression at Merck (CREAM) and Merck online computational chemistry analyzer (MOCCA). CREAM is basically a Python-based modeling tool intended for classification and categorization studies, while MOCCA provides predictive models for toxicity, hepatotoxicity, genotoxicity, ADMET, etc. ANNs play an important role in any such programs by optimizing their training parameters at architectural levels, viz. learning rate, weight decay, batch size, loss-function, signal transfer, and feedbacks, etc.

The importance of any such ANN-based ML methods is that it helps to create newer end-points in the analyzed data and it succeeds to provide better statistical performance compared to conventional methods (Jing et al. 2018). Another example of an ML-based predictive model has been developed by researchers at IBM. The model at IBM helps to predict, represent, and identify which diseases are typically linked to which prominent side effects. The graphical representation included orange and blue dots. An orange dot represents the regions of diseases, while blue dot representations imply the specific side effects associated with them. The IBM models have greatly helped to identify and limit the typical side effects associated with specific diseases.

Recently a newer concept has been proposed by Burden et al., whereby neural networks inversion problem for QSAR studies of dihydrofolate reductase inhibitor was solved by applying GA. A small layered feedforward/back-propagation neural network-based QSAR model was developed and maximum activity on the structure-activity surface was determined using a GA. Such development of hybrid algorithms involving both ANN and GA in ML techniques proves to be a milestone in fast prediction and better evaluation studies in the domain of drug discovery. Proper knowledge and diligent application of each of these ML methods may certainly unveil newer dimensions in multi-millionaire pharmaceutical industrial setups in the near future.

12.10 Conclusions

ML algorithms are playing a game changer role in many sectors. ML approaches are useful in designing and development of new biologically active molecules with desired properties for a drug target. In healthcare sectors, many renowned pharmaceutical companies have made a huge investment in AI companies working on ML as a joint venture to come out with healthcare tools and better drugs. With the rapid advancement in computer processing power and huge datasets availability and synthesis planning, ML algorithms have helped in speeding up drug development. In the near future, ML concepts will permanently change the pharmaceutical industry and the way drugs are discovered.

Competing Interest The authors declare that there are no competing interests.

References

Arif JM, Siddiqui MH, Akhtar S, Al-Sagair O (2013) Exploitation of in silico potential in prediction, validation and elucidation of mechanism of anti-angiogenesis by novel compounds: comparative correlation between wet lab and in silico data. Int J Bioinforma Res Appl 965:336–348

Bakkar N, Kovalik T, Lorenzini I, Spangler S, Lacoste A, Sponaugle K, Bowser R (2018) Artificial intelligence in neurodegenerative disease research: use of IBM Watson to identify additional RNA-binding proteins altered in amyotrophic lateral sclerosis. Acta Neuropathol 135 (2):227–247

Bourquin J, Schmidli H, van Hoogevest P, Leuenberger H (1997) Basic concepts of artificial neural networks (ANN) modeling in the application to pharmaceutical development. Pharm Dev Technol 2(2):95–109

Chen J, Muggleton S, Santos J (2008) Learning probabilistic logic models from probabilistic examples. Mach Learn 73(1):55–85

Chen H, Engkvist O, Wang Y, Olivecrona M, Blaschke T (2018) The rise of deep learning in drug discovery. Drug Discov Today 23(6):1241–1250

Choudhury C, Narahari SG (2019) Pharmacophore modelling and screening: concepts, recent developments and applications in rational drug design. In: Mohan C (ed) Structural bioinformatics: applications in preclinical drug discovery process. Challenges and advances in computational chemistry and physics. Springer, Cham, pp 25–53

Cuperlovic-Culf M (2018) Machine learning methods for analysis of metabolic data and metabolic pathway modeling. Metabolites 8(1):4

DiMasi JA, Grabowski HG, Hansen RW (2015) The cost of drug development. N Engl J Med 372 (20):1972

Douguet DE, Thoreau GG (2000) A genetic algorithm for the automated generation of small organic molecules: drug design using an evolutionary algorithm. J Comput Aided Mol Des 14:449–466

Doyle OM, Mehta MA, Brammer MJ (2015) The role of machine learning in neuroimaging for drug discovery and development. Psychopharmacology 232:4179–4189

Durrant JD, McCammon JA (2010) NNScore: a neural-network-based scoring function for the characterization of protein– ligand complexes. J Chem Inf Model 50(10):1865–1871

Duvenaud DK, Maclaurin D, Iparraguirre J, Bombarell R, Hirzel T, Aspuru-Guzik A, Adams RP (2015) Convolutional networks on graphs for learning molecular fingerprints. In: Advances in neural information processing systems. Curran Associates, Red Hook, pp 2224–2232

Ekins S (2016) The next era: deep learning in pharmaceutical research. Pharm Res 33 (11):2594–2603

Ferrero E, Dunham I, Sanseau P (2017) *In silico* prediction of novel therapeutic targets using gene–disease association data. J Transl Med 15(1):182

Fox T, Kriegl JM (2006) Machine learning techniques for in silico modeling of drug metabolism. Curr Top Med Chem 6(15):1579–1591

Gibson E, Li W, Sudre C, Fidon L, Shakir DI, Wang G, Eaton-Rosen Z, Gray T, Doel R, Hu Y, Whyntie T, Nachev P, Modat M, Barratt DC, Ourselin S, Cardoso MJ, Vercauteren T (2018) NiftyNet: a deep-learning platform for medical imaging. Comput Methods Prog Biomed 158:113–122

Goswami M, Akhtar S, Osama K (2018) Strategies for monitoring and modeling growth of hairy root cultures: an in silico perspective. In: Srivastava V, Mehrotra S, Mishra S (eds) Hairy roots. Springer, Singapore

Gupta MK, Agarwal K, Prakash N, Singh DB, Misra K (2012) Prediction of miRNA in HIV-1 genome and its targets through artificial neural network: a bioinformatics approach. Netw Model Anal Health Inf Bioinf 1:141–151

Gupta CL, Akhtar S, Bajpai P (2014) *In silico* protein modeling: possibilities and limitations. EXCLI J 13:513–515

Hessler G, Baringhaus KH (2018) Artificial intelligence in drug design. Molecules 23(10):2520

Huang G, Li J, Wang P, Li W (2017) A review of computational drug repositioning approaches. Comb Chem High Throughput Screen 20:831. https://doi.org/10.2174/1386207321666171221112835

Huang G, Yan F, Tan D (2018) A review of computational methods for predicting drug targets. Curr Protein Pept Sci 19(6):562–572

Jing Y, Bian Y, Hu Z, Wang L, Xie X (2018) Deep learning for drug design: an artificial intelligence paradigm for drug discovery in the big data era. AAPS J 20(3):58

Kapetanovic IM (2008) Computer-aided drug discovery and development (CADDD): in-silico-chemico-biological approach. Chem Biol Interact 171:165–176

Lavecchia A (2015) Machine- learning approaches in drug discovery: methods and applications. Drug Discov Today 20(3):318–331

LeCun Y, Bengio Y, Hinton G (2015) Deep learning. Nature 521(7553):436–444

Lee EJ, Kim YH, Kim N, Kang DW (2017) Deep into the brain: artificial intelligence in stroke imaging. J Stroke 19(3):277

Leelananda SP, Lindert S (2016) A review of computational methods for predicting drug targets. Beilstein J Org Chem 12:694–2718

Lo YC, Rensi SE, Torng W, Altman RB (2018) Machine learning in chemoinformatics and drug discovery. Drug Discov Today 23(8):1538–1546

Luechtefeld T, Marsh D, Rowlands C, Hartung T (2018) Machine learning of toxicological big data enables read-across structure activity relationships (RASAR) outperforming animal test reproducibility. Toxicol Sci 165(1):98–212

Ma J, Sheridan RP, Liaw A, Dahl GE, Svetnik V (2015) Deep neural nets as a method for quantitative structure–activity relationships. J Chem Inf Model 55(2):263–274

Maltarollo VG, Kronenberger T, Espinoza GZ, Oliveira PR, Honorio KM (2019) Advances with SVM for novel drug discovery. Expert Opin Drug Discovery 14(1):23–33

Mandal AK, Johnson C, Wu F, Bornemeier D (2007) Identifying promisingcompounds in drug discovery: genetic algorithms and some new statistical techniques. J Chem Inf Model 47 (3):81–988

Morris GM, Goodsell DS, Halliday RS, Huey R, Hart WE, Belew RK, Olson AJ (1998) Automated docking using a Lamarckian genetic algorithm and an empirical binding free energy function. J Comput Chem 19:1639–1662

Myint KZ, Wang L, Tong Q, Xie XQ (2012) Molecular fingerprint-based artificial neural networks QSAR for ligand biological activity predictions. Mol Pharm 9(10):2912–2923

Niculescu SP (2003) Artificial neural networks and genetic algorithms in QSAR. J Mol Struct 622 (1–2):71–83

Oquendo MA, Baca-Garcia E, Artes-Rodriguez A, Perez-Cruz F, Galfalvy HC, Blasco-Fontecilla H, Madigan D, Duan N (2012) Machine learning and data mining: strategies for hypothesis generation. Mol Psychiatry 17(10):956–959

Panteleev J, Gao H, Jia L (2018) Recent applications of machine learning in medicinal chemistry. Bioorg Med Chem Lett 28(17):2807–2815

Patra TK, Meenakshisundaram V, Hung JH, Simmons DS (2017) Neural-network-biased genetic algorithms for materials design: evolutionary algorithms that learn. ACS Comb Sci 19 (2):96–107

Pu L, Naderi M, Liu T, Wu H, Mukhopadhyay S, Brylinski M (2019) eToxPred: a machine learning-based approach to estimate the toxicity of drug candidates. BMC Pharmacol Toxicol 20(2):1–15

Rashid MBMA, Chow EK (2019) Artificial intelligence-driven designer drug combinations: from drug development to personalized medicine. SLAS Technol 24(1):124–125

Rashid J, Hodson H (2017) Google DeepMind and healthcare in an age of algorithms. Health Technol 7(4):351–367

Rodrigues T, Werner M, Roth J, da Cruz EH, Marques MC, Akkapeddi P, Werz O (2018) Machine intelligence decrypts β-lapachone as an allosteric 5-lipoxygenase inhibitor. Chem Sci 9 (34):6899–6903

Rojas R (2013) Neural networks: a systematic introduction. Springer-Verlag, Berlin

Sayeed U, Wadhwa G, Jamal QMS, Kamal MA, Akhtar S, Siddiqui MH, Khan MS (2016) MHC binding peptides for designing of vaccines against Japanese encephalitis virus: a computational approach. Saudi J Biol Sci 25(8):1546–1551

Schneider G, Bohm HJ (2002) Virtual screening and fast automated docking methods. Drug Discov Today 7:64–70

Schneider P, Schneider G (2016) De novo design at the edge of chaos: miniperspective. J Med Chem 59:4077–4086

Searls DB (2005) Data integration: challenges for drug discovery. Nat Rev Drug Discov 4(1):45

Segler MH, Kogej T, Tyrchan C, Waller MP (2017) Generating focused molecule libraries for drug discovery with recurrent neural networks. ACS Cent Sci 4(1):120–131

Segler MH, Preuss M, Waller MP (2018) Planning chemical syntheses with deep neural networks and symbolic AI. Nature 555(7698):604

Seifert MH, Kraus J, Kramer B (2007) Virtual high-throughput screening of molecular databases. Curr Opin Drug Discov Devel 10(3):298–307

Shaked I, Oberhardt MA, Atias N, Sharan R, Ruppin E (2016) Metabolic network prediction of drug side effects. Cell Syst 2(3):209–213

Tamaddoni-Nezhad AR, Kakas CA, Muggleton S (2006) Application of abductive ILP to learning metabolic network inhibition from temporal data. Mach Learn 64(1–3):209–230

Terfloth L, Gasteiger J (2001) Neural networks and genetic algorithms in drug design. Drug Discov Today 6(20):102–108

Varnek A, Baskin II (2011) Chemoinformatics as a theoretical chemistry discipline. Mol Inf 30 (1):20–32

Yang JM, Kao CY (2000) Flexible ligand docking using a robust evolutionary algorithm. J Comput Chem 21:988–998

Yang H, An Z, Zhou H, Hou Y (2018) Application of machine learning methods in bioinformatics. AIP Conf Proc 1967:040015. https://doi.org/10.1063/1.5039089

Yosipof A, Guedes RC, Garcia-Sosa AT (2018) Data mining and machine learning models for predicting drug likeliness and their disease or organ category. Front Chem 6:162

Zhong F, Xing J, Li X, Liu X, Fu Z, Xiong Z, Lu D, Wu X, Zhao J, Tan X, Li F, Luo X, Li K, Chen Z, Zheng M, Jiang H (2018) Artificial intelligence in drug design. Sci China Life Sci 61 (10):1191–1204

The manufacturer's authorised representative in the EU is Springer Nature Customer Service Centre GmbH, Europaplatz 3, 69115 Heidelberg, Germany. If you have any concerns regarding our products, please contact ProductSafety@springernature.com

Printed and bound by CPI Group (UK) Ltd, Croydon, CR0 4YY
30/09/2026
02225328-0003